BN 9310877 X

AF841666

WITHDRAWN FROM
UNIVERSITY OF BRIGHTON
LIBRARIES

HISTORICAL CHANGE OF LARGE ALLUVIAL RIVERS: WESTERN EUROPE

FRONTISPIECE The great flood of 25–26 December 1717. 'The great water flood, which was spread over large areas of land by God's hand on the 25 and 26 December 1717 AD in which many people, houses, goods, miserably died and perished. Frm J. M. Fuesslinus, 1717'. Supplied by D. Vischer

HISTORICAL CHANGE OF LARGE ALLUVIAL RIVERS: WESTERN EUROPE

Edited by

G. E. PETTS
Department of Geography,
Loughborough University of Technology

with

H. MÖLLER
Institut für Meereskunde, D-2300 Kiel, West Germany

and

A. L. ROUX
Laboratory of Freshwater Ecology, Université Claude-Bernard, Lyon 1,
F-69622 Villeurbanne Cedex, France

JOHN WILEY & SONS
Chichester · New York · Brisbane · Toronto · Singapore

COVER ILLUSTRATION. The French Upper Rhône at Chautagne contrasts the new man-made canal and the old meandering 'natural' channel which the hydropower works have bypassed. A compensation flow is maintained through the bypassed section which also receives excess flood water

UNIVERSITY
LEARNING
-3 NOV 1994
RESOURCES
BRIGHTON
9310877

Copyright © 1989 by John Wiley & Sons Ltd

All rights reserved.

No part of this book may be reproduced by any means, or transmitted, or translated into a machine language without the written permission of the publisher.

Library of Congress Cataloging-in-Publication Data

Historical change of large alluvial rivers: Western Europe/edited by G. E. Petts with H. Möller and A. L. Roux.
p. cm.
Bibliography: p.
Includes index.
ISBN 0-471-92163-7
1. Alluvial streams—Europe. 2. River channels—Europe.
I. Petts, Geoffrey E. II. Möller, Heino. III. Roux, A. L. (Albert Louis)
GB1281.H57 1989
551.48—dc19 88-31454
CIP

British Library Cataloguing in Publication Data

Historical change of large alluvial rivers: Western Europe.
1. Rivers. Geophysical processes
I. Petts, Geoffrey E. II. Möller H. III. Roux, A. L.
551.48'3

ISBN 0 471 92163 7

Phototypeset by Dobbie Typesetting Limited, Plymouth, Devon
Printed and bound in Great Britain by
Anchor Press Ltd, Tiptree, Essex

List of Contributors

C. Amoros	Laboratory of Freshwater Ecology, Université Claude-Bernard, Lyon 1, F-69622 Villeurbanne Cedex, France
J. Bethemont	Department of Geography, Université de Saint-Etienne, France
M.-C. Borlee	Unite d'Ecologie des Eaux Douces (UNECED), Facultes Notre-Dame de la Paix, rue de Bruxelles, 61, B-5000, Namur, Belgium
G. Braga	Dipartimento di Ingegneria del Territorio, University of Pavia, Italy
J.-P. Bravard	Department of Geography, Université Jean Moulin Lyon 3, 74 rue Pasteur, 69007 Lyon, France
D. Busch	Wasserwirtschaftsamt Bremen, Theodor-Heuss-Allee 21, D-2800 Bremen 1, West Germany
A. De Sostoa	Department of Animal Biology, University of Barcelona, Diagonal 645-647, 08048, Barcelona, Spain
H. Décamps	Centre d'Ecologie des Ressources Renouvelables, CNRS, 29, rue Jeanne Marvig, 31055 Toulouse, France
B. Elvira	National Museum of Natural Sciences, CSTC, Jose Gutierrez Abascal 2, 28006, Madrid, Spain
M. Fortuné	Centre d'Ecologie des Ressources Renouvelables, CNRS, 29, rue Jeanne Marvig, 31055 Toulouse, France
F. Gazelle	Centre d'Ecologie des Ressources Renouvelables, CNRS, 29, rue Jeanne Marvig, 31055 Toulouse, France
S. Gervasoni	Dipartimento di Scienza della Terra, University of Pavia, Italy
J. Hooke	Department of Geography, Portsmouth Polytechnic, Portsmouth, Hampshire, UK
A. Klink	Klink Hydrobiological Consultants, Hydrobiologisch Adviesburo, Boterstraat 28, 6701 CW Wageningen, The Netherlands
J. Lobon-Cervia	Department of Organisms and Systems Biology, University of Oviedo, 33071 Oviedo, Spain
J.-C. Micha	Unite d'Ecologie des Eaux Douces (UNECED), Facultes Notre-Dame de la Paix, rue de Bruxelles, 61, B-5000, Namur, Belgium
H. Möller	Institut für Meereskunde, Dusternbrooker Weg 20, D-2300 Kiel, West Germany

G. Pautou — Laboratoire de Biologie Vegetale, Université de Grenoble 1, F-38402, Saint Martin d'Heres, France

G. E. Petts — Department of Geography, University of Technology, Loughborough, Leicestershire, UK

J.-L. Probst — Centre de Sédimentologie et Géochimie de la Surface, Institut de Géologie, 1 rue Blessig, 67084 Strasbourg Cedex, France

M. C. Rang — Department of Physical Geography, State University of Utrecht, Box 80.115, 3508 TC Utrecht, The Netherlands

C. E. Redmond — Department of Geography, Portsmouth Polytechnic, Portsmouth, Hampshire, UK

P. A. Rincon — National Museum of Natural Sciences, CSTC, Jose Gutierrez Abascal 2, 28006, Madrid, Spain

A. L. Roux — Laboratory of Freshwater Ecology, Université Claude-Bernard, Lyon 1, F-69622 Villeurbanne Cedex, France

M. Schirmer — Arbeitsgruppe Limnische Ökologie, Fachbereich 2, Universität Bremen, Leobener Straße, D-2800 Bremen 33, West Germany

C. J. Schouten — Department of Physical Geography, State University of Utrecht, Box 80.115, 3508 TC Utrecht, The Netherlands

B. Schuchardt — Arbeitsgruppe Limnische Ökologie, Fachbereich 2, Universität Bremen, Leobener Straße, D-2800 Bremen 33, West Germany

H. Smit — Institute for Inland Water Management and Waste Water Treatment, Rijkswaterstaat, NL-8200 AA Lelystad, PO Box 17, The Netherlands

P. Ullrich — Arbeitsgruppe Limnische Ökologie, Fachbereich 2, Universität Bremen, Leobener Straße, D-2800 Bremen 33, West Germany

G. Van Urk — Institute for Inland Water Management and Waste Water Treatment, Rijkswaterstaat, NL-8200 AA Lelystad, PO Box 17, The Netherlands

D. Vischer — Versuchsanstalt für Wasserbau, Hydrologie und Glaziologie (VAW), ETH-Zentrum, CH-8092, Zurich, Switzerland

H. Vivian — Laboratoire de la Montagne Alpine, Institut de Geographie Alpine, rue Maurice Gignoux, 38031 Grenoble Cedex, France

Contents

Historical Change of Large Alluvial Rivers: Western Europe
Edited by G. E. Petts
©1989 John Wiley & Sons Ltd

1

Historical Analysis of Fluvial Hydrosystems

GEOFFREY PETTS

*Department of Geography, University of Technology,
Loughborough, Leicestershire, LE11 3TU, UK*

INTRODUCTION

'At a time when it is acknowledged that global environments may be changing as a result of the impact of human activity, it is important that we are able to learn from the past.'

Sir Malcolm Brown, FRS (1987,ix)

In his forward to a recent volume on developments in palaeohydrology, resulting from the work of the International Geological Correlation Programme, Sir Malcolm Brown emphasizes the importance of the past in understanding the present and predicting the future. Recent decades have witnessed a rapid growth in our understanding of environmental changes over the 10 000–15 000 year timescale (Gregory, 1983; Gregory *et al.*, 1987) and the 10–50 year timescale (e.g. Gregory, 1977; Petts, 1984, 1987a). Substantial efforts have been made to link space and time but greater effort is needed to consider changes over the historical timescale of 100–500 years—the period of intensifying human impact culminating in the creation of truly man-made rivers (Petts, 1984). Knowledge of environmental change resulting from our exploitation of land and water resources over historic timescales is necessary to improve river regulation strategies; to achieve *sustainable* resource development; and to advance environmental conservation.

During the 1980s there have been three major shifts in river regulation. First, there has been a change from river development to river management. This change reflects the increasing recognition of the environmental losses resulting from human impacts; the potential long-term effects of these impacts on Man; and the evolving environmental awareness and needs of societies. Secondly, there has been a move by scientists to tackle problems of large rivers, following two decades of work on experimental catchments and small streams. Thirdly, there has been a development of multidisciplinary approaches to both research and management. As noted by Davies and Walker (1986), new opportunities for both research and management are arising as a result of the recent move away from the reductionist and isolationist philosophies of the previous two decades. In Europe, all three of the above themes are embodied within the 'European Large Alluvial Rivers Network'.

The Network, formed in 1986 in response to an initiative of the Council of Europe to promote scientific and technical cooperation, is concerned with water as a natural resource, and the interaction between that resource and environment. Its purpose is to foster interdisciplinary and international cooperation in the development of environmentally sound management strategies and tools for large rivers in Europe. In order to stimulate research on the sensitivity and responses of hydrosystems, and their components, to human impacts and different forms of management, one early activity of the Network was to review methods

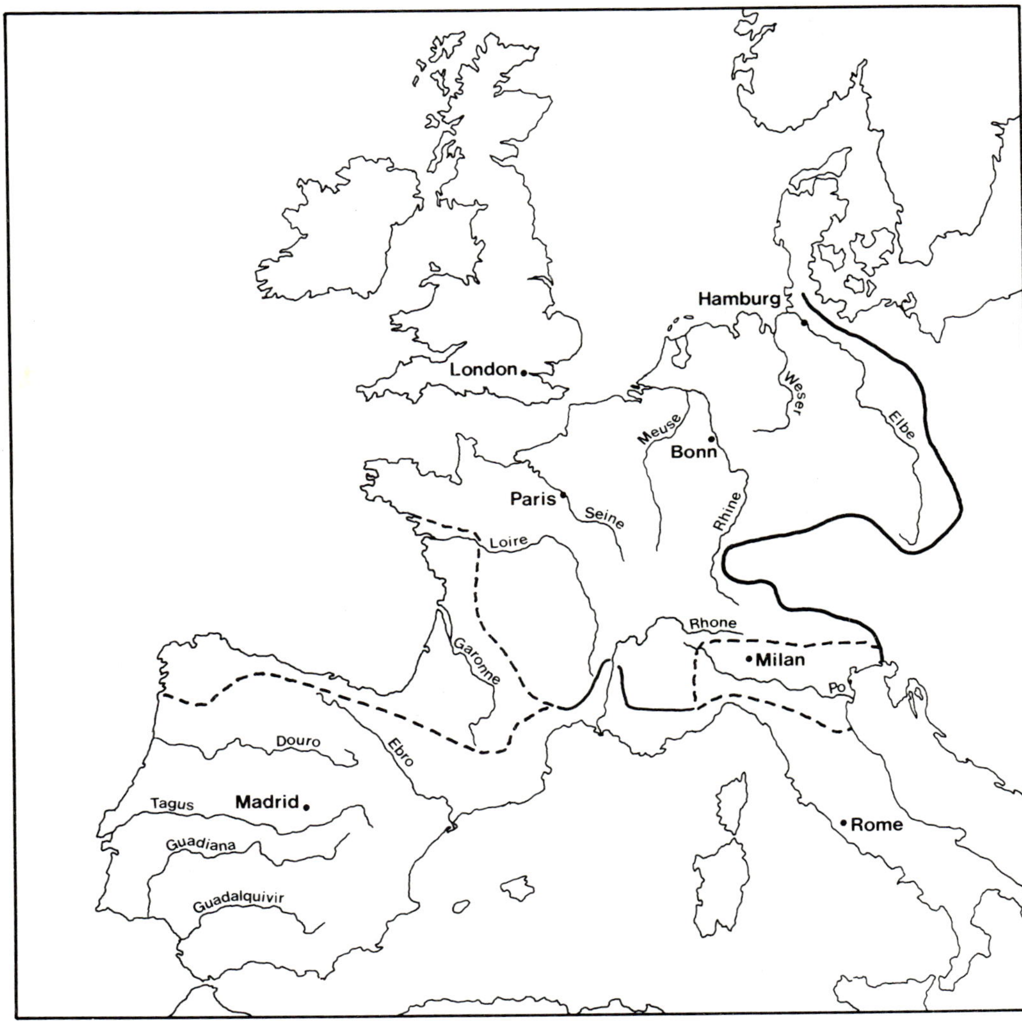

FIGURE 1. The major rivers of western Europe. Included is the watershed separating the rivers draining to the North Sea, Atlantic Ocean, and Mediterranean Sea from those draining to the Baltic Sea and Black Sea. Also shown is the biogeographic divide between the temperate forest zone (mixed broad-leaved and coniferous species) and the evergreen forest, open woodland and meadow of the Mediterranean region

for improving our data base on natural hydrosystems and their historical changes, and case-histories of large rivers in western Europe. This book is one product of this interdisciplinary and international effort.

The river and the alluvial corridor within which it flows may be viewed as a fluvial hydrosystem (Amoros *et al.*, 1987) characterized by longitudinal, lateral and vertical gradients. From headwaters to mouth the hydrosystem changes more or less progressively. The longitudinal gradient has been viewed as a continuum (Vannote *et al.*, 1980) that may be interrupted or rejuvenated by geomorphological discontinuities. The lateral gradient describes the exchanges between the permanent river channel and all the seasonally, or occasionally, inundated and semi-aquatic habitats along its margins. The importance of this dynamic ecotone between the lotic and terrestrial systems for conservation and river management has been the focus of recent attention (Naiman *et al.*, in press). In the mature, lowland reaches of large rivers major alluvial aquifers have been created by aggradation during the Holocene period of rising sea-levels and glacial retreat. The vertical gradient emphasizes the important exchanges that occur between the alluvial aquifer and the river. Within the Catchment Ecobiome, the diversity of terrestrial, semi-aquatic, and aquatic biotopes that comprise the fluvial hydrosystem are determined by the hydrological, water-quality, and geomorphological dynamics.

This book focuses on the large rivers of western Europe (Figure 1). Although none are 'large' in relation to those of eastern Europe or indeed other continents (Table 1) they have experienced a long history of change. The synthesis of information from a wide range of sources has enabled general statements on the history of these rivers to be made. Reference is made to other European rivers to support these statements.

TABLE 1. Major rivers of western Europe. Those in parentheses are not described in subsequent chapters. Data from UNESCO (1974). Threshold drainage area 27 500 km^2

River	Basin	Drainage area to gauging station (km^2)	Climatic class*	Mean discharge (m^3s^{-1})	Mean discharge ($m^3s^{-1}km^{-2}$)	Month of discharge Max.	Month of discharge Min.
Rhine	North Sea	159 680	C	2210	0.014	2	10
Elbe	North Sea	131 950	A	700	0.005	4	9
(Loire)	Atlantic	110 000	A	874	0.008	2	8
Rhône	Mediterranean	95 590	C	1712	0.018	3	8
Douro	Atlantic	91 491	B	496	0.005	2	8
Ebro	Mediterranean	84 230	B	552	0.007	3	8
Po	Adriatic Sea	70 091	D	1480	0.021	5	8
(Guadiana)	Atlantic	60 883	B	202	0.003	2	8
(Tagus)	Atlantic	59 167	B	305	0.005	2	8
Garonne	Atlantic	52 000	D	590	0.011	2	8
(Guadalquivir)	Atlantic	46 995	B	189	0.004	3	8
(Seine)	Atlantic	44 320	A	273	0.006	2	8
Weser	North Sea	37 790	A	305	0.008	2	9
Meuse	Rhine	28 950	C	325	0.011	1	8

*Generalized climatic classes = mean annual rainfall (cm)/July average temperature (°C)
A = $<75/<20$; B = $<75/\geqslant 20$; C = $\geqslant 75/<20$; D = $\geqslant 75/\geqslant 20$

HISTORY OF RIVER REGULATION: EUROPEAN CONTEXT

Traditionally, rivers have been used for water-supply for agriculture, domestic and industrial use; navigation; fisheries; and water-power. Rivers have been viewed not only as resources but also as hazards: floods inundated vast areas; the process of erosion and deposition caused the channels to migrate from one side of the valley to the other; and malaria and other water-associated diseases characterized the floodplain backwaters. The wetland alluvial plains of major rivers were major obstacles to routeways across the valleys. Thus, as noted by Dugdale (1772):

> 'those nations, which be of greatest antiquity, and in chief renown for arts and civility, are also famous for their works of this nature (land-drainage).'

Dugdale considered the drainage of floodplains to be works of 'divine intervention' citing Genesis, Chapter 1, verse 9. He referred to the Egyptians as the 'first masters' in this art of draining and irrigation:

> 'their workmanship about the river Nile, being such that industry surpassed nature . . . through industry it was brought to pass that oftentimes, when nature was defective, there was by the help of trenches and banks, as much ground watered with the smaller floods, as with the greater.'

Today, the large rivers of most regions of the world are regulated, to a greater or lesser degree, by Man. River regulation involves large storage reservoirs, low-head dams and run-of-river impoundments, channel modifications, and water transfers (Kitson, 1984). These actions, together with the effects of pollution, have converted natural hydrosystems to man-made ones. The ecological impacts of river regulation have been dramatic (e.g. Ward and Stanford, 1979; Lillehammer and Saltveit, 1984; Petts, 1984; Craig and Kemper, 1987).

European context

Throughout Europe, large alluvial rivers have experienced a long history of modification by Man such that by the beginning of this century, few reaches of natural river remained. The period of major human impact extends back for at least 200 years: significant developments and selected examples are given in Table 2. Any assessment of the potential effects of human activity on fluvial hydrosystems must take into account the spatial variation of natural hydrosystems which reflects the climatic setting now and in the past. The long-term evolution of European rivers during the Holocene may be subdivided into four regional categories (Starkel, 1987):

1. River systems recently developed on areas formerly covered by ice sheets and affected by isostatic uplift;
2. River systems of the former periglacial zone partly influenced by the ice sheet;
3. River systems of the former periglacial zone with the lower sections influenced by eustatic sea-level changes;
4. River systems of southern Europe, the region of former cold steppe and forest-steppe.

Reconstructions of the climate of the pre-instrumental era, in particular by Lamb (e.g. 1984a,b) have shown that the overall pattern of temperature change since 800 AD has been

TABLE 2. Development of river regulation in Europe

	Historic sequence	Significant developments	Other impacts
1250	Weirs for water power	Stanches widespread in Flanders, Germany, France, Italy and England	
	River improvements for navigation	1398 First summit canal (R. Stecknitz)	Artisanal fishery
		1400 Bertola designed channelization of R. Adda	
		1497 Leonardo designed pound-lock with mitre-gates	
		Verona (R. Adige) and Florence (R. Arno) established river authorities	
		1550 Lupicini designed flood-defences for R. Po	
1500			
	Flood-control and land reclamation	Dredging using endless chain technology developed by 1561	Artisanal fishery
		Pound locks widespread	
		Small rivers channelized e.g. Yevre and Havel	
	Science of regulating rivers established	1577–1643 Castelli (Founder of modern hydraulics)	
		1594 Alicante dam (41 m-high masonry)	
		1692 Completion of Languedoc canal	1616 R. Thames pollution problems
		Guglielmini (1697) and Baratteri (1699) scientific approach to river regulation	
1750			
		Large rivers channelized: e.g. R. Oder	Commercial and artisanal fishery
		R. Guadalquivir—length to Seville reduced by 50 km (40 per cent)	
		Earthbank water-supply dams spread in headwater catchments	Pollution
1850			
	Extensive floodplain reclamation	Major rivers channelized: e.g. Alsatian Rhine and Alpine Rhône	R. Thames severely polluted
		1845 R. Tisza (Theisz) shortened by 340 km, 12.5×10^6 ha reclaimed	
		1849 R. Danube 4×10^6 ha floodplain reclaimed along 230 km reach	
		Masonary headwater supply dams (50 m high) common	
	Water supply dams spread	1898 Hydroelectric power dam at Rheinfelden	Overfishing
1900			
	Hydroelectricity dams	1937 First 1000×10^6 m_3 reservoir: Ivankovo, R. Volga (1120×10^6 m^3)	Severe pollution widespread
		1941 First $25\,000 \times 10^6$ m^3 reservoir: Rybinsk, R. Volga ($25\,400 \times 10^6$ m^3)	
		1950 First 150 m high Dam: Noce–Aldigo, Italy	
	River regulation Dams	1955 First $50\,000 \times 10^6$ m^3 reservoir: V. I. Lenin Dam, R. Volga ($58\,000 \times 10^6$ m^3)	
		1957 First 200 m high Dam: Mauvoisin Dam, Switzerland (237 m high)	
		1961 First 250 m high Dam: Vaiout Dam, Italy (262 m high)	
	Impounded rivers	1962 Grand Dixence 285 m high Dam, Switzerland	Conservation
2000			

characterized by thermal maximum between 1150–1300 AD, followed by a cooling phase leading into a cold epoch, referred to as the 'Little Ice Age', which climaxed around 1550–1700. For example, numerous reports of the freezing of the Thames at London emanate from this period. With regard to rainfall, there was a relatively wet phase, in comparison with 20th century values, around 1200 AD and a dry period about 1600 AD. Summer rainfalls dropped markedly during the medieval warm epoch and since 1300 summer rainfalls have tended to be highest in the second half of each century.

These natural climatic fluctuations would have had hydrogeomorphological effects and ecological implications. However, the effects may have been modified, to a greater or lesser degree, by the activities of Man. These initially affected the catchment—and the river indirectly—and later directly modified the river itself. There is no doubt that early human activity, especially deforestation—to provide fuel and building materials for houses and ships—and the multiplication of grazing animals had the greatest effect on the river valleys of Starkel's fourth region. Such activities would have influenced soil erosion, processes of channel erosion and sedimentation, and the ratio of allochthonous to autochthonous organic matter in streams. Marked changes may have begun in the strongly seasonal regime rivers around the Mediterranean earlier than 2000 BP (e.g. Vita-Finzi, 1969; Davidson, 1980). In general terms, the rivers of the western and central European uplands were affected later, and those of eastern and northern Europe have been affected most recently (Starkel, 1987). Certainly, there is evidence to suggest that the intensification of direct channel works was in some cases at least a response to channel sedimentation or erosion related to land-use changes within the river's catchment.

Prior to the 11th century, river works in Europe were still primitive, but embankments were built for flood control and land reclamation, and the problem of naturally varying river flows for water power had been overcome by the construction of weirs and aquaducts. During the next 700 years the more commercially active countries made improvements to rivers to overcome the obstacles to navigation imposed by shallows and mill weirs, to reclaim alluvial plains, and to control floods.

There were two main centres of technological advance in river regulation. First, in the Netherlands, dredging technology and designs for floodgates, retaining walls, and groynes, were well established by the end of the 16th century. Secondly, in Italy, the problem of *La bonifica* (land reclamation in its broadest sense) stimulated advances in the technology of river training and regulation which was used for the management and systematic control of rivers from the mid-15th century. By the end of the 17th century the 'art' of river regulation had been replaced by a scientific methodology, not least as a result of the work of Castelli, who is recognized as the founder of modern hydraulics. Although the technology of the 17th century was able to master streams and large, slow-flowing rivers, the highly mobile, piedmont rivers were not controlled with any long-term success for another two centuries. Nevertheless, the scale of river engineering works increased dramatically during this period and by 1900 most large rivers had been channelized.

In 1817 Tulla initiated the channelization of the braided Alsatian section of the Rhine and his often quoted statement that 'As a rule, no stream or river needs more than one bed!' became general policy for hydraulic engineers. Possibly the greatest single work of the 19th century was the regulation of the R. Theisz (Tisza) which drains the southern and western Carpathians. Beginning in 1845, 12.5×10^6 ha of floodplain marsh were drained and the river course shortened by 340 km. In part, the rapidity with which engineering works spread during the 19th century reflected three factors. First, there was the need to create more agricultural land along the major routeways to feed the growing populations. Secondly, there was the need to combat malaria by eliminating mosquito habitat. Thirdly, many European valleys were affected by increasing flooding which was perceived to have been caused by deforestation and land drainage:

> 'Recent investigations have shewn that floods and dangerous inundations are on the increase among the rivers flowing through the cultivated lands of central Europe; . . . This tendency is particularly noticeable in the Volga and Dneiper, along whose banks clearings of forest timber has been made to such an extent that the central and lower part of the streams flow through long stretches of treeless country, and the high water mark is always higher than in former times.'
>
> Hermann, Klein and Thome: *Land, Sea and Sky* (*c*. 1890, p. 199) English version translated by J. Minshall and published by Ward, Lock and Co., London.

In part, the increased flooding probably reflected climatic fluctuations. In some locations channel aggradation, sometimes induced by bridge and weir construction or log jams behind narrow-arched bridges, was also a contributory factor.

During the 20th century, river regulation by dams has further contributed to the isolation of rivers from their riparian zones and alluvial plains. Dam-building technology was well established by the end of the 18th century. In a global context, Europe experienced a relatively high rate of dam construction until 1880. An increase in dam building activity occurred between 1900 and 1940 but the peak of dam-building activity was between 1950 and 1970. The size of dams and the reservoirs they created has also grown rapidly. Europe's largest reservoir of $58\,000 \times 10^6\,m^3$, impounded by the V. I. Lenin Dam on the R. Volga, was filled in 1955 and today it is the 16th largest capacity reservoir in the world. The 285 m high Grand Dixence Dam in Switzerland, completed in 1962, was the worlds highest dam and remained so for nearly two decades.

Today, most rivers are truly 'canalized' and their floodplains have been drained for agriculture or plantation forestry. Currently in Europe, large dams greater than 150 m high are being completed at a rate of one every 1.65 years. However, some countries have experienced increased dam-building activity since 1970. This is especially true of Spain where at the present time, two dams over 15 m high are being completed every month (Jalon, 1987). By 1973, 15 per cent of the stable annual runoff was contributed by storage reservoirs (Lvovitch, 1973). Many rivers have been converted to chains of run-of-river impoundments: for example, the Rhône between the Genissiat Reservoir and the Mediterranean—a distance of 470 km—is controlled by a series of twenty low-head hydropower dams.

River regulation in Britain

The long sequence of human impacts on rivers in western Europe may be exemplified by changes in the UK (Petts, 1988), notably those concerning the Thames, Severn, Trent, Yorkshire Ouse and Great Ouse. Over both the Holocene and recent time periods, changes along British rivers closely parallel those of the temperate forest zone of western and central Europe. Many river valleys contain major river terraces dating from the last cold phase of the Pleistocene which ended about 12 000 BP, and many rivers show the legacy of glaciation in the sediments that line the channel and in their slopes. However, the fluvial hydrosystems today relate most strongly to changes induced by Man over the past 5000 years (see Gregory *et al.*, 1987).

Most early settlement took place on the foothills, not on the valley floors which were thickly forested. Deforestation in upland areas spanned a considerable period of time from 5000 to 1300 BP (Chambers, 1981). Floodplain deforestation occurred in two stages 2500–2000 BP and 1500–1000 BP (Wiltshire and Moore, 1983). Wetlands were once common along the river valleys but most were affected by extensive clearance about 2000 years ago. Along the River Severn, wet alder carr dominated the alluvial plain until the Iron Age and the evidence suggests that clearance led to locally rapid channel aggradation and floodplain sedimentation (Brown, 1987). Certainly, by Roman times, much of the lowlands had experienced a long history of cultivation and it is probable that the characteristics of fluvial hydrosystems had been modified by human activity. However, the first major works were undertaken by the Romans who completed drainage schemes (Leach and Leach, 1982) and channelized rivers to aid navigation (Darby, 1983).

By the 9th century, the major rivers were important transport routes. For example, York had become established at the head of a navigable waterway—the Yorkshire Ouse:

> 'From the most distant lands ships did arrive,
> And safe in port lay there, tow'd up to shore.
> Where, after hardships of a toilsome voyage,
> The sailor finds a safe retreat from the sea.
> By flow'ry meads, on each side of its banks,
> The Ouse, well stored with fish, runs through the town.'
>
> Alewin, *c.* 850 (Translated and quoted by Francis Drake 1736 in *The History and Antiquities of the City of York*, p. 227.)

The importance of the Ouse is reflected by the fact that by late Saxon times the city of York ranked second in importance only to London. It has been estimated that since the Roman period the overall width of the channel has been reduced by some 40 m where it flows through York (Y.A.T., 1988), change that reflects the importance of the river for trade.

Weirs to control water levels for mills and for fishing were widespread by the 12th century. The earliest water-mill in England dates from 742 AD and in the Domesday records of 1086 reference is made to more than 5000 water-mills. Medieval bridges with many piers and narrow arches often so obstructed the flow that they created hazardous conditions for boat passage. The extension of trade, partly generated by the mills, led to increasing conflict. Controls on the number of fish weirs on the Yorkshire Ouse were introduced in 1478; flash locks or 'staunches' (navigation weirs) were in use on the Thames as early as 1585; pound locks, a product of the Italian Renaissance, first occurred in England on the River Exe about 1565; and early river engineering works included cutting back the inside of acute bends, dredging out shoals and shallows, piling where the banks were weak, often using alder poles, and weed cutting.

Plans for the systematic 'improvement' of rivers are found from the mid-fifteenth century onwards but there were many local problems. For example, the use of flash locks caused scouring of the channel bed immediately below the weir; the eroded sediment formed a bar below the scour hole and reduced water depths. Indeed, the increasing intensity of river works in the 17th century reflected the increasing trade, larger vessels, and problems of channel siltation. At this time, under royal patronage, leading landowners invested capital in extensive land-drainage works (Sheail, 1988). By 1760, the art of improving rivers for navigation by constructing pound locks and making artificial channels or cuts was well established. However, important reaches of river remained uncontrolled. Notable is the River Trent which due to the numerous shoals remained impassable in dry seasons until about 1800 (Priestley, 1831).

A major increase in navigation occurred from 1750 with the emergence of canals. The need for towpaths led to the clearance of riparian vegetation, although along the Yorkshire Ouse the width of the towpath was restricted to a width of three feet in strict conformity with the provisions of the 1665 Act 'for making divers rivers navigable'. The attitude of politicians and major landowners of the 18th and 19th centuries is reflected by the following statement concerning Otmoor on the River Ray:

> 'such a tract of wasteland in summer, and covered the winter through with water, to remain in such a state, within 5 miles of Oxford and the Thames . . . is a scandal to the national policy'
>
> Arthur Young, *Report of the Agricultural Board*, 1813.

Land-drainage and river engineering works intensified during the 19th century. Although there exist examples of major river canalization and land-drainage works prior to about 1850, they occurred sporadically and in many cases were affected by floods or by silt deposition. However, major training works were complete on most rivers by 1880 and these were supplemented by systematic and regular dredging.

Overfishing had been a problem as early as the 13th century when Edward I restricted salmon fishing. Then, in 1389, the use of fine-meshed nets were banned to protect the fry of salmon and lampreys. However, from about 1600 and particularly after 1750 the rapid deterioration of water-quality along British rivers, associated with the discharge of domestic sewage, had catastrophic ecological effects. Chronic conditions were reached in the 1840s. On the Thames, all commercial fishing in the London area had ceased by 1850. In 1953 anaerobic or near anaerobic conditions persisted for much of the year in a section 50 km long downstream of London Bridge and in 1957 there were no fish in the tidal reaches of the London area.

Impounding reservoirs for water supply were developed from the late 18th century to overcome problems of pollution. Each town developed its own local supply reservoirs, often in the headwaters of the valley in which it was situated. This was particularly true in the highly industrialized area of Lancashire and Yorkshire within the southern Pennines (Petts, 1988). Disputes over water usage resulted from the pressure of mill-owners to maintain discharges and in many cases compensation reservoirs were constructed downstream or in tributary valleys to maintain compensation flows.

The modern era of river regulation is founded in Bills submitted to Parliament during the late 19th century to construct large reservoirs in the Vyrnwy (upper Severn) and in the upper Wye valleys of Wales to supply Liverpool (1880) and Birmingham (1892) respectively (Sheail, 1984). These schemes were important for three reasons: they represented (1) a new phase in dam-building technology; (2) the beginning of large-scale interbasin transfers; and (3) they raised issues of multipurpose river regulation—for water-supply, navigation and ecology. The first significant hydroelectric power development was commissioned in 1896, at Foyers on Loch Ness. However, the major development occurred in Scotland between 1945 and 1965; some 28 schemes were constructed during this period with a capacity of 965 MW (Johnson, 1988). Today, the rivers are regulated by storage behind some 450 large dams (Petts, 1988) and, in England and Wales, 35 500 km of main river is channelized to some degree (Brookes *et al.*, 1983).

Ecological changes

The hydrological and geomorphological characteristics of European rivers have changed dramatically since 1850 and it is likely that anthropogenic changes may have been significant over a longer timescale. However, it is only since about 1975 that scientists have demonstrated the range of ecological effects that may result from river regulation. These effects may be classified as three groups (Petts, 1987b). First, the isolation of the main river from its alluvial plain, eliminating access to backwaters, floodplain lakes and marshes, has had a major effect on both the ecological diversity of the highly productive alluvial corridor (Pautou and Decamps, 1985) and riverine fish populations (Lelek, 1987). Even without the extension of agricultural land into the alluvial plain, succession responding to lowering groundwater levels and the elimination of the flood season causes change usually to a less diverse, and certainly less dynamic, system (Bravard *et al.*, 1986; Amoros *et al.*, 1987). The isolation of backwater and floodplain areas from the main channel has severely impacted fish stocks (Welcomme, 1979; Petts, 1984). Floodplain channels and backwaters provide spawning habitats and refuges during periods of lethal and/or critical water quality. Fish species that spawn on inundated floodplains have been severely affected, for example, in the R. Dnieper (Zalumi, 1970) and R. Volga (Eliseev and Chikova, 1974). On the Rhine, Lelek (1987) noted that the decline of the artisanal fishery coincided with 19th century river channelization works.

Secondly, the introduction of barriers, especially to migration for anadromous fish, has had a widespread impact, but one which is not confined to the large dams of the post-1950 era. On the Dordogne river, *Salmo salar L.* disappeared soon after the first dams were built on the lower reaches between 1842 and 1904 (Decamps *et al.*, 1979). Within eastern European rivers Acipenseridae have been markedly affected (e.g. Zhadin and Gerd, 1963).

Thirdly, changes of instream habitats (involving the interaction of hydrology, water quality and geomorphology) have resulted in major biological changes. Autotrophic production within regulated rivers is often enhanced due not least to the reduced turbidity, flood regulation, and increased water temperatures. Thus, the benthic algal cover increased markedly after closure of the Vir Dam on the R. Svratka (Penaz *et al.*, 1968) and dense algal carpets and drifting filaments made the R. Glama, Norway, less suitable than hitherto for supply and recreational purposes. Similarly, submerged angiosperms have spread within regulated rivers (e.g. Holmes and Whitton, 1977) and 30 km of the R. Dordogne have become infested with aquatic macrophytes (Decamps *et al.*, 1979).

Faunal changes have resulted from a loss of centres of organization, especially spawning habitat and shelter areas, and life-cycle triggers. Henricson and Muller (1979) demonstrated that the type of life cycle largely determines whether a macroinvertebrate species can withstand regulated conditions. Epilimnial, mixed and hypolimnial release dams have caused different changes of the benthos downstream (Armitage, 1978; Scullion *et al.*, 1982; Penaz *et al.*, 1968). Fish species have been particularly affected by changes of the thermal regime below dams, for example, on the R. Elder (Lehmann, 1927) and R. Svratka (Penaz *et al.*, 1968).

Some short- and long-term ecological changes relate to specific events. For example, major fish kills have been caused by the release of anoxic water from some Czechoslovakian reservoirs (Bradka and Rehachova, 1964) and by gas supersaturation below hydroelectric power dams (Pettersen and Mellquist, 1984). Rapid fluctuations of water-level due to supply or power demands can have disastrous effects (Petts, 1984), as can reservoir sluicing and venting operations (Roux, 1984).

Two further impacts that have received relatively little attention in the literature are worthy of note. First is the loss of area of lotic habitat. This relates to the shortening of rivers and the confinement of the flow to a single channel in formerly braided and anastomosed reaches as a result of channelization or channel degradation below dams. Secondly, the loss of catchment integrity has resulted from the introduction of exotic species. In this latter context, the influence of canal construction and interbasin water transfers may be significant.

THE ROLE OF HISTORICAL ANALYSES

River channel changes have provided a focus for geomorphological research for four decades (Gregory, 1977) but the historical analysis of fluvial hydrosystems is a new field of research that has benefited from developments in palaeoclimatology (Lamb, 1982; Flohn and Fantechi, 1984) palaeohydrology (Gregory, 1983; Gregory *et al.*, 1987), palaeoecology (Berglund, 1986) and geoarchaeology (Davidson and Shackley, 1976). Future progress will benefit particularly from links with the last. The riverside has always been the focus of trade and commerce, and the waterlogged soils give human artefacts and fossils excellent preservation potential. Thus, archaeologists value waterfront sites as sensitive barometers of economic, commercial, social and institutional developments (Y.A.T., 1988).

Historical analyses of fluvial hydrosystems are required to meet three needs:

1. To reconstruct the hydrosystems of the pre-industrial era;
2. To determine the rate and magnitude of change of hydrosystems and their components to single and composite impacts; and,
3. To enable the improved prediction of the effects of new impacts and development of ecologically sound management tools.

The results of historical analyses of fluvial hydrosystems are of value in their own right in relation to understanding system dynamics and past environmental change. However, they also have important contributions to make to river management now. It is generally accepted that the existence of a scientific record adds considerably to the value of a site for conservation (e.g. Ratcliffe, 1977) and historical analyses may contribute to an improved policy for deciding whether a river should be preserved, developed or restored. Furthermore, the improved understanding of hydrosystem dynamics resulting from such studies may lead to improved methods for the restoration of damaged systems. Recent research has suggested that rivers can recover, to a greater or lesser degree, from external stresses. Opportunities appear to exist to restore damaged ecosystems or at least some target species or biotopes of particular interest to society (Petts *et al.*, 1989).

One major problem for environmental impact assessment is the timescale required for the fluvial hydrosystem to readjust consequent upon 'control measures' imposed by Man (Petts, 1984, 1987a). Most knowledge relates to changes caused by recent works undertaken during the past forty years. This is problematic because in many rivers ecological adjustments can require a much longer period of time. Biotic responses are usually faster than the abiotic ones and the readjustment of the fluvial hydrosystem depends to a large degree on the magnitude and rate of physical change. The succession of abiotic and biotic changes, along a regulated river (e.g. Petts and Greenwood, 1985) and its alluvial plain (Amoros *et al.*, 1987), form complex temporal sequences within which transient states may dominate for tens of years (Petts, 1987a). Clearly, more information on hydrosystem sensitivity to river regulation, pollution, etc. is needed in order to improve Environmental Impact Assessments and management approaches.

Methods and approaches

Historical analyses of fluvial hydrosystems involve a number of questions, including three primary ones: What changes occurred? When did they occur? And where did they occur? As

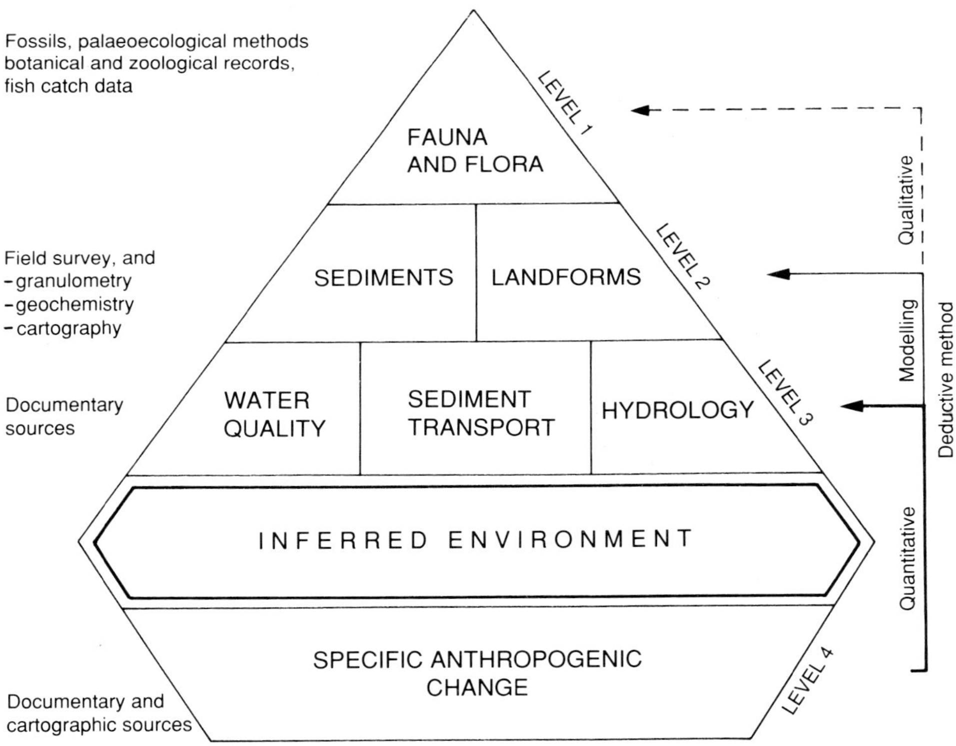

FIGURE 2. The four-level inferential approach to historical analyses of fluvial hydrosystems

in palaeohydrology (Thornes, 1987), at this time the most productive approach to reconstructing the sequence of changes experienced by fluvial hydrosystems over the past three hundred years is inductive (Figure 2). The direction of investigation is the reverse of the direction of causation. The complexity of physical, chemical and biological dynamics and their interactions prohibits the use of a deductive method. Nevertheless, such historical analyses may lead to the development of improved models and to the advancement of deductive methods even if they can never be truly deterministic.

From field observations, documentary and cartographic evidence of former biological populations (Level 1), and fluvial sediments and landforms (Level 2) it is possible to make inferences about the process dynamics of the environment (Level 3). The limitations of this approach are widely appreciated. Reconstructions use information from a number of specific time periods—from maps, plans, photographs and documents, or landforms and stratigraphic units dated by tree-ring analysis, archaeological evidence etc. Of necessity, historical analyses assume uniform change over each period. However, unlike palaeohydrological or geological studies, historical analyses benefit in many cases from more or less detailed documentary evidence concerning the precise nature and timing of anthropogenic changes (Level 4). Such information may be used to derive expected changes in the three primary variables (discharge, sediment transport and water quality) and opportunities exist for the development and application of semideterministic models. Of particular application are sediment transport and channel erosion models, and some water-quality models. With regard to the latter, models to reconstruct thermal regimes and eutrophication are especially promising (Gras, personal communication). It is this fourth level that provides for the

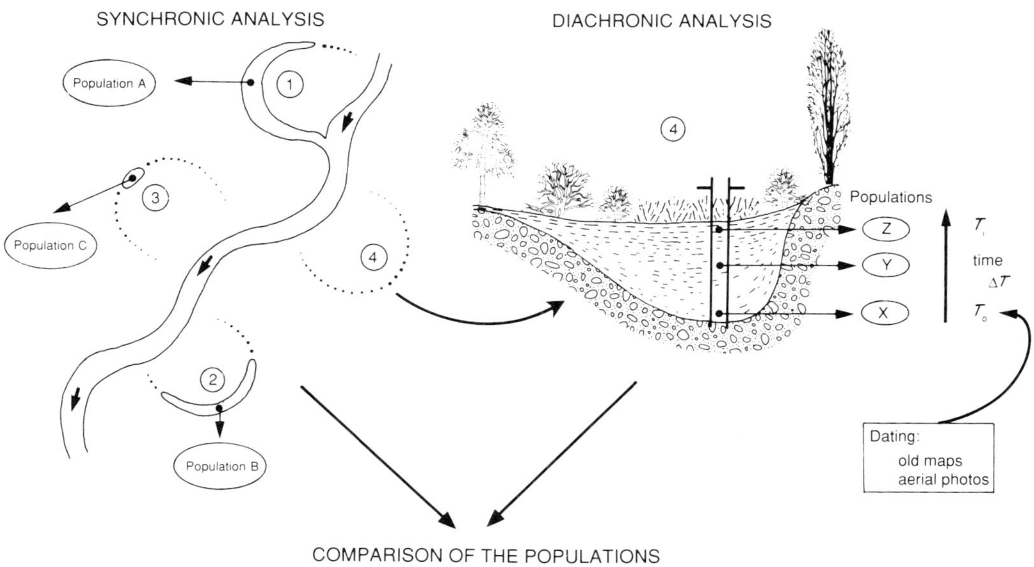

FIGURE 3. A schematic representation of synchronic and diachronic analyses. If the indicator species A, B and C are similar to X, Y and Z respectively, then a succession of changes may be defined. Alternatively, the method may be used to identify historical perturbations to established successions. (From Amoros *et al.*, 1988: reproduced by permission of John Wiley & Sons Ltd)

calibration and refinement of environmental reconstructions based on the inferential approach. As concluded by Thornes (1987) with reference to palaeohydrology, a greater understanding of the dynamics of catchment ecosystems is needed to advance dynamic modelling; historical analyses will contribute to this aim.

A method for applied ecological studies of fluvial hydrosystems has been developed by Bravard *et al.* (1986) and Amoros *et al.* (1987). The methodology involves the combination of *synchronic analysis* and *diachronic analysis* (Figure 3). Synchronic analysis derives successional sequences from a spatial survey of particular geomorphological units of different age. The structure and function of biotope types, such as former channels abandoned at different times, are determined. These are then placed in chronological sequence to develop successional sequences and to highlight complex responses to similar impacts for further investigation. Functional describers, such as indicator species, are identified for each unit and used to derive successions for different primary functional units. Diachronic analysis reconstitutes the succession of development stages from stratigraphic and palaeoecological studies, for example, for infilled abandoned channels. The impacts of human activities, precisely dated, are integrated into the successions. Sedimentological data are used to reconstruct biotope changes and palaeoecological data, using Chironomids or Cladocerans as indicators, are used to describe ecosystem development. Thus, results of such a four-level inferential approach have been used to develop predictive scenarios to assess hydrosystem response to new impacts.

Structure of this book

This introductory chapter is followed by a detailed examination of the sequence of early river engineering works (Visher). However, it is important to set these anthropogenic changes against natural hydrological changes at both the continental (Probst) and river basin (Vivian) scales. The core of this volume includes a series of essentially methodological contributions focusing on Levels 2 and 3 of the inferential approach. Following a discussion of cartographic data analyses and applications to studies of channel change (Hooke and Redmond), cartographic data sources are considered for France (Bravard and Bethemont) and an example of their application to the River Po is given by Braga and Gervasoni. This latter is particularly appropriate as Italy was the centre of map-making worldwide from an early date (Tooley, 1978). Rang and Schouten then demonstrate the use of geochemical analyses and floodplain stratigraphy in deriving pollution histories, and Amoros and van Urk examine the application of palaeoecological methods using fossil invertebrates from alluvial sediments. Interpretations of these data are then exemplified by Smit and van Urk, and Klink for the Lower Rhine.

The third section of the book begins with three chapters illustrating the approaches available for the reconstruction of fish faunas and the historical development of river fisheries of the Elbe (Moller), Douro (Lobon-Cervia *et al.*) and Ebro (Lobon-Cervia and de Sostoa). Moreover, these chapters indicate that, even within western Europe, there is not one single historical model of river development. The final four chapters emphasize this variability: the Garonne (Decamps *et al.*), the Meuse (Micha and Borlee), the Weser (Busch *et al.*) and the Rhône (Roux *et al.*).

PROBLEMS AND DIRECTIONS

Historical analyses are necessary to reconstruct pre-industrial hydrosystems and to explain present-day spatial distributions (of hydrosystems, biotopes, species, etc.). Moreover, they are necessary for the development of both reactive and proactive management. However, it is important that case-histories are multidisciplinary in character and consider an appropriate timescale. The lack of baseline data means that to date changes associated with recent impacts may be over or underestimated. The timescale of fluvial hydrosystem response is such (Petts, 1987a) that during the 20th century at least new impacts have been superimposed upon transient system states—changes initiated earlier this century or even during the last century. Consequently, causal relationships are not easily defined. River regulation in Europe has evolved over more than 200 years and its ecological impacts are masked by the effects of pollution, which became widespread during the 19th century and reached maximum severity on most rivers during the 1960s. Furthermore, physicochemical changes are often confounded with effects of biotic factors (competition, predation, disease), and may effect the ecology indirectly by affecting lower trophic levels or only in combination.

There are two groups of needs which must be addressed if the potential for ecological management is to be realized. First, there is a need for better scientific knowledge of process dynamics and physical, chemical and biological interactions in large rivers. Intrinsic and extrinsic thresholds must be defined and the cumulative effects of a number of small-scale, and individually insignificant, changes of system components for overall system stability must be illuminated. Furthermore, the timescales of system response must be defined, giving due recognition to the occurrence of transient states, and the long-term ramifications of different perturbations must be evaluated. This last includes the biological implications of Complex Response, discussed by Schumm (1973, 1977) to conceptualize the contrasting responses of two similar hydrosystems to the same perturbation.

Secondly, although science can provide information for the formulation of management strategies, the implementation of these strategies requires the necessary social and political will and the appropriate institutional structure. It would be naive to believe that ecological management can restore biological communities completely. Nevertheless, opportunities exist for the restoration of damaged hydrosystems and these will increase as knowledge of species and community requirements is advanced. However, it is unlikely that resources will be made available for the restoration or preservation of large alluvial rivers without an improved awareness of the ecological and societal value of fluvial hydrosystems. Neither will interdisciplinary historical analyses receive due recognition.

This book marks the first step in the development of a European Network of Scientific Cooperation for the improved management of large rivers. We hope that it will encourage and advance the historical analysis of rivers not only in western Europe but worldwide. Such studies will contribute to more rational approaches to river management, to further resource development and to better conservation practice.

REFERENCES

Amoros, C., Roux, A. L., Reygrobellet, J. L., Bravard, J. P. and Pautou, G. 1987. A method for applied ecological studies of fluvial hydrosystems. *Regulated Rivers*, **1**, 17–36.

Armitage, P. D. 1978. Downstream changes in the composition, number and biomass of bottom fauna in the Tees below Cow Green Reservoir and an unregulated tributary, Maize Beck in the first five years after impoundment. *Hydrobiologia*, **58**, 145–56.

Berglund, B. E. (Ed.) 1986. *Handbook of Holocene Palaeoecology and Palaeohydrology*, Wiley, Chichester, 869 pp.

Bradka, J. and Rehackova, V. 1964. Mass destruction of fish in the Slapy Reservoir in winter 1962–63. *Vodni Hospoddarstvi*, **14**, 451–2.

Bravard, J. P., Amoros, C. and Pautou, G. 1986. Impact of civil engineering works on the successions of communities in a fluvial system. A methodological and predictive approach applied to a section of the Upper Rhône River, France. *Oikos*, **47**, 92–111.

Brookes, A., Gregory, K. J. and Dawson, F. H. 1983. An assessment of river channelization in England and Wales. *Science of the Total Environment*, **27**, 97–122.

Brown, A. G. 1987. Long-term sediment storage in the Severn and Wye catchments. In Gregory, K. J., Lewin, J. and Thornes, J. B. (Eds), *Palaeohydrology in Practice*, Wiley, Chichester, 307–322.

Chambers, F. M. 1981. Date of blanket peat initiation in upland South Wales. *Quaternary Newsletter*, **25**, 24–29.

Craig, J. and Kemper, J. B. (Ed.) 1987. *Advances in Regulated River Ecology*, Plenum, New York, 431 pp.

Darby, H. C. 1983. *The Changing Fenland*, C.U.P., Cambridge.

Davidson, D. A. 1980. Erosion in Greece during the first and second millennia BC. In Cullingford, R. A., Davidson, D. A. and Lewin, J. (Eds), *Timescales in Geomorphology*, Wiley, Chichester.

Davidson, D. A. and Shackley, M. L. (Eds) 1976. *Geoarchaeology*, Duckworth, London, 408 pp.

Davies, B. R. and Walker, K. F. 1986. *The Ecology of River Systems*, Junk. Dordrecht.

Decamps, H., Capblanq, J., Casanova, H. and Tourenq, J. M. 1979. Hydrobiology of some regulated rivers in the south-west of France. In Ward, J. V. and Stanford, J. A. (Eds), *The Ecology of Regulated Streams*, Plenum, New York, 273–288.

Dugdale, W. 1772. *History of Imbanking and Drainage*, 2nd Edition, Bowyer and Nichols, London, 461 pp.

Eliseev, A. L. and Chikova, W. M. 1974. Conditions of fish production in the lower reach of the V. I. Lenin Volga hydro-electric station. In Kuzin, B. S. (Ed.), *Biological and Hydrological Factors of Local Movements of Fish in Reservoirs*, Amerind, New Dehli, 193–200.

Flohn, H. and Fantechi, R. (Eds). *The Climate of Europe: Past, Present and Future*, D. Reidel, Dordrecht.

Gregory, K. J. (Ed.) 1977. *River Channel Changes*, Wiley, Chichester.

Gregory, K. J. (Ed.) 1983. *Background to Palaeohydrology*, Wiley, Chichester.

Gregory, K. J., Lewin, J. and Thornes, J. B. 1987. *Palaeohydrology in Practice*, Wiley, Chichester.

Henricson, J. and Muller, K. 1979. Stream regulation in Sweden with some examples from central Europe. In Ward, J. V. and Stanford, J. A. (Eds), *The Ecology of Regulated Streams*, Plenum, New York, 183–200.

Holmes, N. T. H. and Whitton, B. A. 1977. The macrophytic vegetation of the River Tees in 1975: observes and predicted changes. *Freshwater Biology*, **7**, 43–60.

Jalon, G. de 1987. A review of river regulation in Spain. *Regulated Rivers*, **1**, 4, 343–348.

Johnson, F. G. 1988. Hydropower development on rivers in Scotland. In Petts, G. E. and Wood, R. (Eds), *River Regulation in the UK. Special Issue of Regulated Rivers*, **2**, **3**, 277–292.

Kitson, T. 1984. *Regulated River Basins: A Review of Hydrological Aspects for Operational Management*, Working Group on IHP–II Project A.2.2, Unesco, Paris, 122 pp.

Lamb, H. H. 1982. *Climate, History and the Modern World*, Methuen, London, 387 pp.

Lamb, H. H. 1984a. Climate and history in northern Europe and elsewhere. In Morner, N. A. and Karlen, W. (Eds), *Climatic Changes on a Yearly to Millenial Basis*, D. Reidel, Dordrecht, 225–240.

Lamb, H. H. 1984b. Climate in the last thousand years: natural climatic fluctuations and change. In Flohn, H. and Fantechi, R. (Eds), *The Climate of Europe: Past, Present and Future*, D. Reidel, Dordrecht, 25–64.

Leach, R. and Leach, P. 1982. Roman town and countryside. In Aston, A. M. and Burrow, I. (Eds). *The Archaeology of Somerset. A Review to 1500 AD*. Somerset County Council, Taunton, 69–71.

Lehmann, C. 1927. Über den Einfluss der Talspeuen auf die unterhalb liegende Bach und Flussfischerei. *Zeitschrift für Fischerei und deren Hilfswissenschaften*, **25**, 467–76.

Lelek, A. 1989. The Rhine River and some of its tributaries under human impact in the last two centuries. *Canadian Journal of Fisheries and Aquatic Sciences*, in press.

Lillehammer, A. and Saltveit, S. J. 1984. *Regulated Rivers*, Universitetsforlaget, Oslo.

Lvovitch, M. I. 1973. The global water balance. *US Int. Hydrol. Decade Bull.*, **23**, 28–42.

Naiman, D. J., Decamps, H. and Holland, M. (Eds) in press. *Land/Inland-water Ecotones: Strategies for Research and Management*, MAB Series, C.U.P., Cambridge.

Pautou, G. and Decamps, H. 1985. Ecological interactions between the alluvial forests and hydrology of the Upper Rhône. *Archives of Hydrobiology*, **104**, 13–37.

Penaz, M., Kubicek, F., Marvan, P. and Zelinka, M. 1968. Influence of the Vir River Valley Reservoir on the hydrobiological and ichthyological conditions in the R. Svratka. *Acta Sci. natural. Acad. Sci. bohemoslovacae-Brno*, **2**, 1–60.

Pettersen, S. and Mellquist, P. 1984. The effect of supersaturated water at Norwegian water power plants on fish mortality. In Lillehammer, A. and Saltveit, S. J. (Eds), *Regulated Rivers*, Universitetsforlaget, Oslo, 81–86.

Petts, G. E. 1984. *Impounded Rivers*, Wiley, Chichester.

Petts, G. E. 1987a. Time-scales of ecological change in regulated rivers. In Craig, J. and Kemper, J. B. (Ed.), *Advances in Regulated River Ecology*, Plenum, New York, 257–266.

Petts, G. E. 1987b. Ecological management of regulated rivers: a European perspective. *Regulated Rivers*, 363–369.

Petts, G. E. 1988. Regulated rivers in the UK. In Petts, G. E. and Wood, R. (Eds), *River Regulation in the UK. Special Issue of Regulated Rivers*, **2**, **3**, 201–220.

Petts, G. E. and Greenwood, M. 1985. Channel changes and invertebrate faunas below Nant-y-Moch Dam, R. Rheidol, Wales. *Hydrobiologia*, **122**, 65–80.

Petts, G. E., Imhof, J., Maher, J. F. B., Manny, B. and Weisburg, S. 1989. Management of river fish populations. *Canadian Journal of Fish and Aquatic Sciences*, in press.

Priestley, J. 1831. *Historical Account of the Navigable Rivers and Canals throughout Great Britain*, reprinted 1969 by David and Charles, Newton Abbot, 702 pp.

Ratcliffe, D. A. (Ed.) 1977. *A Nature Conservation Review. The Selection of Biological Sites of National Importance to Nature Conservation in Britain*. C.U.P., Cambridge, 2 volumes, 401 and 320 pp.

Roux, A. L. 1984. The impact of emptying and cleaning reservoirs on the physico-chemical and biological water quality of the Rhône downstream of dams. In Lillehammer, A. and Saltveit, S. J. (Eds), *Regulated Rivers*, Universitetsforlaget, Oslo, 61–70.

Schumm, S. A. 1973. Geomorphic thresholds and complex response of drainage basins. In Morisawa, M. E. (Ed.), *Fluvial Geomorphology*, Allen and Unwin, London, 299–310.

Schumm, S. A. 1977. *The Fluvial System*, Wiley, New York, 338 pp.

Scullion, J., Parish, C. A., Morgan, N. and Edwards, R. W. 1982. Comparison of benthic macroinvertebrate fauna and substratum composition in riffles and pools in the impounded River Elan and the unregulated River Wye, mid-Wales. *Freshwater Biology*, **12**, 579–595.

Sheail, J. 1984. Constraints on water-resource development in England and Wales: the concept and management of compensation flows. *Journal of Environmental Management*, **19**, 351–361.

Sheail, J. 1988. River regulation in the U.K.: a historical perspective. *Regulated Rivers*, **2**, **3**, 221–232.

Starkel, L. 1987. The evolution of European Rivers—a complex response. In Gregory, K. J., Lewin, J. and Thornes, J. B. (Eds), *Palaeohydrology in Practice*, Wiley, Chichester, 333–339.

Thornes, J. B. 1987. Models for palaeohydrology in practice. In Gregory. K. J., Lewin, J. and Thornes, J. B. (Eds), *Palaeohydrology in Practice*, Wiley, Chichester, 17–36.

Tooley, R. V. 1978. *Maps and Map-makers*, Batsford, London, Sixth Edition, 140 pp.

UNESCO 1974. *Discharges of World Rivers*, UNESCO, Paris, volume 1.

Vannote, R. L., Minshall, G. W., Cummins, K. W., Sedell, J. R. and Cushing, C. E. 1980. The river continuum concept. *Canadian Journal of Fisheries and Aquatic Sciences*, **37**, 130–137.

Vita-Finzi, C. 1969. *The Mediterranean Valleys*, C. U. P., Cambridge.

Ward, J. V. and Stanford, J. A. 1979. *The Ecology of Regulated Streams*, Plenum, New York, 398 pp.

Welcomme, R. L. 1979. *Fisheries Ecology of Floodplain Rivers*, Longman, London, 337 pp.

Wiltshire, P. E. J. and Moore, P. D. 1983. Palaeovegetation and palaeohydrology in upland Britain. In Gregory, K. J. (Ed.), *Background to Palaeohydrology*, Wiley, Chichester, 433–451.

Y.A.T. 1988. *The Waterfronts of York*, York Archaeological Trust, York, 36 pp.

Zalumi, S. G. 1970. The fish fauna of the lower reaches of the Dnieper: its present composition and some features of its formation under conditions of regulated and reduced river discharge. *Journal of Ichthyology*, **10**, 587–596.

Zhadin, V. I. and Gerd, S. V. 1963. *Fauna and Flora of the Rivers, Lakes and Reservoirs of the USSR*, Israel Prog. Sci. Trans., Jerusalem, 625 pp.

Historical Change of Large Alluvial Rivers: Western Europe
Edited by G. E. Petts
©1989 John Wiley & Sons Ltd

2

Impact of 18th and 19th century River Training Works: Three Case Studies from Switzerland

DANIEL VISCHER

ETH, Zurich, Switzerland

INTRODUCTION

'Ein Sumpf Zieht am Gebirge hin, Verpestet alles schon Errungene; den faulen Pfuhl auch abzuziehn, das letzte wär das Höchsterrungene, eröffn ich Räume vielen Millionen, nicht sicher zwar, doch tätig—frei zu wohnen!'

(aus *Faust II*)

'A swamp follows the mountain base, poisoning all that has already been achieved; to drain the foul slough, that would be the greatest achievement, thereby I would open up space for many millions, to live—not in safety—but free and working!'

(from the drama *Faustus II*)
Johann Wolfgang von Goethe *ca*. 1830

From the point of view of Swiss hydraulics, the 19th century can be viewed as *the* century of river training works. Nearly all the large river training enterprises and erosion abatement schemes in mountain torrents were carried out, or at least begun, at that time. The reasons for this development were fourfold. First, there were an unusually high number of disastrous floods as a result of climatic deterioration and anthropogenic influences, in particular deforestation, and many of these were depicted by artists at the time (e.g. Figure 1). Secondly, malaria and other diseases attributed to wetlands were major problems. Thirdly, there was an urgent need for land as a result of the population growth brought about by industrialization. Fourthly, there was a change in the political structure after the decline of the old federation due to the Napoleonic occupation around 1800.

This chapter describes the training of the River Kander between 1711 to 1714 and which led to an impressive number of other river training enterprises in the 19th century. Examples of these later works include the training of the River Linth and of the rivers at the foot of the Jura Mountains which are also described. In all three projects, one or more natural lakes were integrated into the schemes as retention basins for peak runoff and sediment (Figure 2).

FIGURE 1. Scene of the flooding of the Rhine valley in Canton St. Gallen, autumn 1868, drawing by E. Rittmeier, 1868

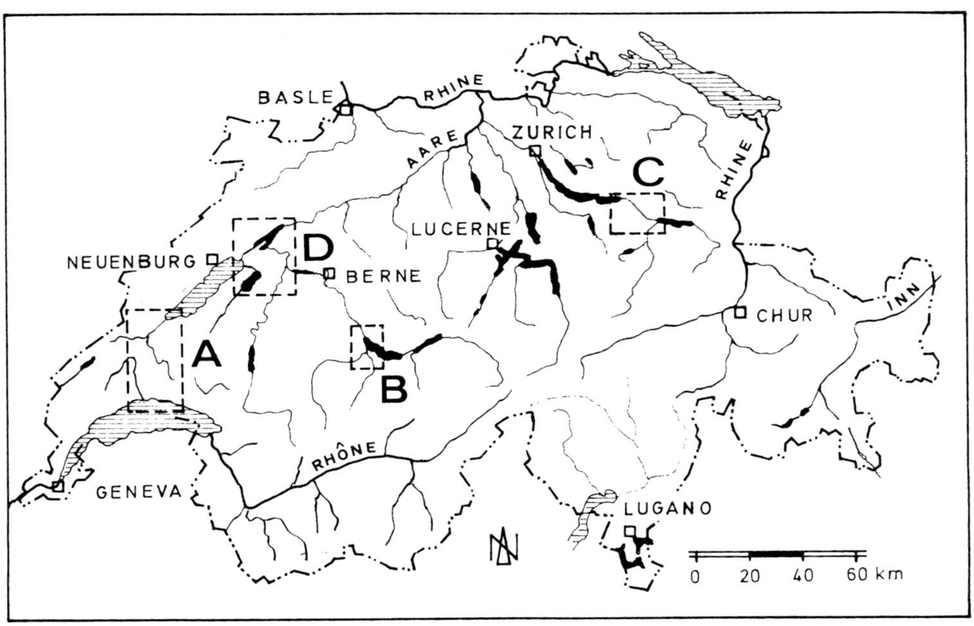

EARLY RIVER WORKS

According to folklore, the earliest river works were undertaken by the monks of Interlaken who diverted the River Lütschine into Lake Brienz making the alluvial fan, the so-called 'Bödeli', available for agriculture. This mountain torrent had originally flowed into the River Aare near Interlaken. Such medieval river training performed the functions of flood abatement and land reclamation simultaneously. Unfortunately, the author was unable to find reliable details concerning this project and its execution. Nevertheless, there is no doubt that the real or imaginary diversion of the Lütschine remained in the minds of the people and even after centuries tempted imitation. It is also referred to in Conrad Escher's advertising pamphlet for the Linth river training published in 1807 (cf. section on Linth).

A second early hydraulic work was the digging of the Canal d'Entreroches between 1638 and 1664 (Figure 3a). However, this canal was not intended as a flood abatement scheme but as a waterway for navigation. It was meant to become a 36 km long connection between Lake Neuenburg and Lake Geneva, crossing the great European water divide between the catchments of the Rivers Rhine and Rhône. The concession was granted to a consortium financed by the Dutch and the citizens of Bern and Geneva. The Dutch interest arose because they were looking for a new link to the Mediterranean ports, which could offer an alternative to the sea-route via Gibraltar past the hostile countries of Belgium and Spain. Supervision of the planning and construction work was undertaken by Elie Gouret-du Plessis, a Huguenot in the service of the Dutch. He also brought hydraulic engineers and carpenters from the Netherlands to build the locks and barges, thereby importing some Dutch hydraulic skills into Switzerland.

The construction work progressed well at the start. The first 16 km as far as Entreroches were completed in only two years; half the waterway followed the small but navigable River Zihl and the other half was excavated in the Orbe plain. The standard profile had a trapezoidal cross-section which was 2.5 m deep, 3 m wide at the base and 5 m wide at the water surface. After this came what was really the summit section. Here the canal had to be partially built up by stones or bricks. Since the climb to the summit only required an altitude difference of 18 m to be overcome, relatively few locks had to be built. However, for the following descent to Lake Geneva 80 m below (following the small River Venoge) 47 locks had to be planned. So the summit and descent sections were technologically complex and extremely expensive. Difficulties delayed building progress. It took eight years to complete the 6 km-long stretch to Cossonay and, around 1648, the consortium ran out of money. In 1664 Bern released it from its obligation to continue work; the plan to construct a waterway of European standing through Switzerland had therefore failed. What remained was a canal of regional importance only. The route from Cossonay to Entreroches was soon taken out of service, but the 16 km long section from Entreroches to Lake Neuenburg remained in operation for 180 years until 1829. It served mainly for the transport of wine and salt from western Switzerland across the Lakes Neuenburg and Biel and on the River Aare to Bern and beyond.

FIGURE 2. (*opposite*) Sketch map of Switzerland. Situation of the project areas: A—Canal d'Entreroches (Figure 3a), B—training of the River Kander (Figure 3b), C—training of the River Linth (Figure 7), and D—training of the Jura waters (Figure 9)

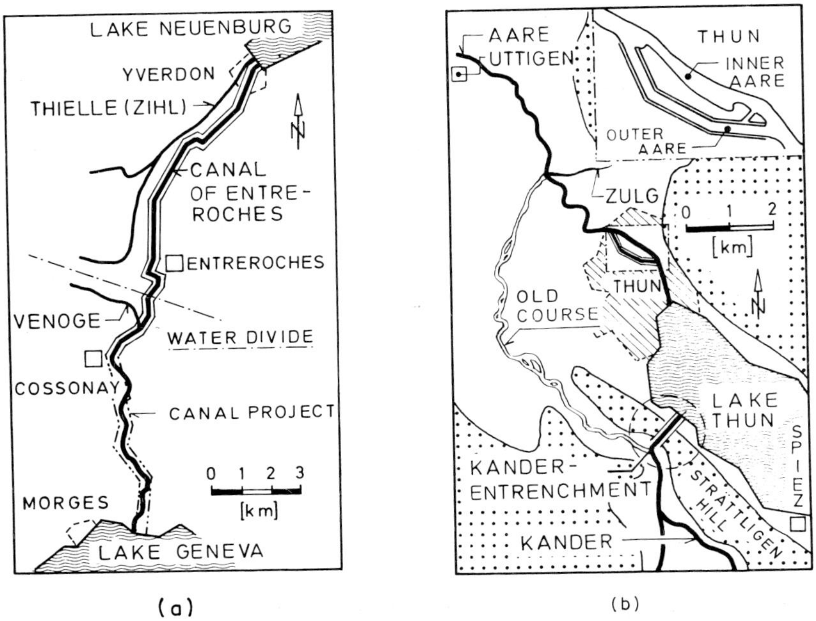

FIGURE 3. Sketch maps showing (a) the Canal d'Entreroches, shipping canal, built 1638–1664 and (b) the training of the River Kander, carried out 1711–1714

THE TRAINING OF THE RIVER KANDER (1711–1714)

The above mentioned early experiences of river works inspired the aristocrats of Berne, the so-called 'Benevolent Masters' who were the first to carry out a major river training effort in Switzerland: the diversion of the River Kander through a channel into Lake Thun in the years 1711 to 1714. This enterprise was, at that time, an extremely bold one, and can be considered to be the forerunner of Swiss river regulation schemes.

The River Kander is a typical mountain torrent which originally flowed past Lake Thun and into the River Aare approximately 4 km downstream from the town of Thun (Figure 3b). The River Zulg also entered the Aare at this point, but from the opposite side. Flood waters regularly inundated the adjoining plains, Thun itself, and the areas around Lake Thun. It was no wonder that the locals approached the Benevolent Masters of Bern. In 1670—here the author follows mainly the descriptions by Bachmann (1983)—a plan was proposed to cut through a range of hills near Strättligen in order to lead the Kander into Lake Thun. At that time, however, the government of Bern had other problems which needed their attention. But when in 1698 the communities were afflicted by still more floods, they pleaded for the Kander to be diverted and even provided labourers and money for the execution of the project; this resulted in the Benevolent Masters organizing a committee to look into the matter. The town of Thun and the other communities along the lake raised opposition immediately but to no avail. In 1700 the investigations of the committee led them to vote in favour of a diversion of the Kander. The decision to go ahead with the work followed in 1711. The responsibility of carrying out the project was given to a newly-formed Kander directorate. The surveyor and artillery lieutenant Samuel Bodmer was to head the planning and direction of the work. He devised a plan to train the River Kander using two interrelated components: (a) an open cut through Strättligen Hill to divert the Kander into Lake Thun and (b) a channelling of the Aare from Thun as far as Uttigen (7 km downstream) in order to increase the discharge from Lake Thun.

The Strättlingen Hill diversion

Only the first component was embarked upon initially. Work started in April 1711. Strättligen Hill was excavated along the projected route, starting from the side of the Kander. Bodmer's plan called for a ditch 950 m long, 85 m wide and up to 80 m deep, and thus required considerable amounts of earth to be moved. The workers were taken to the site daily at 5 a.m. to the accompaniment of a musical band. A working day in those times lasted from 5 a.m. until 7 p.m. Meal breaks were from 7 to 8 a.m. and from 12 noon to 1 p.m.; a working day, therefore, consisted of 12 working hours. Bodmer commanded four categories of personnel:

—12 to 24 foremen and specialized workmen;
—200 to 300 labourers paid at daily rate, and recruited by him for the period from May until October;
—50 to 80 'Ehrtauwner', labourers provided and paid for per day by the communities; and
—60 to 100 poor people, who came voluntarily—beggars, tramps, homeless people and convicts, who were sent by the communities (amongst these were many women and children). With the exception of the convicts, they received a small wage.

It soon became obvious that the problems of moving so much earth had been underestimated. As mentioned before, the planned depth of the trench was 80 m. This required excavation capacities and slope stabilization techniques which were difficult to achieve with the manual methods employed at that time. Therefore, in December 1711, the architect Samuel Jenner, a member of the Kander Committee, suggested a modification to the project; the diversion of the Kander was not to be done entirely by means of an open canal, but should, in its central part, contain a tunnel. However, the second Villmergen War began in early 1712. This civil war between Protestants and Catholics interrupted the work. Bodmer had to lay off a large part of his work force and turn his efforts to fielding fortifications on the Brünig pass. Only in August 1712 could the project be continued along the lines of Jenner's ideas.

Jenner had the tunnel excavated from both the Kander and the lake sides. The profile had a width of 12 m and a height of 4.5 m. These were quite considerable dimensions considering the properties of the loose alluvial and morainic material. It was necessary to proceed with partial excavation followed by the installation of log supports. Despite the geological problems, in December 1713, some of the Kander's water was allowed to follow this new course. This section was then continually enlarged until a development occurred which took the control of the river totally out of the hands of the Kander Directorate.

Along its old bed the River Kander had a slope of 0.5 per cent, whereas the slope in the several hundred metres long tunnel was 7 per cent; therefore, the Kander developed a considerable shear stress on the tunnel bed and walls. The insufficiently-covered tunnel surfaces consequently eroded and widened until, in mid-1714, all the water passed through the tunnel. Continuing erosion finally led to the collapse of the tunnel's arched roof and the slumping of the overlying material. A land slip in July 1714 killed two gentlemen from Bern; they were dragged down and washed away right before the eyes of a group of visitors. The magistrate Emanuel Gross from Thun demanded the Kander be immediately redirected into its old bed. He was too late; the Kander widened the tunnel quickly and unceasingly, forming a gorge whose bed was 25 m lower than planned and so beneath the original bed of the Kander. Inevitably, the eroded material was carried into Lake Thun where it formed a delta larger than half a square kilometre in only two years.

The Aare channelization

By 1715 only the first component of Bodmer's plan had been realized—albeit with nature's own modifications—but the first disadvantages soon became apparent. The diversion of the Kander had, on the one hand, indeed protected the people living adjacent to the River Aare, downstream from Thun, and those along the former section of the River Kander from further catastrophic flooding. On the other hand, Lake Thun now overflowed its shores more frequently, precisely as the people of Thun and those living along Lake Thun had predicted and feared. As a result of this the Benevolent Masters of Bern quickly decided, in October 1716, that the Aare had to have a better outflow from the lake. This, in effect, meant the realization of the second component of Bodmer's scheme. Since this project also set the stage for later Swiss river training works and especially for lake regulation, it is worthy of a brief description here.

At the time of the Kander diversion, the River Aare in the town of Thun consisted of (a) the still-existing Inner Aare and (b) a town moat which branched off the Inner Aare on its south side and re-entered the main river further downstream. Both conduits had weirs at their upper ends which served some mills and which guaranteed a minimum water level for the boats in the lake. At low water levels, water flowed only over the weir on the Inner Aare so that only this branch contained running water. Only at higher water levels did some water enter and flow through the town moat. The total outflow capacity was nevertheless too small at times of high water when the lower-lying houses of Thun were flooded up to the first floor. The situation persuaded the government to first buy the mills at a high price (Nussbaum, 1925) and then have the weirs removed. The result was to accelerate the flow of the Aare through the town. This led to bank erosion and the exposure of bridge foundations; after a short time a bridge and some houses were washed away by the Aare. But the real goal—the prevention of further flooding—was not achieved and further action was initiated based on a project approved in 1720 and drawn up by the magistrate Emanual Gross (Neumann, 1979). First, the town moat was widened and deepened; it was called the 'New' or 'Outer' Aare. This channel can be considered the forerunner of the flood relief channels which have become fashionable again today (an example being the one on the Danube in Vienna). Secondly, the Aare bed was straightened from Thun to Uttigen. Thirdly, the ledges on both sides of the river were replaced by regulating weirs. These were overflowed during high water so that the lake level could be kept low enough to avoid damage. At other times, that is during average or low water levels, they were used to maintain the desirable lake surface level.

In the Outer Aare, an approximately 60 m long weir was fitted by a master carpenter in 1726. He made 10 wooden sliding gates of 4.7 to 5.7 m width, two of which served for

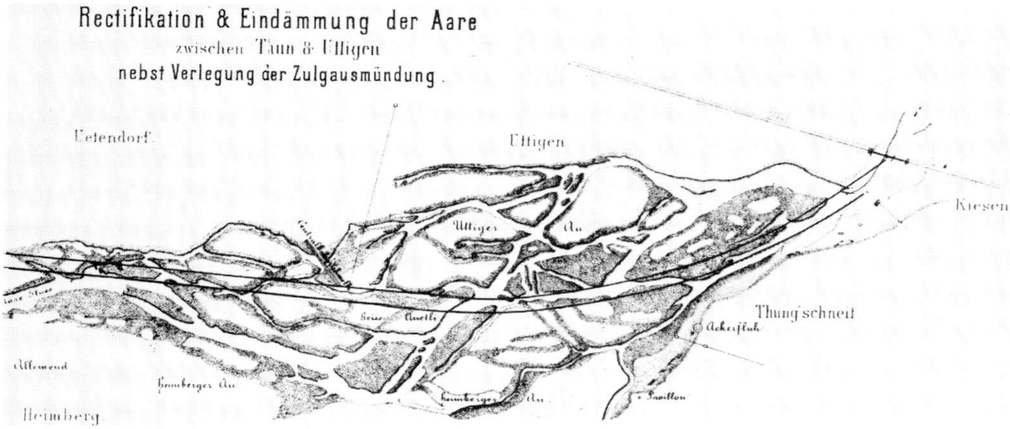

FIGURE 4. Sketch map showing the training of the River Aare following the training of the Kander. Plan from around 1875

the passage of boats. At about the same time the weir in the Inner Aare was fitted with five sliding gates and also a wooden pinion drive (Neumann, 1979). Later, both weirs were restored—the one in the Inner Aare already in 1788 and the one in the Outer Aare in 1818—using plans drawn up by the successful Linth-correction team of Gottfried Tulla and Hans Conrad Escher (cf. the following section on the River Linth).

The scheme which had started with the training of the River Kander could finally be considered completed once the training of the Aare from Thun to Uttigen had also been accomplished (Figure 4) in the years 1782 to 1792 and 1871 to 1873 (Schweizerisches Ober-Bauinspektorat, 1916).

THE TRAINING OF THE RIVER LINTH (1807–1816)

The increasing number of floods in the 18th century, especially the ones in the years 1762 and 1784, caused great problems for the population living along the River Linth and the Walensee (lake, cf. Figure 5). These problems came at a time when other difficulties were already abundant. A poignant picture of this bleak situation is drawn by Becker (1910):

> 'In addition to the devastating effects of the floods in the Linth Valley, which caused damages of millions of francs, an industrial decline took place, resulting in a long period of no income for a population who had just become accustomed to their new way of life as regular wage-earners in the factories. At the same time the potato crops were stricken by disease. What little was left got eaten by the hungry bands of foreign troops who passed through the country in the years 1798 and 1799. The misery was so great that in some communities the children were led to pastures to feed on grasses and roots. They were also taken away in large groups to other regions of Switzerland and even to Swabia; from the Grosstal alone (i.e., the section of the Linth Valley up-river from Schwanden) more than 500 children were led away . . . And even more terrible was the situation in the lowlands. Here the ground was already devastated, the air was

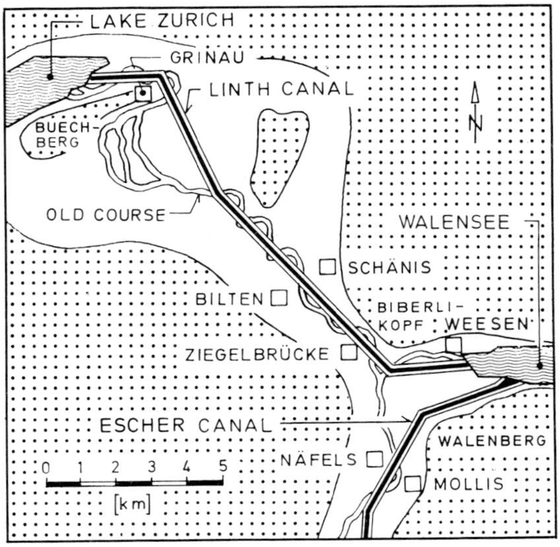

FIGURE 5. Sketch map showing the training of the Linth, carried out between 1807–1816

polluted—whereas at least the air of the mountain valleys was clean—and everything was decaying. One large swamp stretched from Näfels and Oberurnen to the Walensee and to Lake Zürich. The "cold fever" as it was known (malaria in its worst form) became established in the area, ruining people's health and sometimes taking their lives. The inhabitants of the lowlands died a decade earlier than their contemporaries living in higher regions. The whole of the wide valley floor, formerly a pleasant area of meadows, corn and potato fields, was abandoned and became unproductive and uncultivated. The cold fever wrought terrible consequences. A population of more than 16 000 faced physical and moral decline. A greater tragedy had never been witnessed in Switzerland before, the misery lasting not only for years, but for generations.'

It was little wonder that this misery became widely known. In 1738 the 'Landsgemeinde' of Glarus (an annual assembly of all voters in the canton) started to worry about how to drain the swamps. However, little more than minor repair work was initiated or carried out. Later, more extreme measures were called for and finally a bold plan took shape which was mentioned for the first time in the chronicle of 1774 by a local priest:

'One has often talked about suggestions to lower the level of the Walensee and to save the lower areas of the country from swamps. A perceptive mathematician is said to have found a way of diverting the River Linth along the Walenberg to the lake and to give the lake sufficient outflow through the River Maag and several channels. However, such a plan involves so many difficulties and obstacles that we consider it to be but a dream.'

Who might this mathematician have been, whose plan would be put into reality a whole generation later? But the plan remained a dream for the time being. The misery continued and in 1783 the Swiss Diet, the coordinating body of the somewhat loose Swiss Federation, ordered the Bernese Captain Andreas Lanz to draw up a project. Within a year of starting, Lanz produced four different schemes, although he considered only one of these to be a permanent solution to the problems. It called for diverting the Linth into the Walensee. However, the projected costs were so high that his employers were discouraged. In 1792 and 1793 the Helvetic Society took the matter up, and found in the then 26-year-old Hans Conrad Escher from Zürich an extremely attentive and energetic member. Escher subsequently travelled several times through the area of the River Linth and the Walensee and reported his findings in 1796. He fully supported Lanz's plan and gave a lecture on this topic at the last meeting of the Swiss Diet in pre-Napoleonic Switzerland. But soon after, this system of government foundered; the help so desperately needed by the people living near the Linth failed to appear, despite the many stirring, moving letters of complaint which were written. However, at least private donations were directed into the stricken area. The author found, for example, a document dated February 1800, written by a deacon, which had the following title: 'Inventory and account of the voluntary taxes and contributions of noble Swiss men and women for the support of the suffering people in Canton Linth'. The donations, in particular some from many Bernese communities, are recorded not only in units of currency, but also in units of cereals, dried fruit, dried potatoes and in pounds of bacon, cheese and so on.

Only with the new constitution of 1803, the so-called 'Mediationsverfassung', did the necessary peace return to the Swiss Federation, and with it a minimum amount of freedom to act independently. In view of the continuing misery, the Swiss Diet of 1803 set up the first of several committees; this recommended the execution of Lanz's plan, and worked out a second plan in 1804 under Escher's leadership, involving a detailed project with estimated costs and planned financing. In 1805 the decision was made to begin work and the Linth Supervisory Committee was installed with Escher as its president and with the Glarus town councillor Conrad Schindler and a Bernese architect as members, Captain Lanz having died in 1803. But then, public attention was drawn to several unrelated occurrences: in 1805 the international situation made it necessary to station troops on the borders, and in 1806 central Switzerland was struck by catastrophic floods and the landslide at Goldau, the latter causing 400 deaths. The people were affected by these events to such an extent that the commencement of work on the Linth had to be postponed. However, in 1807 those responsible finally became active, foremost among whom was the Swiss President, Reinhard from Zürich.

Since the Swiss Diet had practically no money available, the financing of the scheme was accomplished with the issue of shares and so the first federal shareholding company was founded. In order to advertise the project, Escher and a Bernese dean composed a pamphlet under the title of 'Appeal to the Swiss nation to save the population of the shores of the Walensee and of the Lower Linth Valley who are suffering greatly because of the swamps'. It was published in early 1807. According to Thuerer (1966) it is one of the most remarkable documents in Swiss history combining:

> 'clarity of expertise with an appeal to the heart. As examples it calls to mind the useful practical experience gained in the Bernese Oberland (diversion of the

River Lütschine into Lake Brienz, and later that of the River Kander into Lake Thun), preventing the areas of Interlaken and Thun from becoming swampy, a situation which had now befallen the inhabitants of the Linth area.'

At the end of the advertising pamphlet its authors addressed their fellow citizens with these words:

> 'But remember, help is still available! It lies in your hearts! Let us hurry to help them. Let their sunken ground be freed of those miserable bogs—let their poisoned air be cleansed so that their descendants may be saved from the danger of a slow but unavoidable decline—this is the good deed which they request, and which you cannot refuse them.'

The effect of this advertisement surpassed all expectations; within a few months some 2000 shares—instead of the expected 1600—were subscribed to by cantons, corporations,

FIGURE 6. Johann Gottfried Tulla (1770–1828)

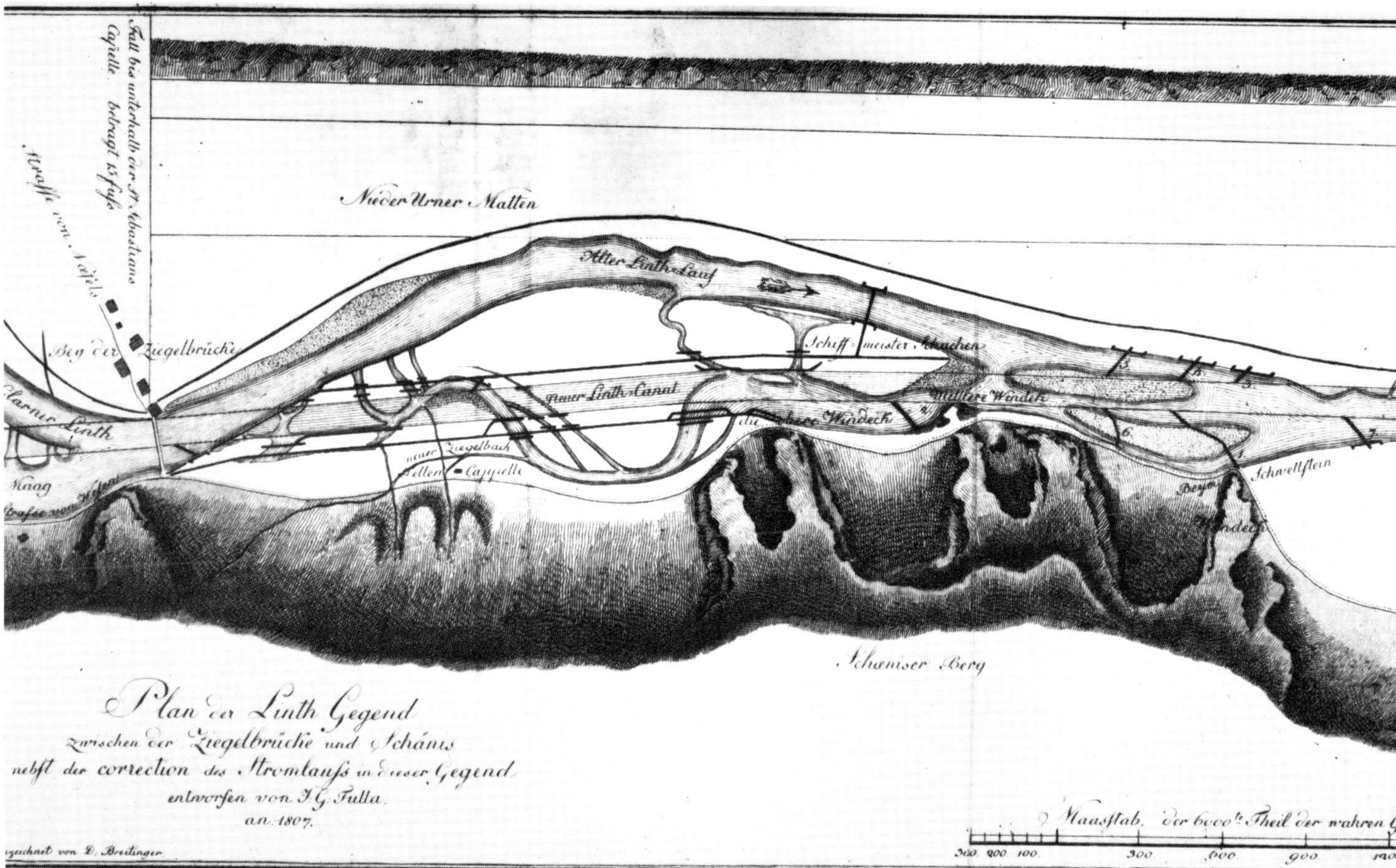

FIGURE 7. Maag–Linth Canal below Ziegelbrucke. Plan by J. G. Tulla, 1807

communities and private people. But where did the technical knowledge for the Linth works come from? Rather surprisingly, the Swiss Diet appointed as 'hydrotect' (project and construction leader) the 'Rheinwuhrinspektor' of Baden, the then Captain, Johann Gottfried Tulla (Figure 6). He was twice granted leave of absence by his sovereign, the Grand Duke of Baden, in order to be able to demonstrate his skill and experience; this happened first from September to November 1807 and then for a few weeks in 1808. During this short time he carried out the hydraulic calculations and the planning (Figure 7) of the main river training sections, based upon geodetic surveys which had been completed before the hydrological measurements made specially for this particular purpose. A contemporary of his, the engineer Colonel Heinrich Pestalozzi (1852), and the subsequent Linth engineer Gottlieb Heinrich Legler (1868), described these efforts in detail confirming that Tulla indeed possessed a great mastery of both theoretical and practical river hydraulics. He received excellent support from his collaborator, the engineer Obrecht, who contributed to the Linth works in 1807 and 1808, each time for five months. From today's point of view it is particularly astonishing how much Tulla already knew at that time about sediment transport in rivers and, correspondingly, how well he calculated the dimensions of the Mollis Canal, later renamed the Escher Canal. It is of no surprise that in 1812 the Swiss Diet sent Tulla (who had advanced to the rank of Major in the meantime), a congratulatory letter from the Swiss President, together with the following section of their minutes (Zier, 1970):

> 'His Excellency the Swiss President has been authorized to declare, by means of a letter written in the name of the Swiss Diet, the compliments, satisfaction and respect of the representatives of the Swiss Federation to the Grand Duke of Baden's engineer—Major Tulla—and through him also to engineer Obrecht; it is a pleasure for them to express their gratitude after having read the carefully written reports of the Linth Supervisory Committee about the great, important service rendered by these distinguished gentlemen in connection with the Linth project.'

Since Tulla was only available for such brief periods, Escher—as president of the Linth Supervisory Committee—was forced to step into his shoes. In addition to his large work load, he had to learn hydraulic techniques. In fact, Escher had never been involved with hydraulics before, but his broad scientific knowledge coupled with a good perception enabled him to become thoroughly trained by Tulla and Obrecht within a few months. The reports contain a letter dated February 1808 from Tulla to Escher in which Escher's success is acknowledged (Pestalozzi, 1852). This speaks for both teacher and student. As from 1808, Escher conducted all the project planning; he shared the duty of construction leader with his colleague, Conrad Schindler, a member of the Supervisory Committee.

From 1807 onwards work progressed briskly at the Linth works. First the channel which was to divert the Linth into the Walensee was dug. This Mollis Canal, as it was called, was 4 km long and had a slope of 3‰. Its cross-section was a so-called double profile consisting of a trapezoidal central channel followed by a levee on each side at a certain distance from the central channel. It is remarkable that the discharge used to calculate the channel dimensions was determined in the natural river bed by means of measurements made by a Woltman propeller. It was the first time that this method was employed in Switzerland. A depth of 2.5 m, a bottom width of 17 m and lateral slope inclinations of

1:1 made for a discharge cross-section smaller than the one up-river. These dimensions compensated for the smaller slope in the canal and (based on Tulla's suggestion mentioned earlier) maintained a sufficiently high shear stress to ensure that the bedload would still be transported all the way to the Walensee.

For the lateral dams, heights of 2.4 and 2.1 m were chosen so that they would provide protection against even the largest flood discharges, such as those which had been observed in the 1760s. The right bank dam was made 0.3 m lower than the left bank one so that it would be overtopped first, and only the right hand, narrow strip of land would be affected. Digging the central channel, building up the levees and protecting the banks—including stone masonry on the sides of the central channel and turfing the dam slopes—was all done by hand. The only supplementary source of energy was the power of draught animals.

The construction of the Mollis Canal represented, however, only the half-way stage to the completion of the Linth works. In addition the outflow of water from the Walensee needed to be improved and the lake surface had to be lowered by about 2 m. For this purpose the old outflow, the small River Maag, and the lower part of the River Linth were replaced by the 15 km long Linth Canal. This canal, having a slope of only 1‰, had at first similar dimensions to the Mollis Canal but considerable bed and bank erosion was expected to take place and this indeed subsequently occurred during the construction period.

Entrenchment was very difficult in certain sections because the canal bed was as much as 5 m lower than the previous bed of the Linth, and depths well below the ground water table of the swampy Linth Plain were reached. Chronicles record the tediousness of work in narrow one-man pits supported by wooden planks in order to dig up loose material and the difficulty of underwater blasting of certain molasse bars using gunpowder. However, the use of fascines as a construction method (Figure 8)—introduced by Tulla and his expert assistants—was a great technological innovation; it allowed the old courses of the Rivers Maag and Linth to be cut off and their waters to be forced into the new conduit. The only 'machine' described in the chronicles is a boring rudder. This was a big iron-coated rudder which was lowered from a boat. It was used to enhance the abovementioned erosion of the channel by causing turbulence and thereby dredging sections rich in clay or other resistant materials.

Particular difficulties were also encountered in connection with the lateral embankments which had been built on top of peat. Considerable settling occurred and this called for repeated rebuilding. Naturally, the Linth discharge had to be guaranteed throughout the construction period, be it through the old river bed or through the newly completed canal sections. In addition, the flourishing trade of timber transportation by water and river navigation had to be maintained.

The actual diversion of the Linth from Mollis to the Walensee took place on May 8, 1811, and the Linth Canal between Weesen and Grinau was opened on April 17, 1816. Now the Linth works were essentially completed and the first positive effects of the enterprise soon became evident. Flooding still occurred in 1817 though, and made some more work necessary. However, the floods also were useful in that the new riverbeds were washed out and widened. Surprisingly, the share holding venture was wound up in 1827; the shares could be cashed in until 1845, less a donation equivalent to roughly 10 per cent of the total costs. Escher did not live to see this final stage of the project; he died on March 9, 1823 at the age of 56. He had become a greatly admired saviour of a region and its thankful people. The Swiss Diet had shown their gratitude several times whilst Escher

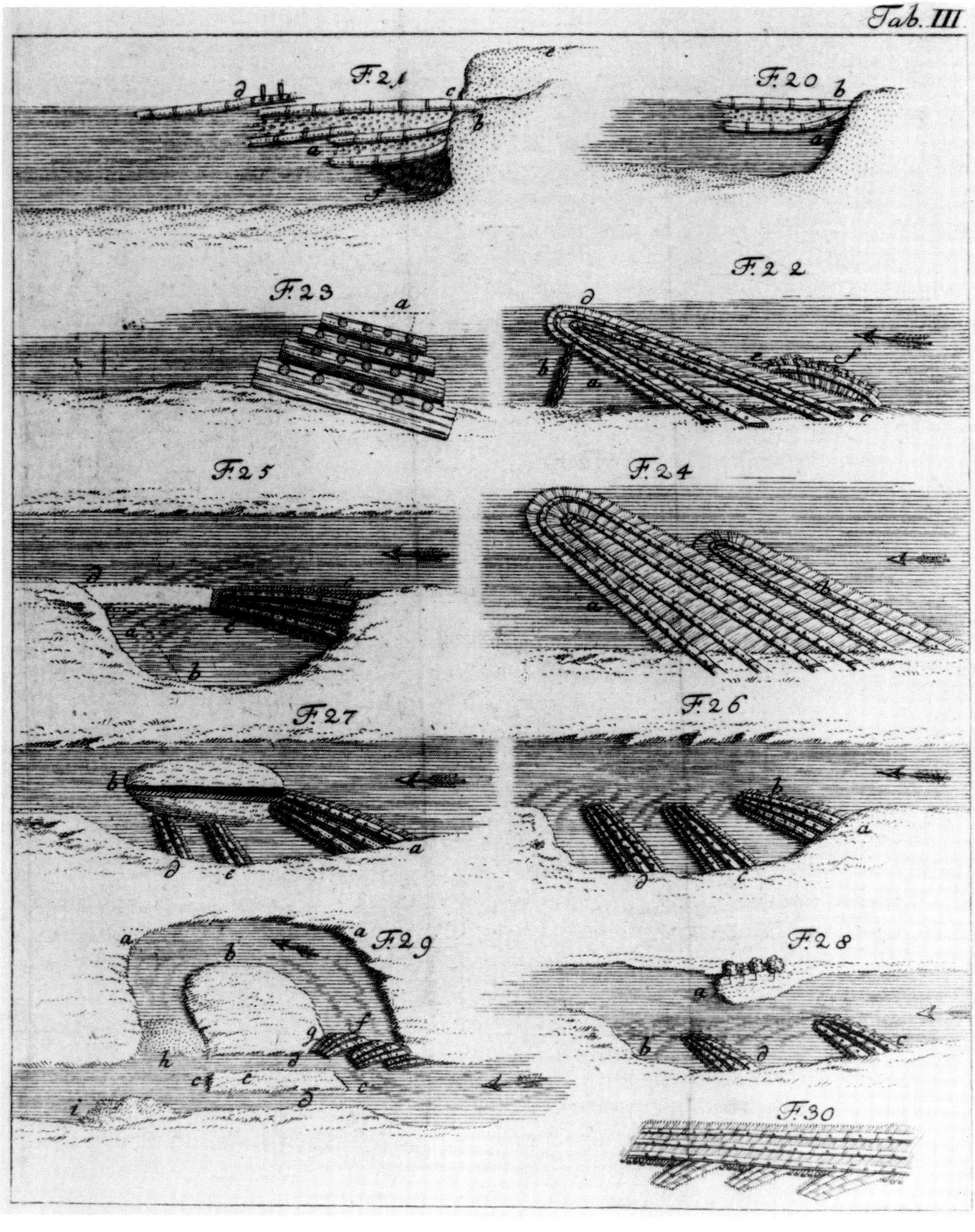

FIGURE 8. Use of fascines to make spurs and groynes in connection with river training works, taken from Silberschlag, 1766

was still alive, but now in June 1823, it decided to bestow on him and his male descendants the honourable title 'von der Linth'. In addition, the Mollis Canal—the true central piece of the project—was renamed Escher Canal. Lastly, it was decided to erect a monument near Ziegelbrücke. This came out simpler than planned but still is impressive enough (Becker, 1910). It consists of a memorial plate mounted on a molasse rock, displaying the inscription:

MD CCC XXXII	MD CCC XXXII
Dem Wohltäter dieser Gegend Johann Conrad Escher von der Linth, geb. den 24. August 1767, gest. den 9. März 1823,	For the saviour of this region Johann Conrad Escher von der Linth, born August 24, 1767, died March 9, 1823,
Die Eidgenössische Tagsatzung,	The Swiss Diet,
Ihm danken die Bewohner Gesundheit, Der Fluss den geordneten Lauf, Natur und Vaterland hoben sein Gemüt, Eidgenossen! Euch sei er ein Vorbild!	The inhabitants owe to him their health, The river its orderly course, Nature and fatherland lifted his spirit, Citizens! May he be an example to you all!

As previously mentioned only the most essential sections of the Linth works had been completed by 1816; modifications and refinements were needed after Escher's death. The Escher Canal had to be lengthened through the delta which the Linth had formed in the Walensee. Likewise the Linth Canal downstream from Grinau had to be lengthened as far as Lake Zürich. Becker (1910) reports:

> 'Especially the backwaters now had to be regulated. They needed another discharge channel since the dams along the Linth Canal prevented the backwaters discharging into the main channel. To conduct this work and to administer all Linth affairs, a permanent commission was installed, the so-called Linth Police Commission (Linthpolizeikommission). It consisted of a federal representative, one representative from each of the Cantons Schwyz, Glarus, St. Gallen and Zürich, and a Linth engineer. The first Linth engineer, from 1827 to 1863, was Colonel La Nicca from Canton Graubünden. He was a hydraulic and traffic engineer, who well deserved the respect shown to him throughout the whole of Switzerland. The man appointed as his assistant in 1844, and who was to become his successor in 1863, was Gottlieb Heinrich Legler from Canton Glarus, who carried out his responsibilities with true devotion until his own death in 1895.'

We are indebted to Legler for his excellent descriptions of the Linth works and many other river training projects in Switzerland.

THE FIRST CORRECTION OF THE JURA WATERS (1868–1891)

After leaving the hilly countryside at Aarberg (Figure 9), the original course of the River Aare—nowadays called the old Aare—followed the edge of the great plain, the so-called Grosses Moos, between Lake Murten, Lake Neuenburg and Lake Biel. After this it meandered its way through the wide, flat area between Büren and Solothurn. The low slope of this almost 30 km-long stretch caused bedload sedimentation and narrowing of the riverbed in several places. This situation obstructed high-water discharge and led to frequent flooding. Based on old chronicles it appears that the frequency of these adverse events increased around the middle of the 16th century. At this time, the settlements in the marginal zones suffered increasingly from flood damage; the sown fields and crops were repeatedly destroyed in the cultivated plains and the ground gradually turned into a swamp. Accordingly, the population's cries for help became louder and louder until, starting in 1704, the Bernese government—the author follows here mainly the reports by Stambach (1970) and Ehrsam (1974)—contracted in succession a total of a dozen experts to devise a correction plan.

One of these was Johann Gottfried Tulla. His basic assumption was that the source of all the problems lay in the bedload. He gave a clear assessment of the hydraulic situation and suggested a river correction in the area of the confluence of the River Aare with the River Zihl which drained Lake Biel. Both river courses were to be shifted and straightened in such a way that their confluence would be 10 km further downstream. Tulla also considered an even more comprehensive correction of the Aare as far as the entrance of the River Emme downstream from Solothurn, including the removal of the delta of the Emme which was an obstacle to the Aare's discharge. In addition, he considered a correction of the section of the River Broye leading from Lake Murten to Lake Neuenburg and the section of the River Zihl leading from Lake Neuenburg to Lake Biel. Furthermore, he proposed a diversion of the Aare into Lake Neuenburg or Lake Biel as an interesting alternative.

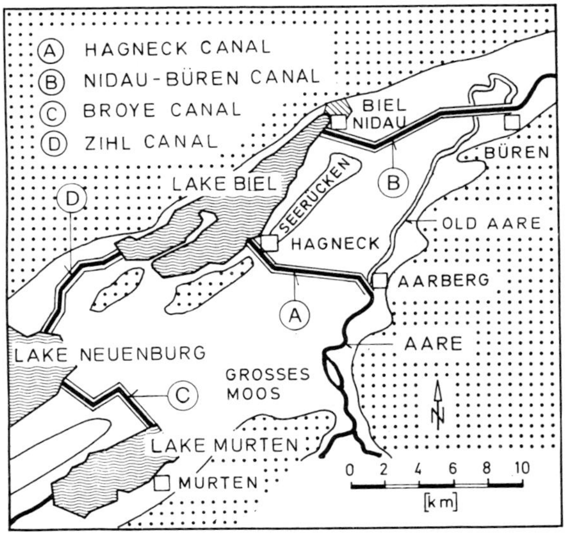

FIGURE 9. Sketch map showing the correction of the Jura waters, carried out 1868–1891

For the time being the projects remained in plan only until further floods underlined their urgency. The population's experiences during one of the flood catastrophies were described in a moving way by Johann Rudolf Schneider, a doctor who had grown up in Meienried:

> 'Indeed a sad and terrible view, so many thousands of acres of fertile land and all their produce now under water! Misery is immense. Lost, totally lost are the fruits of this diligent population's untiring energy. The three large Lakes Murten, Neuenburg, and Biel seem to form *one* large water basin. The houses in Landeron and Nidau stand out like islands in the middle of it. A traveller told me today that even the upper River Broye and the River Orbe (leading into Lake Murten and Lake Neuenburg) had left their beds, and that the plains of Chablais, of Orben and of Iferten (above Lake Neuenburg) were also under water. The view yesterday must have been terrible when the plain between Jensberg and Solothurn was also flooded by the Aare, which, as you can see now, has already retreated considerably. The Aare rose with unexpected swiftness the night before, carrying away the weirs and dams—which had been constructed with so much effort over large distances. Its destructive effect was greatest from Kappelen (near Aarberg) as far downstream as Meienried (Schneider's home near Büren). When I woke up yesterday morning the waves were already lapping at the door of my house; near Andern the water entered the houses, even reaching the rooms where the people were still asleep. The towns of Schwadernau, Scheuren, Meienried, Reiben, Staad and Altreu (villages near Büren) were also under water. Near Meienried the Aare rose 21 feet 8 inches above its lowest-ever level. Most of our meadows had been mown, but because of the continuous rain nothing could be collected. Not only the hay fell prey to the floods, but also a father and a child were led to early graves. Our wheat fields are covered by mud, sand and gravel. In a few days—especially if, as it seems, hot weather should set in—we shall not have a single healthy ear of corn left. The potato crop has been lost completely, the villages have been filled with washed-up debris and the dwellings have become a refuge for all sorts of vermin . . .'

The floods were followed on each occasion by epidemics and mortality was high in the areas around the lakes. Catastrophic floods of 1831 and 1832 led to a meeting of a group of men who formed a committee under the leadership of Schneider. This committee studied the possibilities of fighting the Aare and it called for action by means of newspaper articles, pamphlets, brochures and a public meeting. Its success was marked by the fact that it was given the status of a 'preparatory association' by a decree of the Cantons Bern, Freiburg, Waadt and Neuenburg. In 1840 it entrusted the chief engineer of Canton Graubünden, and the already-mentioned first Linth engineer Richard La Nicca with the preparation of a solution.

In March 1842 La Nicca presented his first project. It contained the following suggestions for corrections:

—Diverting the Aare from Aarberg into Lake Biel;
—Conducting the combined waters of the Aare and the Zihl from Lake Biel to Solothurn;

—Correcting the Broye section leading from Lake Murten to Lake Neuenburg and the Zihl section leading from Lake Neuenburg to Lake Biel; and
—Draining the swamps of the Grosses Moos and of the adjacent areas.

Most radical was the proposal to divert the Aare from Aarberg through the Grosses Moos using an artificial cut in the Seerücken at Hagneck into Lake Biel, and initially a great many people opposed it including experienced technicians. Many were unable to understand how it could be possible to lower the level of Lake Biel by diverting the Aare into it, thereby adding even more water. To understand the sceptics one needs to recall the initially negative outcome of the diversion of the River Kander, and that the second correction of the River Aare (which cured these problems) from Thun to Uttigen was not completed until 1873. The only positive precedent at their disposal was the largely completed correction of the River Linth. The preparatory association therefore had to make large efforts to explain the project so that the population came to understand that at times of flooding much less water needed to flow out of Lake Biel, and that simultaneously all of the Aare's bedload could be deposited in this lake. A total of 26 years had to pass before the first shovelful of earth could be turned and construction work begun! Mainly to blame for this delay was the political situation at the time (e.g. the confusion of a civil war, the 'Sonderbundkrieg') and also the lack of a legal foundation. Considerable delay was also caused by the fact that the correction involved five cantons; negotiations between these were frequently most tedious. On top of all this came difficulties of financing the project, recurring doubts about its feasibility, paltry hesitations, obstinacy and petty jealousy. The battle for and against La Nicca's project was fought bitterly.

Only the federal constitution of 1848 brought the legal foundation for a realization of the project. Article 21 states: 'The federal government has the right to execute public works at the expense of the confederation, also to support such execution and even to assert the right to expropriate for this purpose'. But now more and more competing projects emerged. This caused the Federal Council to have a series of technical evaluations done. However, in December 1863 the Federal Assembly declared that the correction of the Jura waters according to La Nicca's plan fell under Article 21, and it granted the project a considerable amount of financial support from the federation.

Spring 1865 again saw extreme flooding which affected practically all the areas around the lakes, joining Lake Murten, Neuenburg and Biel into one large water body. With this last disaster in their minds, all five cantons decided in July 1867 to proceed with the correction works. The federal government raised its contribution somewhat, with the condition that the correction be conducted according to a modified version of La Nicca's plan, produced by the experts appointed by the Federal Assembly, and that the following works had to be included:

—Diversion of the Aare from Aarberg into Lake Biel through the Hagneck Canal;
—Draining the combined waters of the Aare and Zihl from Lake Biel through the Nidau–Büren Canal to Büren;
—Correction of the upper Zihl between Lake Neuenburg and Lake Biel;
—Correction of the lower Broye between Lake Murten and Lake Neuenburg; and
—Execution of correction works along the 30 km stretch downstream from Büren to the entrance of the Emme, as far as they should prove necessary.

The plan to drain the swamps of the Grosses Moos and adjacent areas as suggested in La Nicca's first project had been abandoned for financial reasons. The Nidau–Büren Canal

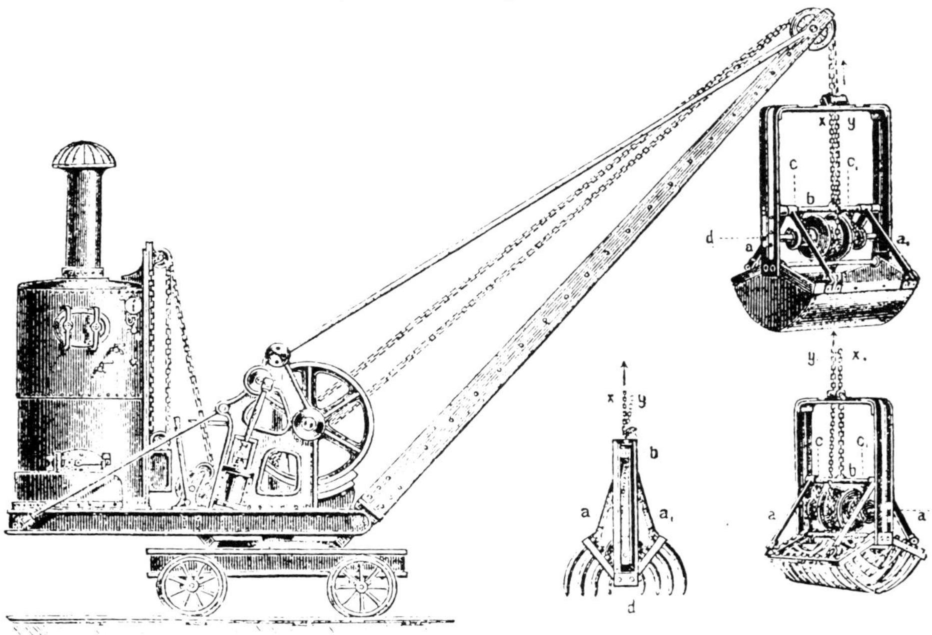

FIGURE 10. Steam dredger on rails, taken from Franzius, 1890

was to be completed in seven years and the Hagneck Canal in ten. The correction of the Aare between Büren and the mouth of the Emme as well as those of the lower Broye and the upper Zihl were to be done within the three years following the completion of the Nidau–Büren Canal and lowering of the average level of the Lake Biel.

The conditions put forward by the federal government were accepted by the cantons and execution of the correction works commenced. Each had to create its own organization since they could not agree on a joint directorate in charge of construction work. Canton Bern was responsible for diverting the Aare from Aarberg into Lake Biel through the Hagneck Canal and the carrying of the combined Aare and Zihl waters through the Nidau–Büren Canal to Büren. Engineer Gustav Bridel from Biel was appointed chief engineer; later, he was succeeded by Karl von Graffenried from Bern. State (that is cantonal) supervision was the duty of the so-called 'Entsumpfungsdirektion' (swamp drainage directorate). The engineers Fraisse from Lausanne and La Nicca acted as federal inspectors. The construction programmes were formulated, plans were published, land acquisition was organized and the stock was produced. This last consisted of two steam dredgers (Figure 10), two steam cranes, 24 transport boats, 122 tipping boxes, 60 goods wagons, two small steam locomotives and 4 km of rail with the necessary sleepers. The town of Nidau provided the old town hall as a base for the work's management and a telegraph office was installed. A large repair workshop was also set up in Nidau in which, later on, several iron bridges were constructed. In addition the enterprise had to ensure the supply of about 300 000 m^3 of limestone for the protection of river banks and the supply of energy necessary for construction work. The dredgers, cranes and the repair workshop consumed roughly

2200 kg of coal every month. When it became impossible to obtain a regular supply of coal after the outbreak of the German–French war and especially after the occupation of the Saarland by German troops, work had to be interrupted for several weeks in the months of February and April 1871. Coal prices rose more than 70 per cent at that time.

In order to avoid a repetition of the already-mentioned negative aspects of the diversion of the River Kander, it was necessary to complete most of the Nidau–Büren Canal (which was to drain Lake Biel) before the Aare could be diverted into Lake Biel. The first sod was turned in August 1868. When the canal was almost finished, the discharge from Lake Biel increased so rapidly that the lake level dropped almost below that foreseen by La Nicca. This caused certain problems which are described later.

To divert the River Aare, a small canal of only 5 to 10 m width was dug at first from Aarberg to Hagneck, with the intention of letting the Aare's waters scour a wider course. In order to stop these waters from entering the new canal prematurely, the end of the canal in Aarberg was deliberately kept closed. Accordingly the Aare maintained its old course for the time being, as far as Büren. The opening of the lead canal was planned for August 17, 1878, but during a flood on August 16, it overflowed the barrier and some of the Aare flowed into Lake Biel. At first, the water which flowed at a rate of approximately 100 m^3 per second from the Aare into the new canal did little erosion. The desired erosional process was considerably slower than had been hoped for. In 1882 only about 40 per cent of the Aare's water entered Lake Biel via the Hagneck Canal. It took until 1887 for the diversion to be complete and the canal to reach its planned dimensions. Nevertheless, by then almost 2 million m^3 of gravel and earth had been transported into Lake Biel. Unfortunately erosion in the Hagneck Canal progressed even further, necessitating some considerable construction work in the cut through the Hagneck hill. This was carried out in 1888. But finally, in 1891, both the Hagneck and the Nidau–Büren Canals were inaugurated.

As mentioned above, the opening of the Nidau–Büren Canal eventually led to a considerable lowering of the level of Lake Biel. This effect was only partially compensated by the additional inflow of the Aare water. The exceptionally low lake level had adverse effects on the stability of its shores; together with other factors, it led to a slumping of the shores and land slides at several places. Damage had to be repaid by the enterprise. A weir was installed in the Nidau–Büren Canal downriver from Nidau as counter measure. In addition, the building of a lock at Nidau was considered at the time. At high water, however, the lowering of Lake Biel corresponded to all expectations and thus the danger of the lake overflowing had been removed. The lowering also exposed a large area of new land along the shore which could be sold to interested parties. This was a much needed source of income for the enterprise. Similar sales of shore terrain also took place elsewhere, especially in the neighbouring cantons where the lake lowering produced similar positive effects.

Although the federal government's ruling of July 15, 1867 permitted further correction work only after the completion of the Nidau–Büren Canal, the Bernese government had already asked the Cantons Freiburg, Waadt and Neuenburg in October 1872 to start work on the correction of the upper River Zihl. The reasoning was that Lake Biel had been lowered enough so that no further flooding occurred in the area of the upper Zihl. The invited cantons agreed upon a common directorate to lead the construction work and chose Henri Ladame from Neuenburg and, later on, Borel to take on the responsibilities as chief engineers. As in the case of the Bernese correction works, the federal technical experts were again La Nicca and Fraisse. The stock of building tools which had become available in Canton Bern was acquired immediately; this action was naturally advantageous to both parties alike. Any missing stock was produced at a local workshop.

Correction works on the River Broye consisted of excavations and three large cuts. Material extracted from dry trenches was used to fill adjacent, lower-lying areas. Excavations from submerged areas, on the other hand, were loaded onto barges and dumped into the lake. Rocks required for the protection of canal banks and for the piers in Lake Murten and Lake Neuenburg were brought from quarries along Lake Neuenburg. In 1878 the main work was completed. After the protection of the banks had been shown to be too weak for navigation in the canal, it was improved further until 1883. Dedication of the Broye Canal took place in 1886.

The work in the Zihl Canal also consisted of excavations and several cuts. As with the Broye Canal all the dredged material was dumped into the lake. The rocks needed for bank protection and the piers came not only from quarries along Lake Neuenburg but also from those along Lake Biel. After completion of the main work in 1878 the bank protection again proved insufficient. In addition, the piers in Lake Neuenburg became too short, because of the lowering of the lake, and so additional work became necessary.

Lowering Lake Murten and Lake Neuenburg in connection with the canal construction did not lead to any slumping of the shore or landslides. However, as expected, the harbours and several boat jetties became dry. The enterprise again was responsible for the necessary modifications.

The correction works along the Aare section from Büren to the mouth of the Emme were not carried out. This gave rise to certain disagreements and complaints addressed to Canton Solothurn which should have undertaken these works. Indeed these corrections were not entirely necessary; the training works already carried out had decreased the peak flow level in the Aare at Solothurn to the extent that the river profile was now sufficiently large. In 1891 the first training of the Jura waters was complete. The draining of the wetland adjacent to the Jura waters, not included in this project, was left to the cantons. Of course there were some after-effects; the continuing erosion in the Hagneck Canal has already been mentioned. Subsidence of the banks in the Hagneck cut did not cease for quite some time, and eventually it became necessary to rebuild the Hagneck bridge. In order to finally produce a stable situation, the hydroelectric power station Hagneck was built between 1897 and 1900 at the entrance of the canal into Lake Biel. It dammed the canal over a distance of 4 km and thereby stopped its erosion.

The provisional weir below Nidau in the Nidau–Büren Canal could not permanently fulfill the demands made of it. Changes and improvements were made several times. Nevertheless it could barely cope with the exceptional high water of 1910. Further renovations were also only of a provisional nature, so that between 1936 and 1940 Canton Bern had a modern lock installed. The regulation of Lake Biel and, hence, that of the connected Lakes Neuenburg and Murten, was now guaranteed.

All in all the first correction of the Jura waters had been a great success. The correction meant, according to Ehrsam (1974) that

> 'the Lake Country (Seeland) was freed from centuries of flooding (with the inevitable swamps and other disturbances), which had caused the population to resort to a poverty-stricken, idle existence since neither healthy agriculture

nor trade could develop. Some areas could barely be traversed before the establishment of the inland waterway network. Now, after systematic soil treatment and fertilization, the same terrain bore witness to the most beautiful vegetable fields; low quality grass had been replaced by valuable grain, sugarbeet, potatoes and vegetable crops of all kinds. In the picturesque villages of the lake country a healthy population prospered. The thankful inhabitants of this area erected a memorial in 1908 to honour—at the very site of their first successful work—the main patrons of the Jura water correction, namely the doctor and politician Johann Rudolf Schneider and the ingenious engineer Richard La Nicca from Graubünden.'

The fact that 70 years after the completion of the first correction of the Jura waters a second one became necessary (it was carried out from 1962 to 1973) does not speak against the farsightedness of these men. Even in river engineering there is no measure which is satisfactory forever; *not only do the waters themselves and the land surrounding them undergo changes, but so do the requirements of the local inhabitants.*

REFERENCES

Bachmann, Gottfried 1983. Die Ableitung der Kander und die Juragewässerkorrektion (the diversion of the River Kander and the training of the Jura waters). In *Die Geschichte der Gewässerkorrektionen und der Wasserkraftnutzung in der Schweiz*, Volume 9E, 9, Internationale Fachmesse Pro Aqua-Pro Vita, Basel.

Badische Wasser- und Strassenbaudirektion 1929. *Johann Gottfried Tulla, sein Leben und sein Wirken* (Johann Gottfried Tulla, his life and his work), published by the author, Karlsruhe.

Becker, Friedolin 1910. *Das Linthwerk und seine Schöpfer* (The Linth works and their creators), Jahresbericht der Geograph.-Ethnographischen-Gesellschaft, Zürich.

Ehrsam, Emil 1974. *Zusammenfassende Darstellung beider Juragewässerkorrektionen* (A summary describing both trainings of the Jura waters), Internationale Baukommission der II. Juragewässerkorrektion, published by the author, Bern.

Legler, Gottlieb Heinrich 1868. *Hydrotechnische Mitteilungen über Linthkorrektion, Runsenbauten, Zürichseeregulierung usw*, (Hydrotechnological information about the Linth correction, constructions in ravines, the regulation of Lake Zürich etc.), published by the author, Glarus.

Neumann, Kornel 1979. Projekt wie die Innundation zu Thun und dortigen Seeörtheren zu verhindern (Project to prevent the flooding of Thun and other communities along Lake Thun), *Zeitschrift Wasser, Energie, Luft* Nr. 9, Vol. 71, Baden.

Nussbaum, F. 1925. *Kleine Heimatkunde des Kantons Bern; Lehrbuch für das V. Schuljahr* (A pocket local history of Canton Bern: textbook for the fifth grade), Kantonaler Lehrmittelverlag, Bern.

Pestalozzi, Heinrich 1852. Das Linthwerk in hydrotechnischer Beziehung (Hydrotechnological aspects of the Linth works). Appendix to: Hottinger, J.J.: *Hans Conrad Escher von der Linth; Charakterbild eines Republikaners*, Orell-Füssli, Zürich.

Schweizerisches Ober-Bauinspektorat 1916. *Flusskorrektionen der Schweiz; Aare zwischen Thuner-See und Bieler-See* (River training works in Switzerland: the Aare between Lake Thun and Lake Biel), Bern.

Stambach, Ernst 1970. Die Juragewässerkorrektion (The training of the Jura waters). *Schweizerische Technische Zeitschrift*, 24/25, Zürich.

Thuerer, Georg 1966. *Das Linthwerk Hans Conrad Eschers* (The Linth works of Hans Conrad Escher), Neue Zürcher Zeitung, October 30th, Zürich.

Zier, Hans Georg 1970. Johann Gottfried Tulla, ein Lebensbild (A portrait of Johann Gottfried Tulla). *Zeitschrift Badische Heimat*, Nr. 4, Vol. 50, Freiburg i.B.

Historical Change of Large Alluvial Rivers: Western Europe
Edited by G. E. Petts
©1989 John Wiley & Sons Ltd

3

Hydroclimatic Fluctuations of Some European Rivers since 1800

JEAN-LUC PROBST

Centre de Sédimentologie et Géochimie de la Surface, Institut de Géologie, 1 rue Blessig, 67084 Strasbourg Cedex, France

INTRODUCTION

The water cycle may be investigated at different spatial (global, regional or local) and temporal scales. Over a long time period, the three main components of the water cycle are precipitation (P), evapotranspiration (E) and runoff (R) which results from the difference between the two first components ($R=P-E$). The runoff may be positive as in continental areas (except deserts) or negative as in ocean areas. River discharges may be considered as the runoff component of the water cycle in a natural system, represented by the drainage basin. Considering large rivers in Europe, the water cycle can be studied most appropriately at regional scale. River discharge depends on several factors but it may be viewed as an integral of the different hydroclimatic factors which have affected the drainage basin. Several authors have already used the long period streamflow fluctuations to characterize different climatic periods (Hoyt and Langbein, 1944; Riehl and Meitin, 1979; Faure and Gac, 1981; Hassan, 1981; Degens and Spitzy, 1983; Liebscher, 1983: Kuusisto quoted by Zuidema, 1983; Probst and Tardy, 1985, 1987; Tardy, 1986). The fluctuations of river discharge are cyclic and some authors have used knowledge of the periodicity of such cycles in order to manage the regional water resources and even to forecast the recurrences of dry and humid periods (Rozhdestvenskiv, 1968; Solyankin, 1978; Faure and Gac, 1981). Other authors have used the river discharge as a representative hydroclimatic factor in order to determine the influence of extraterrestrial factors on the discharge cycles (Dyer, 1978; Fairbridge, 1984; Smirnov, 1974). At present, it is important to follow Smirnov (1969, 1972, 1973), Hirst and Hastenrath (1983) and Probst and Tardy (1987) to determine the relationships between the general atmospheric circulation and the river discharge fluctuations.

For geologists and geomorphologists who study the continental erosion in drainage basins and the riverine transport of dissolved and suspended matter, it is also very important to better understand the changes of river discharge in order to calculate the mass balance fluctuations and to place the study periods in a wider hydroclimatic context (Gac and Kane, 1986; Kattan *et al.*, 1987; Nkounkou and Probst, 1987).

The main purpose of this chapter is to determine the major hydroclimatic periods which have affected large European rivers. Fourteen rivers have been selected from the different regions. The river discharge fluctuations in these different regions are compared following latitudinal and longitudinal transects.

DATA SOURCES AND METHODOLOGY

The discharges of most European rivers have been intensively controlled since 1900. But some river discharges have been recorded since the beginning of the nineteenth century; such is the case, for example, of the Vanern Gota in Sweden which has been gauged since 1807. For the Garonne river (SW France), the discharges have been reconstructed for the

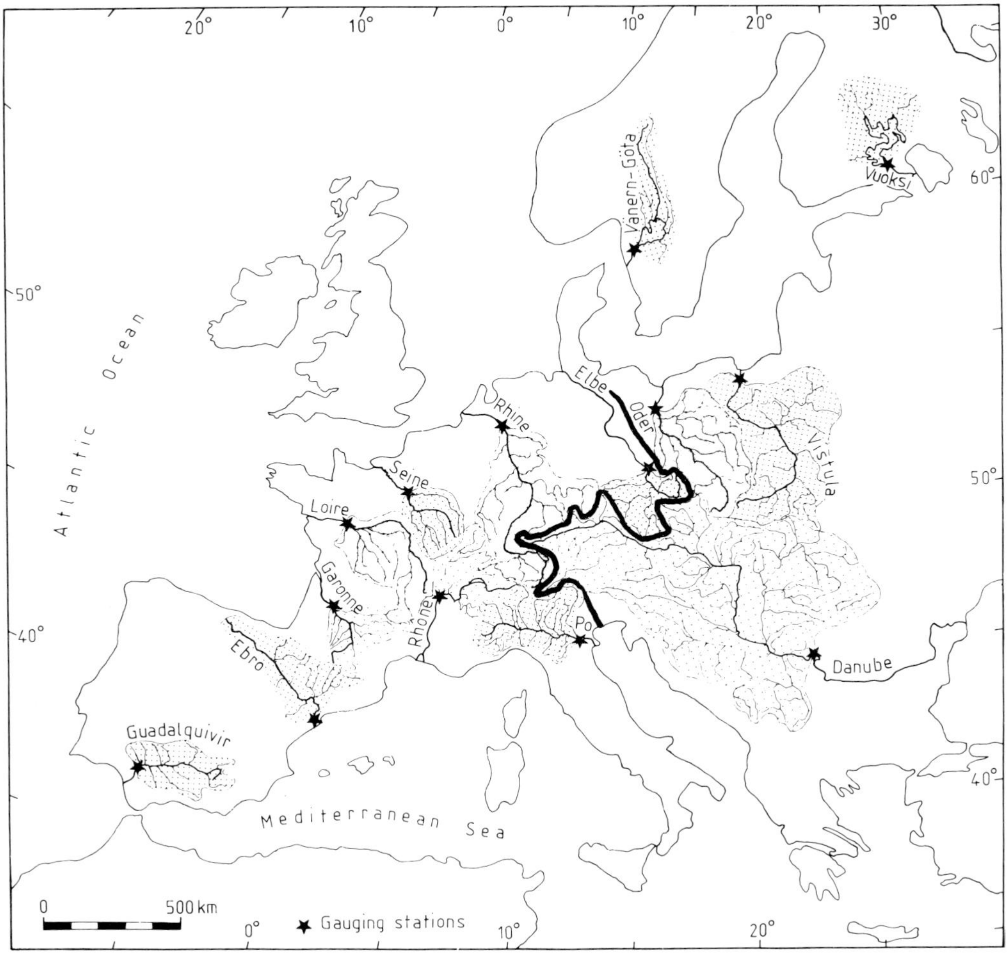

FIGURE 1. General location map of the fourteen selected river basins and gauging stations. The watershed between western and eastern Europe is highlighted. However, note that the hydrological characteristics of the Elbe place it in the central European hydrological region

TABLE 1. Characteristics of the different selected rivers and gauging stations. Q_m, mean annual discharge; Q_{max}, mean annual maximum discharge; Q_{min}, mean annual minimum discharge; σ, standard deviation; CV, variation coefficient. *, some lags in the data series

Rivers	Gauging stations	Country or region	Basin areas (10^3 km^2)	Data periods	Q_m ($m^3 s^{-1}$)	Q_{max} ($m^3 s^{-1}$)	Q_{min} ($m^3 s^{-1}$)	σ ($m^3 s^{-1}$)	CV
Vanern–Gota	Vanesborg	Sweden	46.8	1807–1983	532	748	225	97	0.181
Vuoksi	Imatra	Finland	61.3	1847–1975	590	954	246	116	0.196
Loire	Montjean	France	110.0	1863–1974*	814	1960	276	281	0.345
Garonne	Mas d'Agenais	France	52.0	1832–1978*	632	1020	188	186	0.295
Seine	Paris	France	44.3	1928–1982	273	460	101	103	0.379
Rhône	Lyon	France	50.2	1900–1982	1051	1644	475	249	0.236
Rhine	Lobith	W. Europe	160.0	1901–1983	2215	3149	1095	475	0.214
Elbe	Decin	Czechoslovakia	51.1	1851–1975	300	697	152	90	0.301
Oder	Gozdowice	Poland	109.3	1901–1978*	532	895	287	134	0.251
Vistula	Tczew	Poland	193.8	1901–1978*	1041	1540	598	221	0.213
Danube	Orsova	Hungary	578.3	1840–1975	5447	8050	3340	961	0.176
Po	Pontelagoscuro	Italy	70.1	1918–1975	1448	2620	905	398	0.267
Ebro	Tortosa	Spain	84.2	1913–1974*	1374	2475	350	534	0.389
Guadalquivir	Alcala del Rio	Spain	49.9	1913–1972*	508	1306	46	323	0.636

period 1832–1859 by Baumgarten (1848) and Fargue (1868) using the water heights recorded at Tonneins and Langon stations. For some other rivers, the discharges have been recorded since the mid nineteenth century: Danube (1840), Vuoksi (1847) and Loire (1863). Most of the discharge data are available in UNESCO (1971, 1974, 1979, 1985). For French rivers, the data have been kindly provided by the Services Hydrologiques Centralisateurs (S.H.C.).

Discharge fluctuations have been studied using mean annual flow data which are the most appropriate for determining the long-term trends and the major hydroclimatic phases over long time periods. For shorter time periods, the mean monthly discharges should be used (e.g. Liebscher, 1983). Summer seasonal flows might be also used rather than annual totals in order to study drought events (Beran and Rodier, 1985).

The mean annual flow data vary widely and have to be filtered in order to eliminate the yearly variation 'noise' and to determine clearly the interannual hydroclimatic periods. Different filtering methods have been used by hydrologists, as described by Probst and Tardy (1987), but the methods based on the moving average techniques are the most commonly employed. In this study, the data have been filtered by a simple moving average technique based on the five year arithmetic mean:

$$\bar{X}_i = (X_{i-2} + X_{i-1} + X_i + X_{i+1} + X_{i+2})/5$$

in which X_i is the value for year i and $\bar{X}_i$, the average filtered value.

In order to compare more easily the discharge variations of the different rivers, the data have to be standardized first by calculating the weighted deviation ($d/\bar{X}$) in which d is the deviation ($X_i - \bar{X}$) from the mean interannual value, $\bar{X}$. Among the available data for the European river discharges, fourteen large rivers have been selected (Figure 1) in the different countries according to the importance of the drainage basin and to the length of record, taken as long as possible. The characteristics of the selected rivers are reported in Table 1.

LONG-TERM RIVER DISCHARGE FLUCTUATIONS

As described by Probst and Tardy (1987) in a recent study, the long range streamflow fluctuations of fifty world major rivers appear to be synchronous in regional areas or even in small continental areas. In Europe, this is the case, for example, for the Aquitaine Basin (SW France) where the main rivers (Garonne, Lot, Dordogne, Adour and Gaves) present synchronous fluctuations (Probst and Tardy, 1985). The same synchronism has been described by Probst and Tardy (1987) between the discharge fluctuations of the four main French rivers (Garonne, Loire, Rhône and Seine). As seen in Figure 2a, one can add to this group the Rhine which follows the same fluctuations. The correlation coefficients between these five river discharge fluctuations are very good (Table 2, group II).

In large continental areas or between two different continents, Probst and Tardy (1987) have shown that river discharge fluctuations might be contrasting, displaced or even synchronous. One can cite for example the contrast between North America and Europe, and the nine year lag between Africa and Europe. In the Euroasian continent, the river discharge fluctuations are displaced progressively in time going from the west part to the east part, i.e. from Loire (France), Elbe (Czechoslovakia) to Ural (W USSR) and Yenissei (Siberia) (Probst and Tardy, 1987). This lag can be correlated with both meridional and

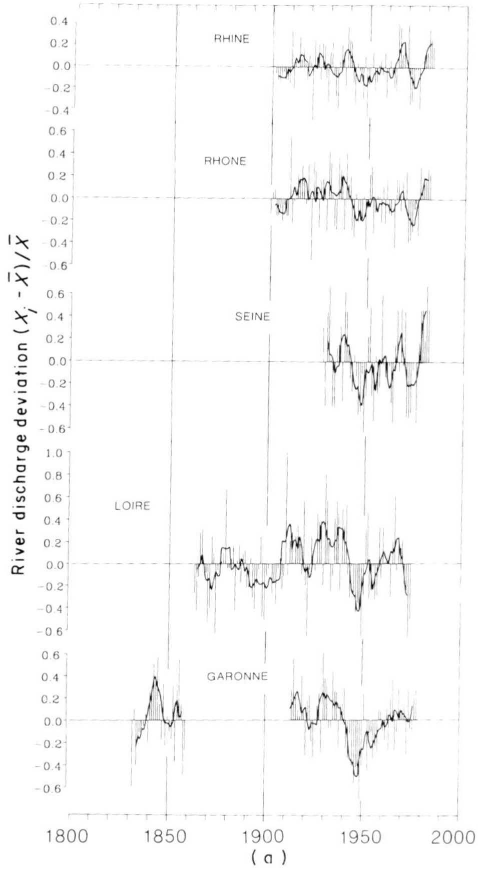

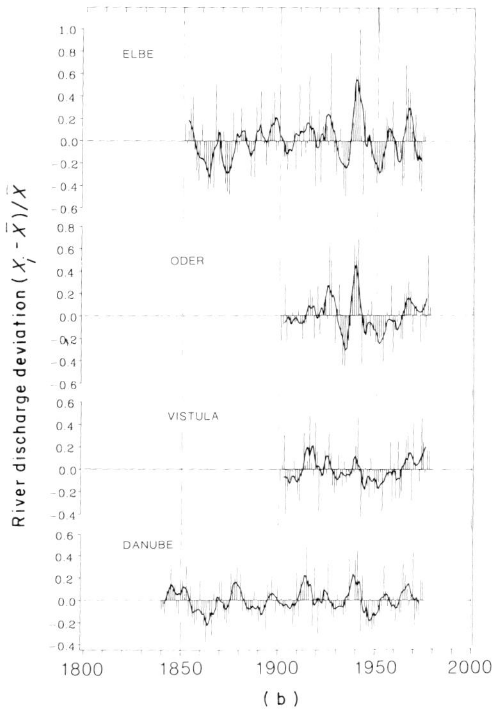

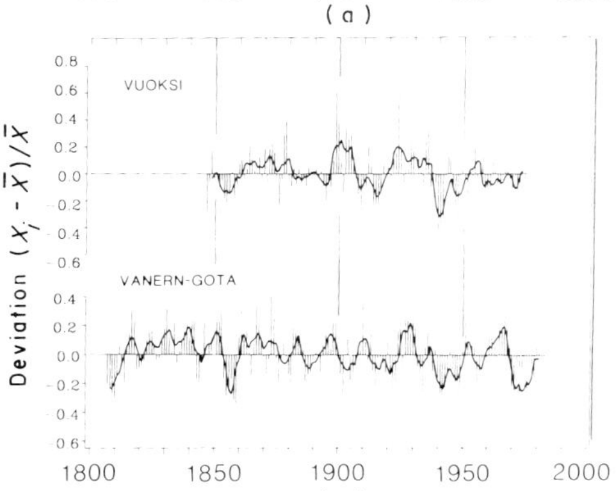

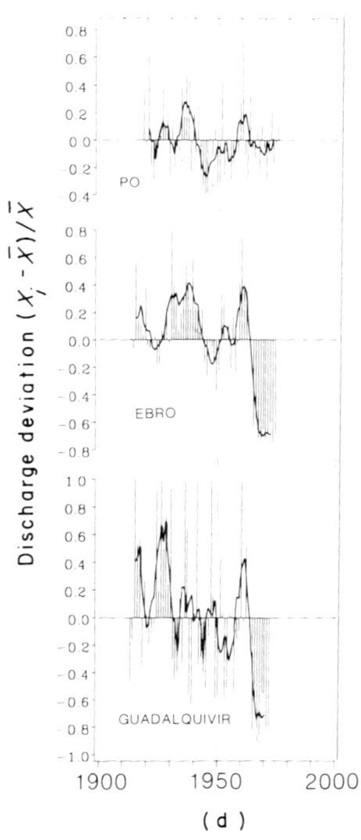

FIGURE 2. Long-term discharge fluctuations for all the selected rivers, plotted using the standardized data $((X_i - \bar{X})/\bar{X})$. Straight lines, unfiltered data; curves, 5-year moving average. (a)—Western Europe rivers, (b)—Central Europe rivers, (c)—Northern Europe rivers, (d)—Southern Europe rivers

TABLE 2. Pearson's correlation coefficients calculated for all river discharge fluctuations, for individual pairs. I—rivers of Northern Europe; II—Western; III—Central; IV—Southern, Va, Vanern Gota; Vu, Vuoksi; Lo, Loire; Ga, Garonne; Se, Seine; Ro, Rhône; Ri, Rhine; El, Elbe; Od, Oder; Vi, Vistula; Da, Danube; Po, Pô; Eb, Ebro; Gu, Guadalquivir

	Va	Vu	Lo	Ga	Se	Ro	Ri	El	Od	Vi	Da	Po	Eb	Gu
		(I)												
Va	1													
Vu	0.401	1												
Lo	0.287	—	1											
Ga	—	—	0.749	1		(II)								
Se	0.430	—	0.841	0.597	1									
Ro	0.309	—	0.833	0.696	0.853	1								
Ri	0.329	—	0.702	0.500	0.842	0.851	1							
El	—	−0.179*	0.425	0.275	0.524	0.502	0.727	1		(III)				
Od	—	−0.151*	0.366	0.249	0.518	0.416	0.580	0.819	1					
Vi	—	—	0.266	0.280	0.407	0.335	0.473	0.558	0.642	1				
Da	—	—	0.477	0.437	0.533	0.564	0.711	0.701	0.499	0.593	1		(IV)	
Po	0.172*	—	0.369	0.444	—	0.299	—	—	—	—	—	1		
Eb	—	—	0.443	0.308	0.232	0.371	—	—	—	−0.185*	—	0.373	1	
Gu	—	—	0.368	0.279	—	0.346	—	0.225*	—	—	—	0.476	0.596	1

Significant level, $p < 5\%$

$*5 < p < 20\%$

—no significant correlation $p > 20\%$

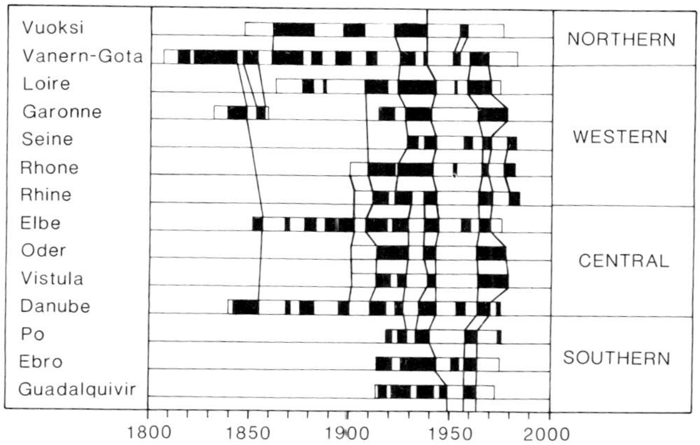

FIGURE 3. Mean hydroclimatic periods for all the European rivers, determined using the 5-year moving average curves. Black periods: discharge in excess; white periods: discharge in deficit (both in relation to long-term mean)

zonal propagation of the two dimensional pressure anomaly pattern recently described by Krishnamurty *et al.* (1986).

If one compares in detail the discharge fluctuations of the five western European rivers with the other European rivers, one can detect some differences between the successions of dry and humid periods, and between the duration and the intensity of these periods. The discharge fluctuations of the fourteen selected rivers are plotted in Figure 2, using the standardized data. In order to summarize all these fluctuations, in Figure 3 is shown, for all the rivers, the periods which have been in deficit or in excess from the long-term mean annual discharge. The limits of these periods have been determined using the filtered data curve.

As seen in Figures 2 and 3, it is very difficult to compare the discharge fluctuations during the nineteenth century because only six rivers have been gauged. Nevertheless, from the Danube, Garonne and Vanern Gota records, one can observe a humid period during the years 1840–1850 which may be common throughout central, western and northern Europe. This humid period is particularly well defined for the Garonne river; the discharge deviates from the mean by up to 60 per cent and this period is on average the most humid which has affected the Garonne basin for the available data period. It is interesting to remark that during this period, the sediment discharge measured by Baumgarten (1848) indicates a specific transport of about 100 t $km^{-2} yr^{-1}$. This specific transport is 4 to 5 times more important than the values (20–30 t $km^{-2} yr^{-1}$) given by Probst and Bazerbachi (1986) for the period 1970–1980 which is on average relatively moderate.

Following this humid period, the late 1850s–early 1860s was a dry period which began earlier for Vuoksi and Vanern Gota, and ended later for Elbe and Danube. There was a return to humid conditions during the late 1870s. The humid maximum occurred in 1880 on the Loire, Elbe, Danube and Vuoksi; but these contrast with the Vanern Gota which experienced low flows at this time.

During the 20th century, it is easier to determine the succession of dry and humid periods for all the selected rivers. As shown in Figure 3, five main hydroclimatic periods can be distinguished:

1. A dry period around the years 1900–1910;
2. A humid period until 1940;
3. The drought of 1940–1950;
4. A humid period around the years 1960–1970; and
5. A new dry phase after 1970.

In detail, one can observe some lags or even contrasts between the different rivers and some shortening or lengthening of the periods. The dry period of 1930–1935 only affected the rivers of Central Europe and to a lesser degree, Vanern Gota, Rhine, Seine, Po and Guadalquivir. The beginning of the great dry period from 1940–1950 seems to have developed progressively from northern Europe to western and central regions. This period is reduced for the southern rivers and more particularly for Ebro and Guadalquivir. Some rivers (Loire, Seine, Rhône, Elbe, Danube and Ebro) also show a return of a short humid period between 1950 and 1960.

The humid period which follows the European drought of 1940–1960 started earlier than 1960 for Vanern Gota, Loire and for the rivers of southern Europe. The dry period of 1970 started earlier for rivers of northern and southern regions but the Garonne, Oder and Vistula were not affected by this dry phase. For the rivers of western Europe, this drought is confined to the period 1970–1976, whereas for Vuoksi and Vanern Gota, the drought extended until 1980. During the same years, the western rivers have returned to a humid period. It is important to note that this last drought phase was particularly pronounced on the Ebro and Guadalquivir; the deficit of these rivers reached about 85 per cent from the long-term mean annual flow.

PRINCIPAL COMPONENT ANALYSIS

Principal component analysis has been used by hydrologists to determine long-term synchronous or asynchronous river discharge fluctuations in regional areas and to group rivers into homogeneous areas where the river discharges present the same fluctuations (Smirnov, 1972 and 1973; Bartlein, 1982; Lins, 1985). Matalas and Reiher (1967) state a clear preference for using the principal component approach in problems related to reducing the variability and dependency of multivariate data. In this study, a principal component analysis has been performed using Factor procedure (SAS Institute Inc., 1985) on the mean annual discharge fluctuations of the fourteen selected rivers for a common time period of 34 years (1936–1962 and 1965–1971). The analysis output includes all the eigenvalues and the pattern matrix for eigenvalues greater than one. Three factors have been retained by the mineigen criterion (Table 3). The cumulative explanation of these three factors reaches 75.85 per cent. The final communality estimates, which are the squared multiple correlations for predicting

TABLE 3. Principal component analysis results: eigenvalues and percentage variance explained by the three retained factors

Factor	1	2	3
Eigenvalue	5.99	2.59	2.03
% variance explained	42.81	18.54	14.51
Cumulative %	42.81	61.34	75.85

16

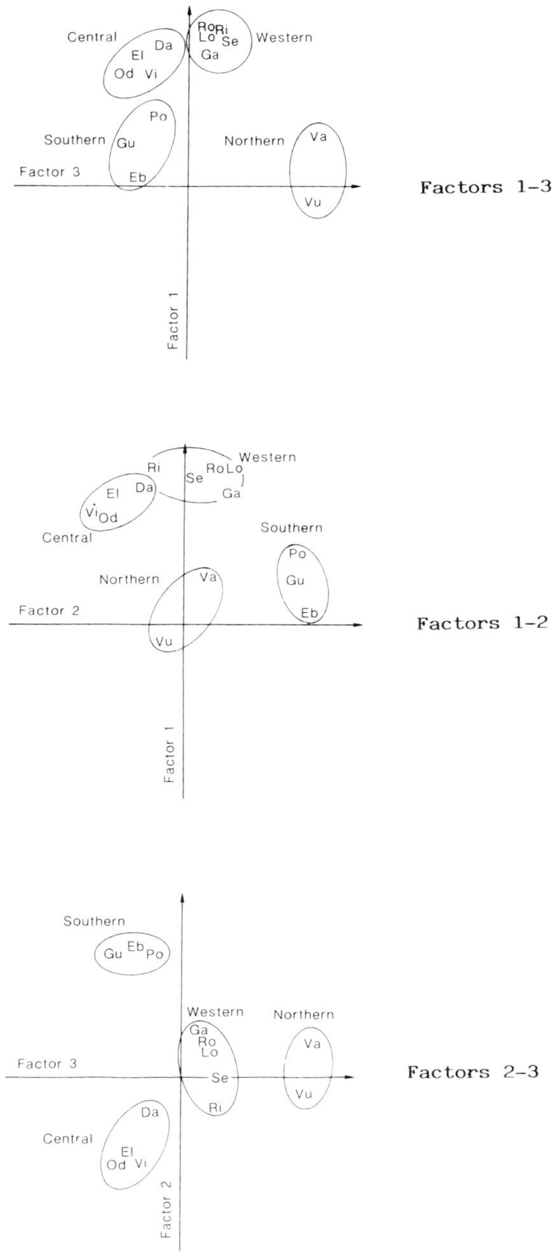

FIGURE 4. Principal component analysis: plots of factor patterns. Key: Va, Vanern Gota; Vu, Vuoksi; Lo, Loire; Ga, Garonne; Se, Seine; Ro, Rhône; Ri, Rhine; El, Elbe; Od, Oder; Vi, Vistula; Da, Danube; Po, Pô; Eb, Ebro; Gu, Guadalquivir

the variables from the estimated factors, show that all river discharge fluctuations are well accounted for by three components, with final communality estimates ranging from 0.637 for the Vuoksi to 0.918 for the Loire.

The factor patterns have been plotted in Figure 4. The first component has positive loadings for all rivers (except Vuoksi), and more particularly for rivers in western and central European regions. This component indicates that the most frequently occurring pattern of annual streamflow is one where most of Europe uniformly experiences above or below mean conditions. This pattern explains 42 per cent of the total annual streamflow variance.

The second component contrasts the central region and the southern region. The third factor defines a north–south contrast in components loading; large positive loadings for Vanern Gota and Vuoksi (northern region) and negative loadings for rivers of central and southern regions.

Finally, the principal component analysis allows the classification of the fourteen large European rivers into four groups which correspond to four main regions where the annual river discharges fluctuate synchronously:

1. Northern: Vuoksi and Vanern Gota
2. Western: Rhine, Rhône, Seine, Loire and Garonne
3. Central: Elbe, Vistula, Oder and Danube
4. Southern: Ebro, Guadalquivir and Pô.

LONG-TERM RUNOFF BALANCE FLUCTUATIONS FOR EUROPE

The total mean annual runoff for the whole European continent is estimated to 3×10^{12} m^3. This mean annual value varies according to the computation methods used by the different authors (Table 4). The purpose of this section is not to recalculate the runoff balance of Europe but to reconstitute the runoff fluctuations for a long time period. Similar works have been already realized by UNESCO (1978) for the period 1918–1967 and more recently by Probst and Tardy (1987) for the period 1863–1974.

The runoff balance fluctuations of UNESCO (1978) have been estimated using the data provided by stream gauging stations located near the mouths of large European rivers. The ungauged areas have been computed by geographical interpolation with the nearest gauging station and the long-term observations have been reconstituted using relations between annual river runoffs. The estimates of Probst and Tardy (1987) are based on multiple regressions between the amount of European river discharges calculated for a common period of time (34 years) and the discharges of two representative European rivers

TABLE 4. Annual water balance of Europe. 1—Baumgartner and Reichel (1975); 2—UNESCO (1978); 3—Shiklomanov and Sokolov (1983)

Areas (10^3 km^2)	Precipitation (km^3)	Precipitation (mm)	Evaporation (km^3)	Evaporation (mm)	Runoff (km^3)	Runoff (mm)	Runoff Coef. (RP^{-1})	Ref.
10 025	6587	657	3761	375	2826	282	0.43	1
10 500	8290	789	4940	470	3210	306	0.39	2
10 495	8290	790	5320	507	2970	283	0.36	3

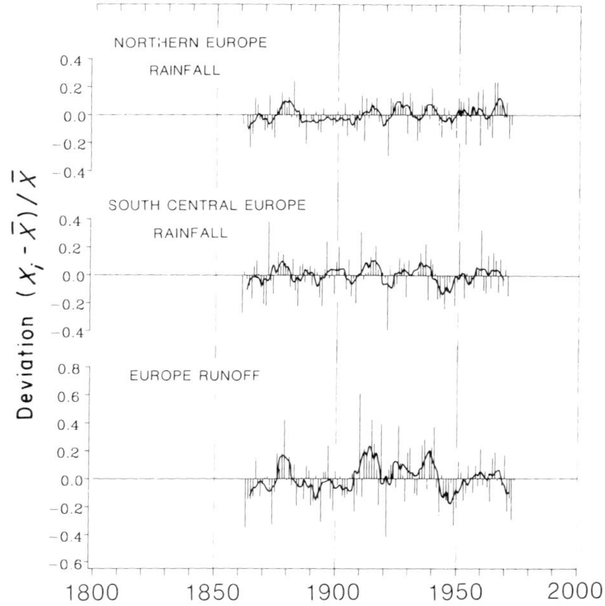

FIGURE 5. Comparisons of long-term fluctuations of Europe runoff calculated by Probst and Tardy (1987) and rainfall patterns in north and south central Europe calculated by Vines (1985). Straight lines, unfiltered data; curves, 5-year moving average

(Loire and Danube), recorded for a longer period of time (112 years). The regression equation given by Probst and Tardy (1987) is as follows:

$$\Sigma Q_i = 0.172\ Q_{Danube} + 6.099\ Q_{Loire} \quad (1)$$

where Q_i are the discharges of each individual European river, *i*.

These two approaches for estimating runoff fluctuations in Europe produce comparable data for the common time period (1918–1967) and the correlation coefficient between these two data sets is fairly good ($r = 0.703$).

The sequence of runoff fluctuations derived by Probst and Tardy (1987) are plotted in Figure 5 using the annual runoff deviations from the long-term mean. At this regional scale of analysis, Europe runoff fluctuates rather less than the individual river discharges. The coefficient of variation calculated for the period 1863–1974 is only 0.181 and this contrasts with those of river discharges (Table 1). As seen in Figure 5, the 5-year moving average curve exhibits a succession of dry and humid hydroclimatic periods which have affected the European continent as the whole:

1. A dry period from 1863 to 1875 with only one year (1867) which presents a significant discharge excess from the long-term mean;
2. A short humid period (1876–1882);
3. A long dry phase around 1900 (1883–1910) with some recurrences of humid years;

4. A very long humid period (1911–1941) with several annual discharge excesses which reach 30 to 60 per cent, interrupted only by a short drought around the year 1920; and
5. The European drought of 1950 (1942–1956) with some humid recurrences during 1951–1952. It is interesting to note, like Probst and Tardy (1985), that this period coinsides with the world thermal maximum (1945) calculated by Mitchell (1963) and to the warming culmination of springs (1947 to 1949), summers 1942–1943) and autumns (1942 to 1951) described by Le Roy Ladurie (1983);
6. A moderate humid period (1957–1968) with some dry recurrence years; and
7. The dry period of the 1970s.

The available discharge data of some rivers, suggests that a new humid period has established since 1977.

There is good agreement between the periods defined above and the fluctuations of the drought index calculated by Beran and Rodier (1985) for west and central Europe, using summer seasonal river flows rather than annual discharges. This drought index expresses the proportion of studied rivers experiencing conditions drier or wetter than the 5-year return period level.

TABLE 5. Pearson's correlation coefficients between European runoff and rainfall patterns. R, Europe runoff; Psc and Pn, rainfall patterns in south–central and northern Europe

	R	Psc	Pn
R	1		
Psc	0.614	1	
Pn	0.577	0.610	1

It is interesting now to compare in Figure 5, the runoff fluctuations of Europe with the rainfall fluctuations in northern and south–central Europe calculated by Vines (1985) using the data of Tabony (1980, 1981). The two rainfall patterns are synchronous. As seen in Table 5, the correlations between the rainfall and runoff patterns are very good and the correlation coefficients are highly significant. Then, it may be possible to express the European discharge variations as a function of rainfall variations in south–central and northern regions. The regression equation is as follows:

$$R_d = 0.507\ P_{n_d} + 0.578\ P_{sc_d} \qquad (2)$$

with $r = 0.663$ for $n = 109$ years.

where R_d, P_{n_d} and P_{sc_d} correspond respectively to the weighted deviations from the long-term mean $((X_i - \bar{X})/\bar{X})$ of runoff, northern precipitation and south–central precipitation. As seen in Figure 5, the two rainfall patterns fluctuate less than runoff. The variation coefficients of northern (0.112) and south–central (0.132) rainfalls are lower than that of runoff fluctuations (0.181) and, the annual deficits or excesses are sometimes more important for runoff than for rainfall. These differences may be attributed to the complexity of the transfer function between precipitation and runoff. The runoff results from the difference between precipitation and evapotranspiration which increases, for example, when the temperature increases. The annual storage variations in ground water may also contribute to these differences.

CONCLUSIONS

The long-term discharge fluctuations of large rivers may be viewed as a regional 'barometer' that allow the identification of the major hydroclimatic periods which have affected, on average, the river basins. In a small continental area, such as Europe, the river discharges fluctuate more or less synchronously over long time periods. The succession of major dry and humid periods is the same for all European rivers, except the Vuoksi in the north. However, in detail, over a short time period, some hydroclimatic changes may be delayed or even reversed. These short periods allow the classification of the fourteen selected rivers into four groups which correspond to four hydrological regions (northern, western, central and southern). In each region, the river discharge fluctuations are synchronous and the correlation coefficients between the different river discharge variations are highly significant. The progressive lags of the hydroclimatic periods, going from west to east and from south to north, have not yet been completely elucidated, but may be related to the general atmospheric circulation and more particularly to the zonal and meridional propagation of the two-dimensional pressure anomaly pattern. Our present investigations are focused on this relationship. European runoff fluctuations, reconstituted by Probst and Tardy (1987) for a long time period using regression analysis between river discharges, appear to be in phase with the European rainfall patterns calculated by Vines (1985).

This study has focused on the European continent but it is important to place the runoff fluctuations in a wider geographical context and to compare these fluctuations with those of other continents, described by Probst and Tardy (1987). Runoff fluctuations in Europe have been shown to contrast with runoff variations for rivers in North and South America. If one compares Europe with Africa, one can detect an average 9-year lag. Europe is also positively correlated with Asia but this correlation reflects the similar negative long-term trends of these two continents, i.e. in both Europe and Asia runoff patterns decrease significantly as a function of time for the period 1910–1975. Globally, these two continents present a more humid regime during the first half of the 20th century than during the second half. Finally, it is important to point out that when Europe experiences a drought event, this is generally not a worldwide tendancy. At such times, somewhere in the world, other regions or continents are affected by humid regimes.

ACKNOWLEDGEMENTS

The author thanks all researchers and public authorities who sent the needed information concerning the discharge data of European rivers: Dr R. Castensson (Linköping University, Sweden), Dr M. Fosse (Service de la Navigation de la Seine, Paris), Dr W. Froehlich (Institute of Geography, Krakow, Poland), Dr C. H. Van der Weijden (Rijksuniversiteit Utrecht, Netherlands), the Rijkswaterstaat, Directie Gelderland (Netherlands) and the Services Hydrologiques Centralisateurs (SHC Seine, Loire, Garonne and Rhône).

The author is also very grateful to Dr R. G. Vines (CSIRO Division of Forest Research, Canberra, Australia) for providing the basis data on European rainfall patterns which have been calculated by him in a recent study.

REFERENCES

Bartlein, P. J. 1982. Streamflow anomaly patterns in the USA and southern Canada 1951–1970. *Journal of Hydrolology* (Elsevier, Amsterdam), **57**, 49–63.

Baumgarten, M. 1848. Navigation fluviale, Garonne. *Annales des Ponts et Chaussées, Mémoires et Documents*, Paris, no 197, 2, **XVI**, 1–157.

Baumgartner, A. and Reichel, E. 1975. *The World Water Balance*. Elsevier, Amsterdam, 179 pp.

Beran, M. A. and Rodier, J. A. 1985. Hydrological aspects of drought. *UNESCO/WMO, Paris, Studies and Reports in Hydrology*, **39**, 149 pp.

Degens, E. T. and Spitzy, A. 1983. Paleohydrology of the Nile. In *Proc. SCOPE/UNEP Workshop, Transport of carbon and minerals in major world rivers*, Part 2. *Mitteilungen aus dem Geologish–Paläontologischen Institut der Universität Hamburg, SCOPE/UNEP Sonderband.*, Heft, **55**, 1–20.

Dyer, T. G. J. 1978. On the 11-year solar cycle and river flow. *Water South Africa, ZAF*, **4**, 4, 157–160.

Fairbridge, R. W. 1984. The Nile floods as a global climatic/solar proxy. In Morner, N. A. and Karlen, W. (Eds), *Climatic Changes on a Yearly to Millenial Basis*, D. Reidel Publishing Company, Dordrecht, 181–190.

Fargue, 1868. Etude sur la configuration du lit moyen de la Garonne. *Annales des Ponts et Chaussée, Paris*, **I**, 34–92.

Faure, H. and Gac, J. Y. 1981. Will the Sahelian drought end in 1985? *Nature*, **291**, 5815, 475–478.

Gac, J. Y. and Kane, A. 1986. Senegal river. Part I — Water budget and flux of carried load through the estuary. *Sciences Géologiques Bulletin, Strasbourg*, **39**, 1, 99–130.

Hassan, F. A. 1981. Historical Nile floods and their implications for climatic change. *Science*, **212**, 1142–1145.

Hirst, A. C. and Hastenrath, S. 1983. Diagnostics of hydrometeorological anomalies in the Zaire (Congo) basin. *Quarterly Journal of Royal Meteorological Society*, **109**, 881–892.

Hoyt, W. C. and Langbein, W. B. 1944. The yield of streams as a measure of climatic fluctuations. *The Geographical Review*, **34**, 218–234.

Kattan, Z., Gac, J. Y. and Probst, J. L. 1987. Suspended sediment load and mechanical erosion in the Senegal basin. Estimation of the surface runoff concentration and relative contributions of channel and slope erosion. *Journal of Hydrology (Elsevier, Amsterdam)*, **92**, 59–76.

Krishnamurty, T. N., Sho-Hang, Chu and Iglesias, W. 1986. On the sea level pressure of the southern oscillation. *Archives for Meteorology, Geophysics and Bioclimatology (Springer–Verlag, Berlin), A*, **34**, 385–425.

Kuusisto, E. 1983. *Variations in the water balance of Finland during the last fifty years*, International Association of Hydrological Sciences, Wallingford U.K. (IAHS), Hamburg Symp., unpublished communication, quoted by Zuidema (1983).

Le Roy Ladurie, E. 1983. *Histoire du Climat Depuis l'An Mil*, Ed. Flammarion, Paris, 2 vol., 287 pp. and 245 pp.

Liebscher, H. J. 1983. The use of long term river level and discharge records in the study of climatic variations in the Federal Republic of Germany. In Street-Perrott, A. *et al.* (Eds), *Variations in the Global Water Budget*, Oxford Symp., D. Reidel Publishing Company, Dordrecht, 173–184.

Lins, H. F. 1985. Interannual streamflow variability in the United States based on principal components. *Water Resources Research (American Geophysical Union)*, **21**, 5, 691–701.

Matalas, N. C. and Reiher, B. J. 1967. Some comments on the use of factor analyses. *Water Resources Research (American Geophysical Union)*, **3**, 213–223.

Mitchell, J. M. 1963. On the world-wide pattern of secular temperature change. In *Changes of climate*, Proc. UNESCO/WMO Symp., Rome 1961, UNESCO Paris, 161–181.

Nkounkou, R. R. and Probst, J. L. 1987. *Hydrology and geochemistry of the Congo river system.* In Degens, E. T., Kempe, S. Wei-bin, G. (Eds), *Transport of Carbon and Minerals in Major*

World Rivers, Part 4, Mitteilungen aus dem Geologisch-Paläontologischen Institut der Universität Hamburg, SCOPE/UNEP Sonderband, Heft 64, 483–508.

Probst, J. L. and Bazerbachi, A. 1986. Solute and particulate transports by the upstream part of the Garonne river. *Sciences Géologiques Bulletin, Strasbourg*, **39**, 1, 79–98.

Probst, J. L. and Tardy, Y. 1985. Fluctuations hydroclimatiques du Bassin d'Aquitaine au cours des 70 dernières années. *Revue de Géologie Dynamique et de Géographie Physique (Paris)*, **26**, 1, 59–76.

Probst, J. L. and Tardy, Y. 1987. Long range streamflow and world continental runoff fluctuations since the beginning of this century. *Journal of Hydrology (Elsevier, Amsterdam)*, **94**, 289–311.

Riehl, H. and Meitin, J. 1979. Discharge of the Nile River: a barometer of short period climate variation. *Science*, **206**, 4423, 1178–1179.

Rozhdestvenskiv, A. V. 1968. Conversion of annual streamflow to a long period. *Soviet Hydrology: selected papers* (GGI, Sbornik rabot po gidrologii, No. 8, 1968, 3–13), Issue no. **4**, 365–374.

SAS Institute Inc. 1985. *SAS User's Guide: Statistics, Version 5 Edition*, Cary, N.C.: SAS Institute Inc., 956 pp.

Shiklomanov, I. A. and Sokolov, A. A. 1983. Methodological basis of world water balance investigation and computation. *International Association of Hydrological Sciences (IAHS), Wallingford, UK, Publ. no. 148*, Proc. Hamburg Symp., 77–92.

Smirnov, N. P. 1969. Causes of long-period streamflow fluctuations. *Soviet Hydrology: selected papers* (Izvestiya VGO, vol. 101, no. 5), Issue no. **3**, 308–314.

Smirnov, N. P. 1972. Asynchronous long-period streamflow fluctuations in the European USSR. *Soviet Hydrology: selected papers* (Izvestiya VGO, vol. 104, No. 1, 1972, 59–63), Issue no. **1**, 46–49.

Smirnov, N. P. 1973. Spatial patterns of long-period streamflow fluctuations in the European USSR. *Soviet Hydrology: selected papers* (Vodnyye Resurcy, no. 2, 1973, 21–32), Issue no. **2**, 112–122.

Smirnov, N. P. 1974. Solar activity and river runoff fluctuations in the USSR. *Soviet Hydrology: selected papers* (Izvestiya VGO, No. 3, 1974, 211–216), Issue no. **3**, 205–210.

Solyankin, E. V. 1978. Problème concernant le calcul de la fluctuation à long terme du débit de la Volga. (text in Russian) *Vodnyye Resurcy, SSSR*, **2**, 74–82.

Tabony, R. C. 1980. *A Set of Homogeneous European Rainfall Series*, Met. 0-13. Branch Memorandum 104, Met. Office Library, Bracknell (UK).

Tabony, R. C. 1981. A principal component and spectral analysis of European rainfall. *Journal of Climatology (Royal Meteorological Society)*, **1**, 283–294.

Tardy, Y. 1986. *Le cycle de l'eau. Climats, paléoclimats et Géochimie globale*. Masson, Paris, 338 pp.

UNESCO 1971, 1974, 1979 and 1985. *Discharge of Selected Rivers of the World*, Volume III, parts I, II, III and IV, UNESCO Press, Paris.

UNESCO 1978. World water balance and water resources of the earth. *Studies and Reports in Hydrology*, **25**, UNESCO Press, Paris (English translation, printed in the USSR), Order no. 3853, 663 pp.

Vines, R. G. 1985. European rainfall patterns. *Journal of Climatology (Royal Meteorological Society)*, **5**, 607–616.

Zuidema, F. C. 1983. General report on global, regional, local water balances and long term forecasts. *International Association of Hydrological Sciences (IAHS, Wallingford, UK), Publ.*, **148**, Proc. Hamburg Symp., 113–123.

Historical Change of Large Alluvial Rivers: Western Europe
Edited by G. E. Petts
©1989 John Wiley & Sons Ltd

4

Hydrological Changes of the Rhône River

H. VIVIAN

Laboratoire de la Montagne Alpine, Institut de Geographie Alpine, rue Maurice Gignoux, 38031 Grenoble Cedex, France

INTRODUCTION

With an area of some 100 000 km^2, the Rhône drainage basin is one of Europe's largest catchments and its hydrological regime is influenced markedly by the Alps and Jura mountains. The Rhône was important for ancient civilizations and today is a heavily populated and industrialized basin. The history of the river blends with the history of Man who readily exploited its various resources while seeking to protect himself from flood and erosion hazards. It is not surprising that the flow of the river, which today is used for navigation, irrigation and hydroelectric energy production, has changed as a result of regulation and diversion. However, the hydrological changes also relate to climatic change recorded since the beginning of the observation period (19th century). This paper examines the changing hydrological regime of the Rhône in relation to both climatic and anthropogenic influences.

Major man-made changes have affected both the Rhône river and its main tributaries (Figure 1). The most obvious of these has been the hydroelectric developments undertaken during the second half of the 20th century by the Compagnie Nationale du Rhône (C.N.R.) on the main river, and by Electricite de France (E.D.F.) and Grand Dixence in the mountain basins (Jura and Alps). The C.N.R. operates nineteen multipurpose developments from Genissiat to Arles (twenty power plants, fourteen sluices and twenty-four dams on a total of 481 km of regulated river) of which 200 km of river is bypassed by canals. Flood protection, irrigation and agricultural land drainage schemes have accompanied these developments including approximately 100 000 ha of actual and projected irrigated land and 41 220 ha of land protected from floods (C.N.R., 1981). In the southern part of the catchment, the Durance is used for the production of hydroelectricity by E.D.F. and for irrigation by the Societe du Canal de Provence. Between 1959 (the date of construction of the Serre-Poncon dam) and today, ten plants have been put into operation on the middle section of the Durance. On the lower section, downstream from Cadarache, 80 per cent of the annual discharge is diverted into the Berre pond for irrigating the Provence and to drive five additional plants. When the Durance flows into the Rhône at Avignon, only residual stream flow remains (Juramy and Montfort, 1987). In the upper basins, Grande Dixence and E.D.F. have installed numerous dams to regulate the river's highly variable

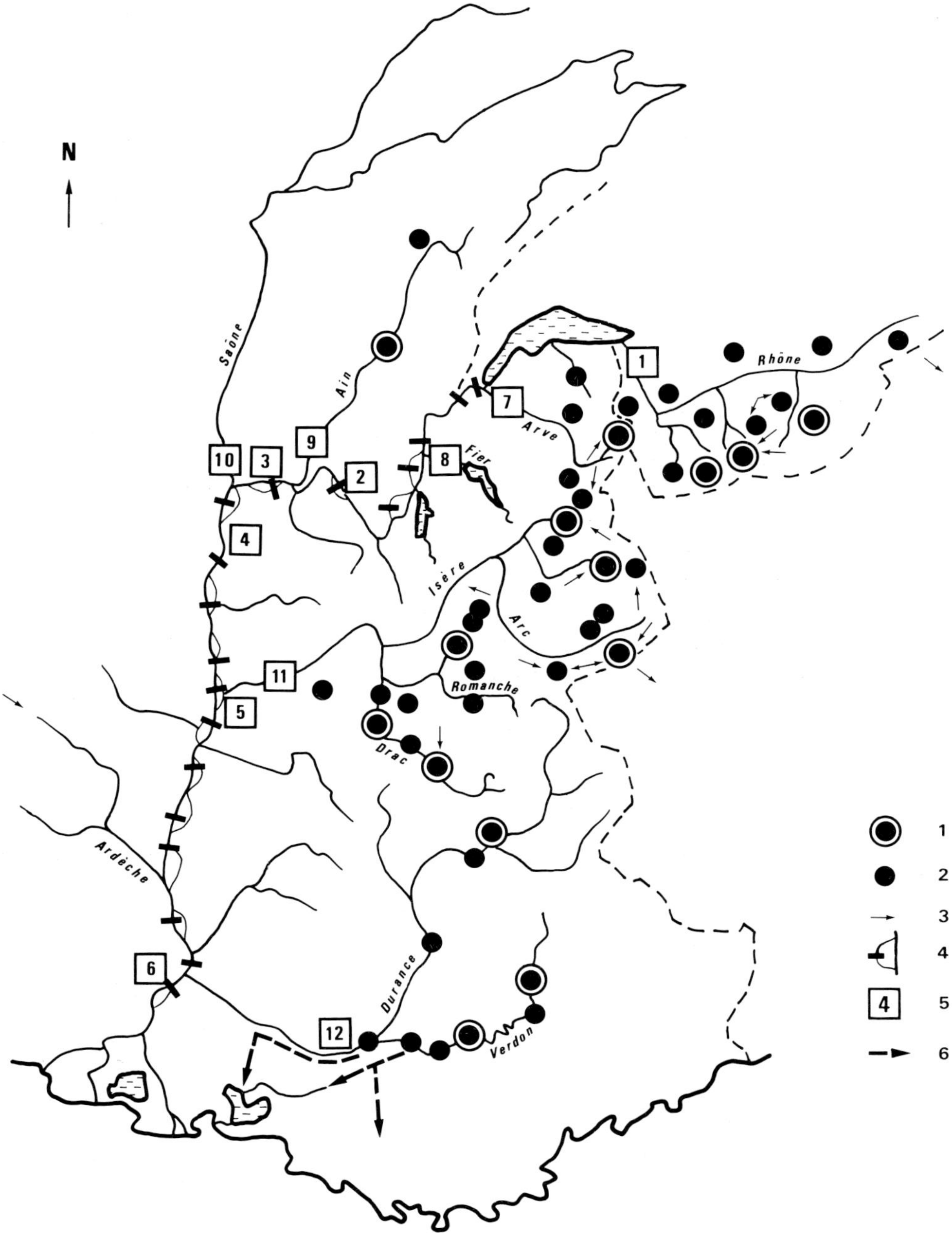

FIGURE 1. Distribution of the major water works in the Rhône drainage basin.
Key: 1. Reservoirs with a capacity >100 hm_3; 2. Reservoirs with a capacity <100 hm_3; 3. Interbasin transfer; 4. Rhône river developments (bypass channel diversion from the river); 5. Gauging station used in Table 1; 6. Abstraction from the Durance and Verdon river (irrigation)

regime (summer floods and winter low flows) in order to adapt the electrical production to the demand (summer storage during low demand, winter production during high demand). The largest dams were constructed after the 1950s and are filled by capturing runoff from numerous torrents outside the natural catchment of the reservoirs yielding a total capacity of 4500 million m^3 which corresponds to 8 per cent of the average Rhône discharge (Edouard and Vivian, 1982, 1984; Bezinge, 1983).

The natural parameters controlling the water supply may be grouped into two categories. The first category is the change in runoff pattern resulting from glacial dynamics. In the Valaisan Alps (Switzerland) the glacial surface area has decreased by 26 per cent between 1867 and 1968, resulting in a decrease of summer discharge by 600 million m^3 at Porte du Scex (Kasser, 1969). In France, this decrease has been estimated to be 24 per cent with a similar reduction of summer flows on the Arve and Isere (Vivian, 1975). Secondly, is the change of the rainfall regime since the beginning of the century (Martin *et al.*, 1981; Vivian, 1984). The average annual rainfall in Southeast France decreased between 1941 and 1979 and this is reflected by the extension of the period of summer low flows from June and July through to September–October. Furthermore, winter rainfall has been lower than the long-term average since 1950, especially in January and February.

Whilst describing the changes of the Rhône's regime, this study will attempt to define the relative influence of natural and anthropological factors. By using the method of accumulated normalized deviations, the discharge and rainfall trends are described in order to detect any behavioural similarity or difference between the two variables. A similarity indicates a link between climatic factors and discharge. But a difference does not necessarily indicate the effects of anthropological influences, although the independent time series of the two variables may signify that there is only a tentative and indirect relationship between precipitation and discharge. This may be due to the fact that the Rhône is supplied by various sources (Vivian, 1984).

FLUCTUATIONS OF THE ANNUAL YIELD

The method of accumulated normalized deviations was used to analyse both annual and monthly discharges in chronological series established from C.N.R. records of 1920 to 1979 (Vivian, 1983). A time series plot of the accumulated normalized deviations is a simple method, the only objective of which is to identify data trends. These may be compared to the mean of a long series and hence for successive sequences of years it is possible to define the magnitude of fluctuations as either a positive or negative deviation from the mean.

The C.N.R. gauging stations used in this study, from upstream to downstream, are Sault Brenaz (15.380 km^2), Ternay-Givors (51.080 km^2), Valence (66.450 km^2) and Beaucaire (81.360 km^2). Interpretation of the data is complicated by the variable climatic regimes within different parts of the basin: downstream to Lyon, the flow regime reflects the mountainous climate partly regulated by snow melt; from Lyon to Valence an oceanic pluvial influence dominates; at Valence with the confluence of the Isere river, the alpine influence reappears; and finally, from Valence to Beaucaire the influence is pluvial with a mediterranean tendency. Nevertheless, Figure 2 shows a striking similarity in the general pattern of the data. The four graphs show that data oscillate only moderately about the O axis. However, the curves may be combined into two pairs: Sault Brenaz–Beaucaire, and Ternay–Valence.

At the widely separated Sault Brenaz and Beaucaire gauging stations, the curves show that oscillations in relation to the 1920–1979 average, have little amplitude. Nonetheless, for

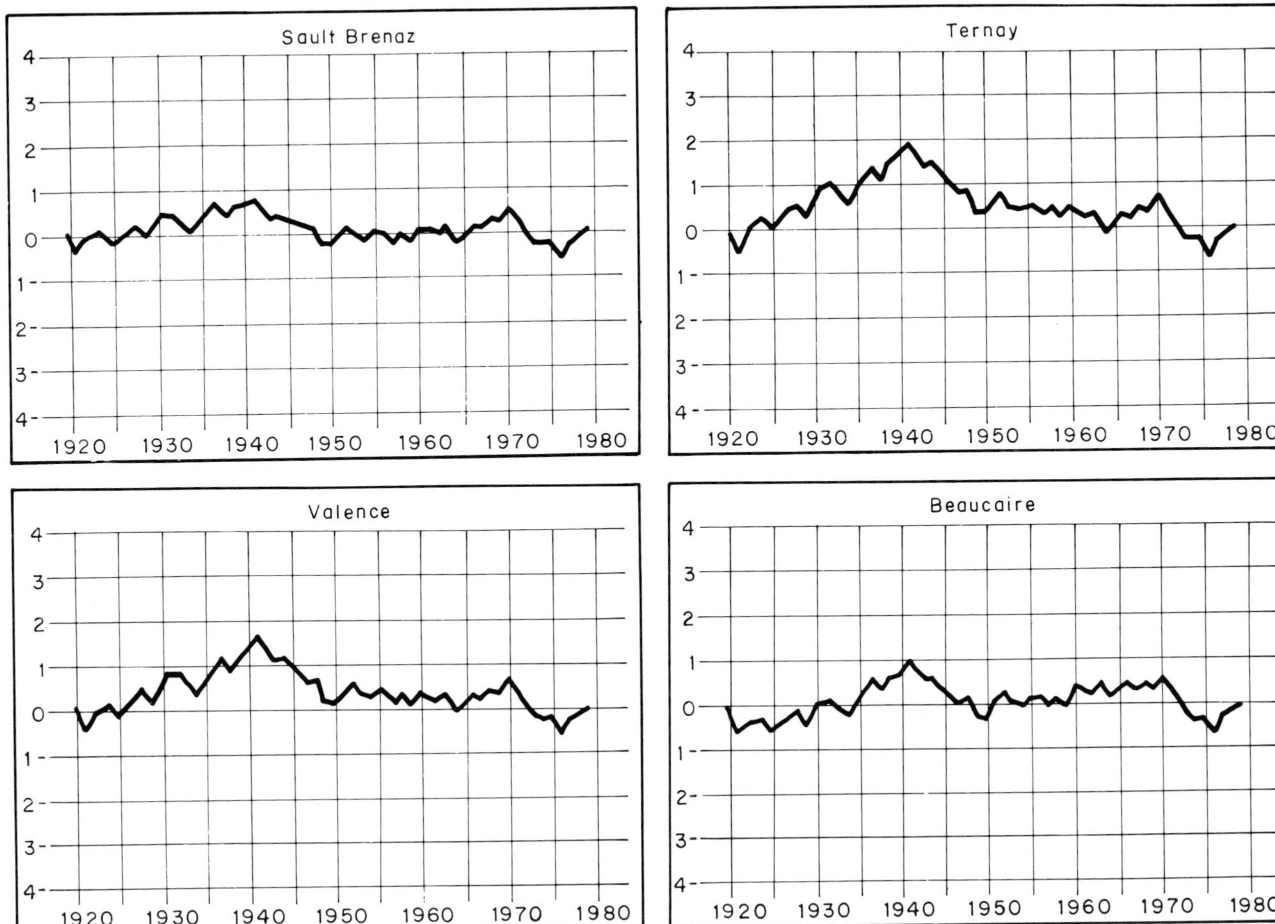

FIGURE 2. Graphs of the accumulated normalized deviations–average annual discharge rates (1920–1979)

the series 1921–1941, they both have an upward tendency (+6 per cent and +8 per cent respectively) followed by a characteristic decrease for the dry decade 1941–1950 (−11 per cent and 16 per cent). The discharge values then vary little until 1970 and then decrease progressively until 1976 (average deficit 1970–76: −16 per cent and −17 per cent). After this date, an upward trend has been recorded which persists until today.

At the Beaucaire station, the graph shows that despite the marked reduction of the Durance's discharge during the 1960s, an average of $-100\,m^3\,s^{-1}$ being abstracted annually, there has been little change of the overall flow pattern. The effect of the Durance's abstractions may be recognized by comparing the mean discharges for the periods 1877–1936 ($1818\,m^3\,s^{-1}$) and 1951–1981 ($1695\,m^3\,s^{-1}$).

The data for Ternay and Valence show the same trends for the period as in the preceding case, but with greater amplitudes. The average increase since the beginning of the century, which starts in 1921 reaches +11 per cent at Ternay and +10 per cent at Valence. A decrease is then recorded from 1941 to 1964 (−8 per cent at Ternay, −7 per cent at Valence) with a maximum decrease for the series 1941–1950 (−16 per cent, −15 per cent). A short increase from 1964 to 1970 follows, and then a decrease from 1970 to 1976 (−19 per cent at both Ternay and Valence). Thereafter the present trend towards increases (+15 per cent) is evident.

For the four gauging stations, there is no clear change after 1950 to reflect the important anthropological influences after this date. The succession of oscillations detected at Ternay and Valence faithfully reflect the variations of the rainfall and the increasing influence of the Saône's oceanic pluvial flow regime. The Saône's discharge rate increased steadily (+24 per cent) from 1921–1941 and decreased thereafter (−26 per cent). This trend persisted until 1964 (just as at Ternay and Valence) and accelerated between 1970 and 1976 (average decrease −22 per cent). The overall pattern is similar to that described by the Rhône data and the increase during the last four years is identical to that recorded at the Rhône stations. At Sault Brenaz, above the Rhône–Saône confluence, this influence is not demonstrated and the influence of the Saône tends to progressively decrease downstream. In order to identify any anthropogenic influence it is necessary to consider the monthly and daily timescale: annual volumes remain practically unchanged due to compensation schemes.

CHANGES OF THE SEASONAL REGIME

To detect changes in the distribution of monthly mean discharges two approaches are used. The first method involves comparison of the monthly graphs of the accumulated normalized deviations for the chronological sequence of maximum and minimum flows with the monthly mean of 1920–1979. The second method compares two or several averaged flow regimes corresponding to chosen periods. This comparison leads to a simple graphic representation in the form of histograms for each chosen gauging station. Monthly mean discharge changes from Porte du Scex to Beaucaire. Tables 1 and 2 show the discharge values from the beginning of the century (Parde, 1942) and those from the 1951 to 1981 period based on data from C.N.R. and E.d.F. using the gauging stations of the Rhône and its major tributaries. The monthly discharge series presented in Figure 3 show two types of 'natural', historical, seasonal regime. The first major type of regime (e.g. the Upper Rhône, Isere) is typical of mountain rivers, supplied by summer glacier and snow melt whilst undergoing mass storage during the winter. These regimes have summer maximum

TABLE 1. Average monthly discharges of the Rhône river at the beginning of the century and today (in $m^3 s^{-1}$)

Stations	J	F	M	A	M	Jn	J	A	S	O	N	D	Annual discharge	Maximum/Minimum	Basin (km^2)
Porte du Scex															
1914–38	57	50.1	57.8	90.7	224	403	449	375	236	123	82.5	65	185	8.8	5220
1951–81	90.5	95.3	99.7	119.7	195.5	332.2	341.4	294.2	200.5	128	105.7	94.3	174.7	3.8	5230
(1)															
Sault Brenaz															
1877–1936	344	335	406	513	580	677	720	644	525	458	412	386	500	2.1	15381
1951–81	395	462	468	469	487	571	584	507	405	345	373	405	456	1.7	15380
(2)															
Lyon St Clair															
1877–1936	500	500	610	705	730	775	800	715	600	590	585	570	640	1.6	20500
1951–81	582	681	674	631	653	669	661	584	504	467	535	600	603	1.5	20300
(3)															
Givors–Ternay															
1877–1936	1160	1140	1230	1220	1100	1020	985	870	780	905	1120	1190	1060	1.6	51400
1951–81	1279	1455	1296	1126	925	918	837	745	718	745	991	1243	1023	2	50400
(4)															
Valence															
1877–1936	1380	1370	1550	1640	1690	1670	1480	1210	1075	1255	1470	1550	1440	1.6	65800
1951–81	1580	1810	1703	1577	1459	1492	1298	1072	1007	1054	1397	1553	1409	1.8	66450
(5)															
Beaucaire															
1877–1936	1800	1740	1970	2100	2185	2085	1640	1340	1280	1700	2020	1960	1818	1.7	95500
1967–85	1983	2380	2070	1917	1997	1740	1360	1043	1090	1395	1545	1821	1695	2.3	81360
(6)															

TABLE 2. Average monthly discharges of the major tributaries (in $m^3 s^{-1}$)

Stations	J	F	M	A	M	Jn	J	A	S	O	N	D	Annual discharge	Maximum/Minimum	Basin (km^2)
Arve Genève															
1905–38	46.3	42.7	59.8	88.3	129	137	128	110	83.3	64.4	59.8	56.3	84	3.2	1983
1951–81	45.7	54.7	64.4	86	110	133	117	96.1	73.9	54.4	61.9	50.9	79	2.9	1976
(7)															
Fier-Vallières															
1911–39	46	47	62	76	72	41	34	31	34	44	56	57	50	2.4	1385
	40.3	55.9	60.3	66.9	61.3	44.6	32.6	24	27.1	28.8	44.7	49.8	44.7	2.8	1376
(8)															
Ain Chazey															
1877–1936	140	150	190	185	135	90	75	68	72	125	160	170	130	2.8	3672
1951–1981	151	194	185	149	110	79	58	61.5	79.5	95.5	151	184	125	3.3	3680
(9)															
Saône–Lyon															
1877–1936	644	625	606	503	360	237	180	152	177	308	522	606	410	4.2	29908
1951–81	695	769	871	492	318	245	175	163	212	273	451	642	442	5.3	29908
(10)															
Isère B. Monteux															
1904–1938	204	206	272	380	615	665	485	331	264	260	270	247	350	3.3	11799
1963–81	253	310	339	399	482	536	427	285	260	264	272	265	341	2.1	11800
(11)															
Durance Mirabeau															
1892–1950	133	135	182	240	323	331	184	113	106	174	206	179	193	3.1	11917
1961–80	10.4	15.2	13.9	6.5	21	29.7	6.9	2.9	2	4	16.6	4.5	11	14.8	—
(12)															

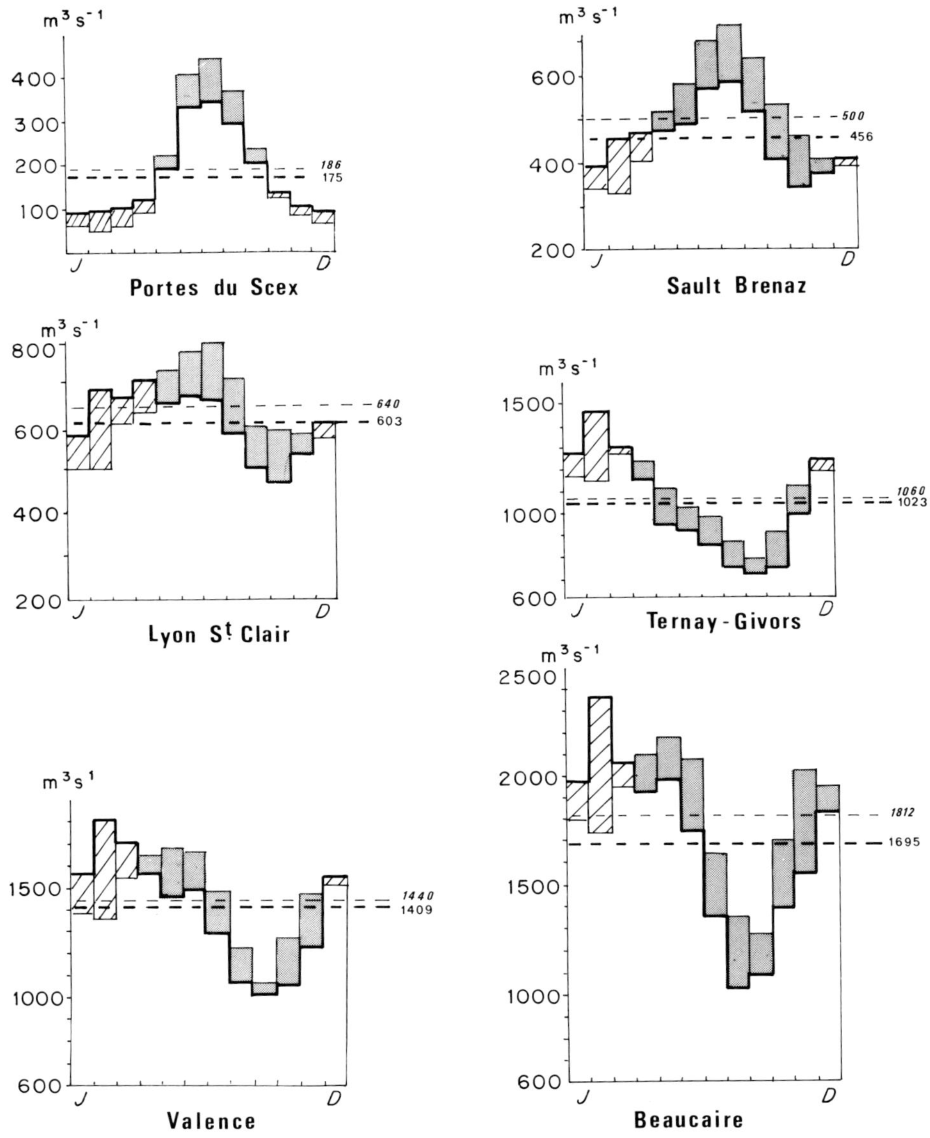

FIGURE 3

flows and winter minimum flows e.g. Rhône at Porte du Scex, at Sault Brenaz, at St. Clair and the Isere at Beaumont Monteux). The second major type is an oceanic or mediterranean pluvial regime with maximum flows from November to March. Such regimes have summer minimum flows and cold season maximum flows (e.g. the Rhône at Ternay, the Saône at Lyon and the Ain).

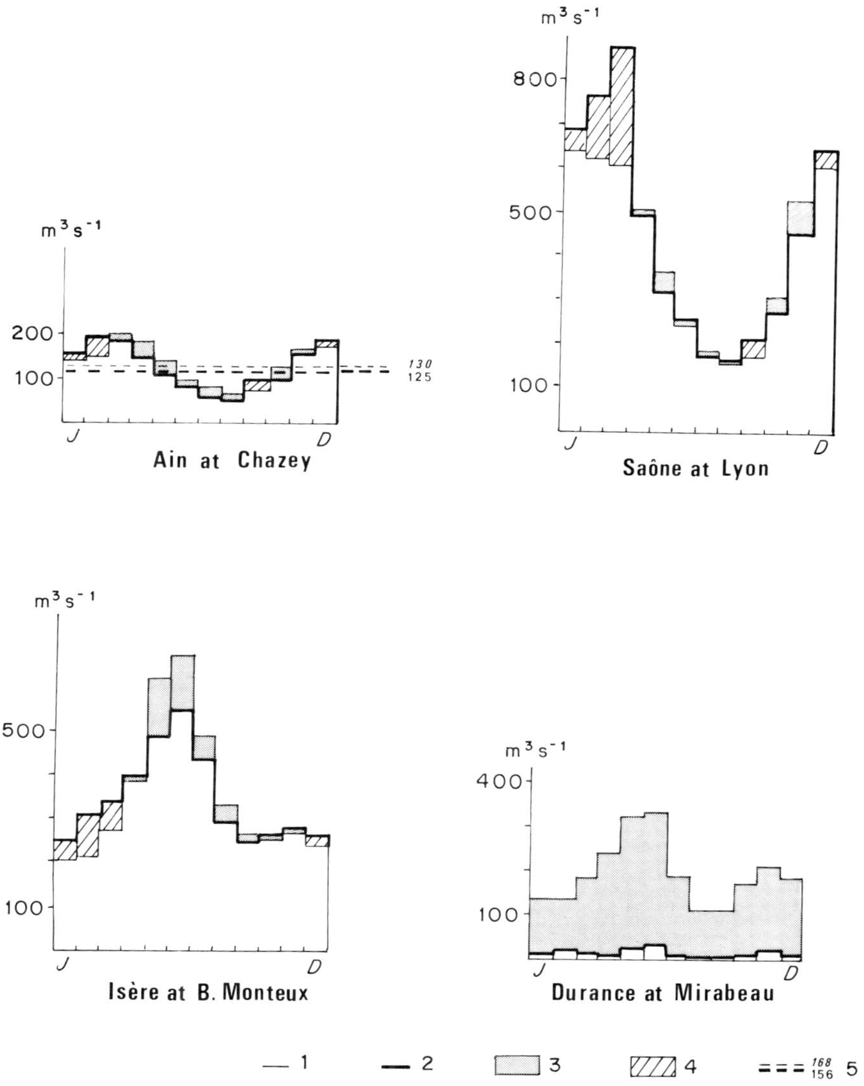

FIGURE 3. Monthly flow regimes for gauging stations on the Rhône and its major tributaries. Key: 1. Discharge rate at the beginning of the century; 2. Actual discharge rate; 3. Loss; 4. Increase; 5. Average annual discharge rate (beginning of the century and actual)

Between these two major types of simple regimes are the transition types: the Rhône at Valence with maximum flows in the spring (May) due to the influence of the Isere river; the Rhône at Beaucaire, experiencing the subdued influence of both the Rhône at Valence, and of the Durance (with maximum spring flows from the upper basin and a summer mediterranean drought). Given the sizes of these drainage basins, these regimes are moderate with intermonthly coefficients of 1.6 to 3.3, except for the Rhône at Porte du Scex with its exaggerated alpine characteristics reflected by the intermonthly coefficient of 8.8.

In contrast to these it is possible to define actual 'man-modified' regimes characterized by intermonthly regularity resulting from a lowering of the summer high flows (15 per cent to 24 per cent) and an increase in the winter low flows. Flows in May and August are reduced the most (–20 per cent to –30 per cent); maximum flow increases have occurred in January, February and March (as much as +90 per cent!). These regimes are those of the upper Rhône, from Porte du Scex to Lyon St. Clair and on the Isere. The month of minimum flow may be displaced in relation to the former minimum, in this way the low flows at Sault Brenaz and Lyon are not only inferior to those at the beginning of the century, but the winter minimum is in the fall (October)! The hydrological regime has become an oceanic type . . .

At other stations on the Rhône downstream from Lyon and on the Saône and Ain rivers, there has been a simple exaggeration of the natural regime. There has been an increase in the usual months of high flow (especially February, from +25 per cent to +44 per cent) while the summer minimum is even lower (–14 per cent to –22 per cent). The month of maximum flow at Ternay, Valence and Beaucaire is February, and the flow is double that recorded at the beginning of the century. The month of minimum flow remains the same (July or September) but with a slight decrease (from –11 per cent to –18 per cent).

The present regime of the Saône differs only from the one described by Parde (1942), by the large increases to the February and March flows (+23 per cent to +44 per cent). The Durance at Mirabeau shows the most dramatic change, not in its seasonal rhythm, but in the considerable decrease of flow throughout the year (–91 per cent to –98 per cent) yet retaining the same monthly proportions! The data clearly show a transformation of the Rhône's monthly discharge values, that is equally true for its major tributaries with the Saône showing the least change. Generally speaking, these regimes remain moderate with intermonthly coefficients ranging from 1.5 to 5, except for the Valaisan Rhône, which becomes increasingly moderate (8.8 instead of 3.8), and the residual flow of the Durance which becomes extremely high (14.8)! These comparisons which are carried out on averages calculated from two series separated in time, may be reinforced and complemented by examining the chronological variation of the monthly discharge rates (graphs of accumulated normalized deviations).

Chronological variation of mean monthly discharges from 1920 to 1979

Discharge rates presented in Tables 1 and 2 show that only five or six months have undergone changes characterized by percentages greater than 10 per cent (in relation to the reference series) for all gauging stations. Of the winter months, data for January and February show the greatest differences between the first and second halves of the century.

Using the accumulated normalized deviations, in January, two phases may be distinguished in the series with the opposed phases centered around 1954. From 1920 to 1954 there is a downward trend, up to –10 per cent in relation to the 1920–1979 mean, and from 1954 to 1979 an increase of up to +10 per cent. The increase from 1954 to 1969 is particularly consistent at all four stations (*c.* 17 per cent). For February data from the Sault Brenaz and Beaucaire stations, 1954 again separates the two major periods, the first decreasing (–12 per cent) and the second increasing (+17 per cent) (Figure 4). At the gauging stations

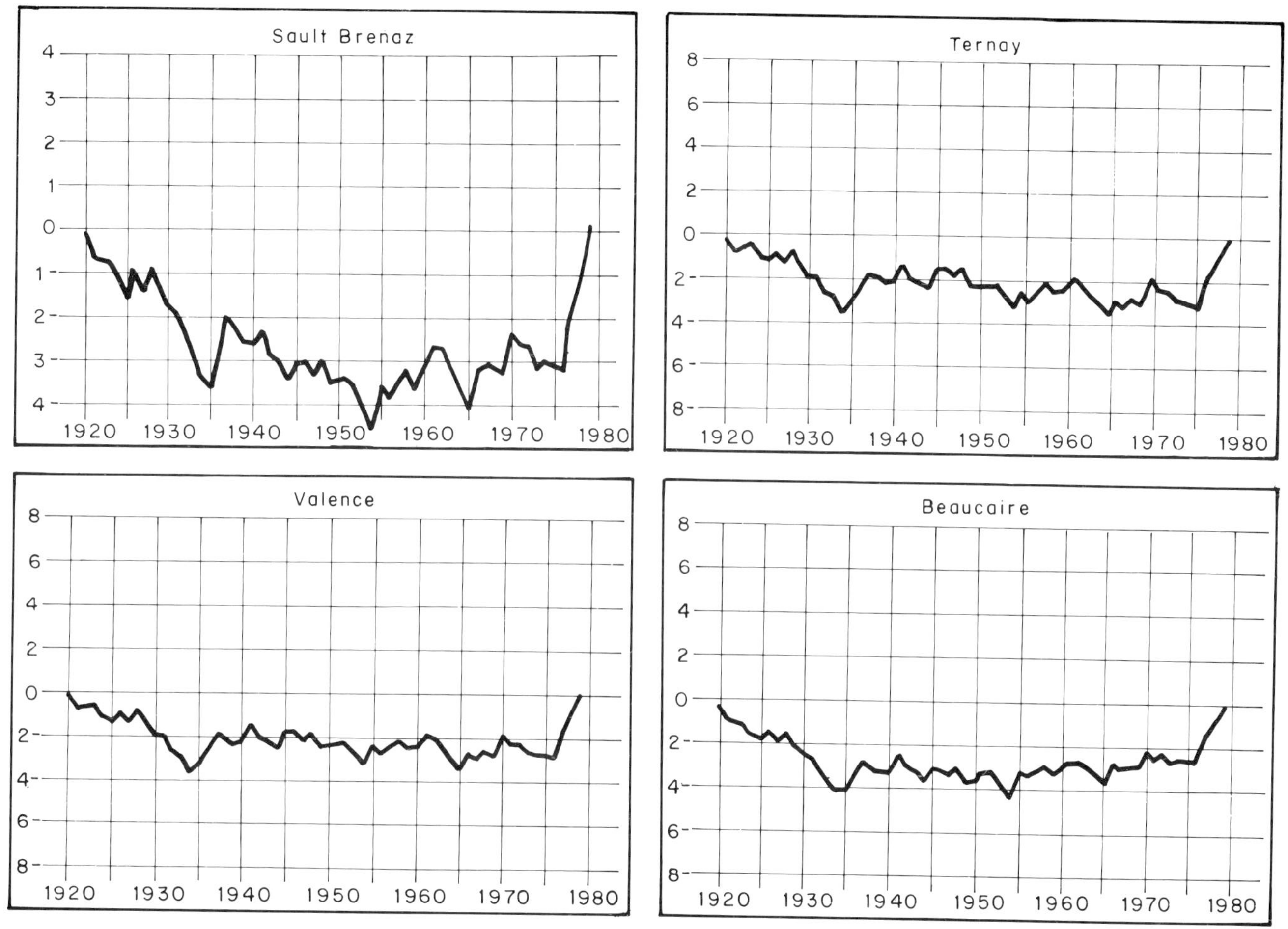

FIGURE 4. Graphs of the accumulated normalized deviations–average discharge rates for February (1920–1979)

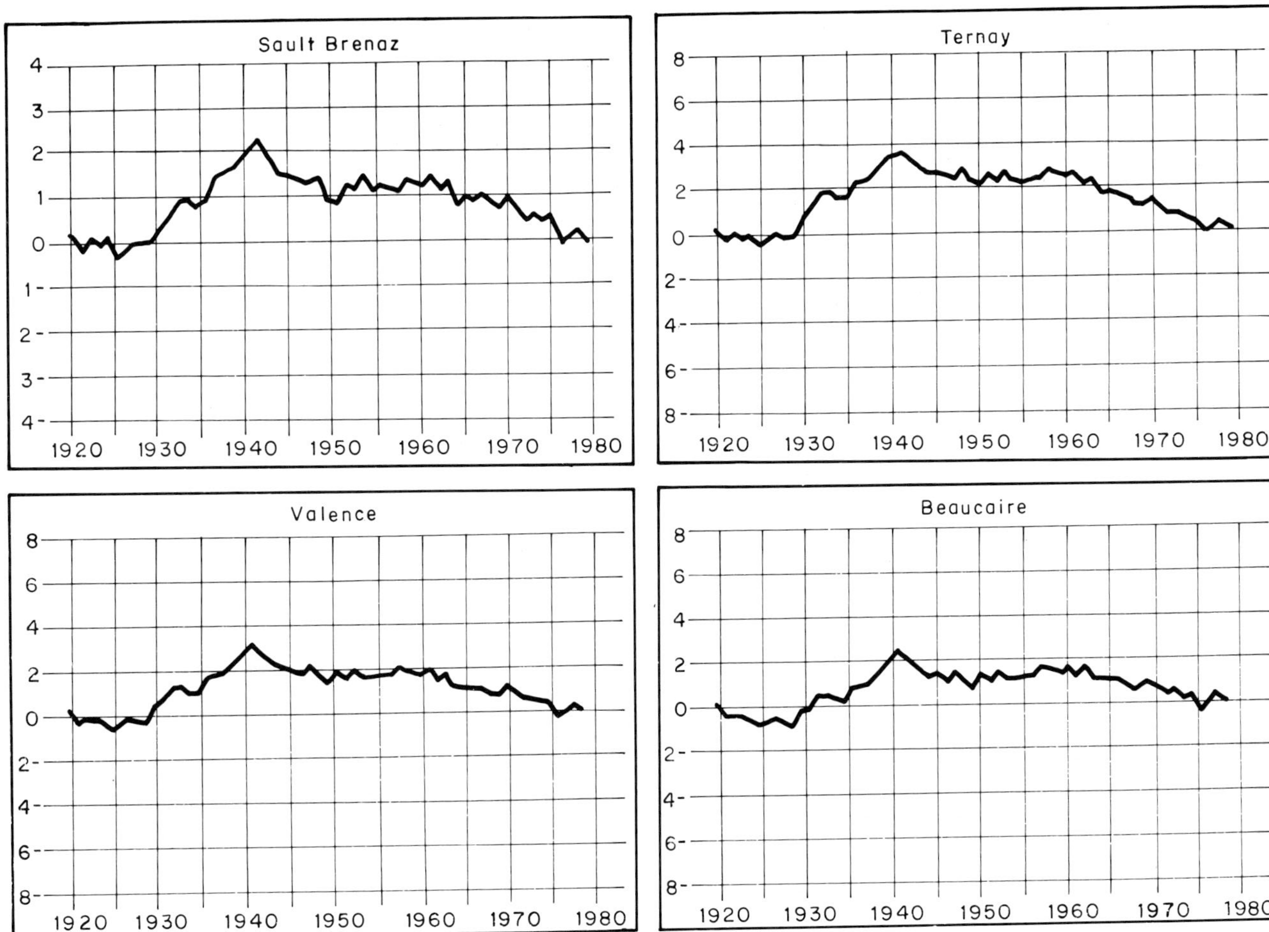

FIGURE 5. Graphs of the accumulated normalized deviations–average discharge rates for July (1920–1979)

of Ternay and Valence the sharp decrease stops in 1934 (−21 per cent) and the increase actually starts in 1965 and reaches +20 per cent in 1979. These fluctuations, which are similar from the north to the south of the drainage basin, are consistent with the assertions made by comparing the average discharge rates in Table 1. The increasing/decreasing phenomenon sometimes extends into March, as at Sault Brenaz, where after recording a decrease from 1920 to 1956 (−7 per cent), the series for 1956–1979 is 11 per cent greater than the 1920–1979 average.

For the summer months (June, July and August) the data (Figure 5) show that 1941 is the turning point of the two opposing trends with an increase or relative stability from 1920 to 1941, followed by a decrease until 1976. The most recent period 1976–1979 greatly exceeds the 1920–1979 average. The decrease from 1941–1976 was recorded equally at the four stations with average deficits ranging from −8 per cent to 11 per cent. This deficit may also be seen in October with identical values of −8 per cent to −12 per cent. The fall season everywhere seems to have been the driest during the second half of the century as is shown by the rainfall record at St. Genis Laval near Lyon (Vivian, 1984). The summer deficit may also be partly due to a decrease in glacier melt. However, summer deficit (and winter surplus) may also be caused by the storage and release of water from the large reservoirs. The role played by the alpine tributaries that are heavily influenced by hydropower operations, as well as the Grande Dixence development of the Valais, must be influential (see Bezinge, 1983).

VARIABLE CONTRIBUTIONS FROM MAJOR TRIBUTARIES

When considering that 80 per cent of the Rhône's mean annual discharge at Beaucaire is supplied by its major tributaries, it becomes evident that a significant change to one of them, between 1877 and 1981, could markedly affect the main river (see Figure 1).

From Table 1 by calculating the average deficits for the two seasons, six months of the cold season and six months of the warm season, it is possible to determine the contribution of the alpine part of the Rhône and its major tributaries to the average annual discharge at Lyon and Beaucaire at the turn of the century and after 1951. The data indicate that the individual and cumulative percentage contributions by the Arve, the Fier, the Ain and the Saône, changed little from 1877 to 1981 in relation to the Rhône's mean annual discharge at Givors (64 per cent before 1951–67 per cent after). Due to the abstractions from the Durance on the lower Rhône, the percentage contribution of the tributaries between Lyon and Beaucaire is reduced (33 per cent before 1951, 24 per cent after). If the Valais Rhône at Porte du Scex is added to the modified French tributaries the percentages remain quite constant except at Beaucaire. The significance of the Saône river, which has increased during the recent period, provides a more faithful reflection of climatic influences than the Isere or Durance rivers.

Seasonal variations

With reference to the cold season from October to March, as with the previous calculation, the percentage of the mean annual discharge of the Rhône at Givors and Beaucaire represented by the mean discharge rate for the cold season, of the upper Rhône and its tributaries, yields the following results: Givors 84 per cent before 1951, 96 per cent after. The tributaries influenced by hydroelectric developments show an increase during the winter, but so does the more 'natural' Saône. The weakened influence of the Durance's flow causes a reduction in the percentage of the middle and lower Rhône's major tributaries represented by the mean discharge rate at Beaucaire (27 per cent before 1951, 23 per cent after); however, an increase in the winter flow since 1951 is evident throughout the drainage basin (76 per cent to 81 per cent).

Contrary to the winter series, in the warm season (April–September) the trend is characterized by a clearly decreasing influence of tributary flow during the year. It should be noted that the part played by the Saône contrasts with the changes caused by hydroelectric storage since during the winter the natural and anthropological changes counter balance one another. The hydroelectric and irrigation schemes of the Isere and Durance rivers further modify the flows of the lower Rhône. The contributions of the Isere, Ardèche and Durance between 1951–1981 to the Rhône at Beaucaire was 434 $m^3 s^{-1}$ (26 per cent) compared with 710 $m^3 s^{-1}$ (39 per cent) at the beginning of the century. When considering all the tributaries of the Rhône (Swiss and French) in relation to the annual discharge at Beaucaire, the discharge during the warm season is equal to 85 per cent of the total flow before 1951 and 70 per cent thereafter.

Variations in the alpine Rhône and its tributaries is undeniable as far as an explanation of the seasonal variations of the river between the Valais and the delta is concerned; these variations being the consequences of the changing climatic and anthropogenic influences. In order to determine the relative importance of each influence and the extent to which each affects the monthly discharge rates, a simple but demonstrative comparison was carried out.

NATURAL AND ANTHROPOLOGICAL INFLUENCES ON THE RHÔNE DISCHARGE

This section summarizes the research published by Vivian (1984, 1986a, 1986b). For several gauging stations the discharge records for the period of 1920 to 1979 have been analysed as three groups:

- —Slightly changed discharges, before 1950, called 'natural' discharges;
- —Artificial discharges, measured at the gauging stations during the entire 1950–1971 period;
- —'Naturalized' discharges which use a correction factor to take into account the hydroelectric utilization of water, yielding a hypothetical value that excludes the effect of upstream developments.

The 20th century series was chosen for reasons of data homogeneity and was obtained from the C.N.R. Although the reference period, 1920–1949, includes some minor development, it is certain that at this time flow variations were more the result of climatic fluctuations than the influence of water projects. This study is based upon the discharges recorded at the gauging stations of the alpine Porte du Scex and Sault Brenaz, whose positions upstream from Lyon are close to the influence of hydroelectric reservoirs, while remaining under the influence of the alpine and oceanic climates. Lastly, Valence, situated at the junction of the Isere and Rhône, represents a climatic transitional zone, namely, oceanic–mediterranean. The naturalized discharges for the recent period are taken to represent the climatically

TABLE 3. Natural influence (*NI*) and anthropological influence (*AI*) for the 1950 to 1979 period (reference period 1920–1949) expressed in a positive or negative percentage of the reference discharge rates

Station	J	F	M	A	M	Jn	J	A	S	O	N	D
Porte du Scex												
1920–49	54.7	51.4	71	67.2	206	390	439	374	244	121	87	62
1950–79 *QI*	83.9	87.7	114.1	109	190.3	327.2	336.4	284.9	190.8	119.2	99.7	88.7
1950–79 *QRN*	47.7	48.5	77.6	84.2	198.5	383.6	409.7	352.9	203.6	110	73.7	54.7
NI	−12.8	−5.6	+9.3	+25.3	−3.6	−1.6	−6.7	−5.6	−16.6	−9.0	−15.3	−11.8
$\frac{QI-Qr}{Qr}$	+53.4	+70.6	+60.7	+62.2	−7.6	−16.0	−23.4	−23.8	−21.8	−1.5	+14.6	+43
AI	*+66.2*	*+76.2*	*+51.4*	*+36.9*	*−4.0*	*−14.4*	*−16.7*	*−18.2*	*−5.2*	*+7.5*	*+29.9*	*+54.8*
Sault Brénaz												
1920–49	341	360	393	436	475	591	607	553	451	339	391	338
1950–79 *QI*	383	455	450	458	486	565	571	502	404	324	389	398
1950–79 *QRN*	317	390	395	454	544	673	640	559	407	303	359	354
NI	−7.0	+8.3	+0.5	+4.0	+14.5	+13.9	+5.4	+1.0	−9.7	−10.6	−8.2	+4.7
$\frac{QI-Qr}{Qr}$	+12.3	+26.4	+14.5	+5.0	+2.3	−4.4	−5.9	−9.2	−10.4	−4.4	−0.5	+17.7
AI	*+19.3*	*+18.0*	*+14.0*	*+1.0*	*−12.2*	*−18.3*	*−11.3*	*−10.2*	*−0.7*	*+6.2*	*+7.7*	*+13.0*
Valence 1920–49	1525	1511	1541	1538	1556	1638	1387	1159	1091	1073	1490	1321
1950–79 *QI*	1612	1788	1614	1563	1462	1477	1249	1058	998	987	1357	1526
1950–79 *QRN*	1496	1680	1550	1569	1567	1667	1380	1223	1030	947	1253	1421
NI	−1.9	+11.2	+0.6	+2.0	+0.7	+1.8	−0.5	+5.5	−5.6	−11.7	−15.9	+7.6
$\frac{QI-Qr}{Qr}$	+5.7	+18.3	+4.7	+1.6	−6.0	−9.8	−9.9	−8.7	−8.5	−8.0	−8.9	+15.5
AI	*+7.6*	*+7.1*	*+4.1*	*−0.4*	*−6.7*	*−11.6*	*−9.4*	*−14.2*	*−2.9*	*+3.7*	*+7.0*	*+7.9*

Note: *Qr* : reference discharge (1920–49)
QI : influenced discharge
QRN : natural or reconstituted natural discharge
NI : natural influence
AI : anthropological influence

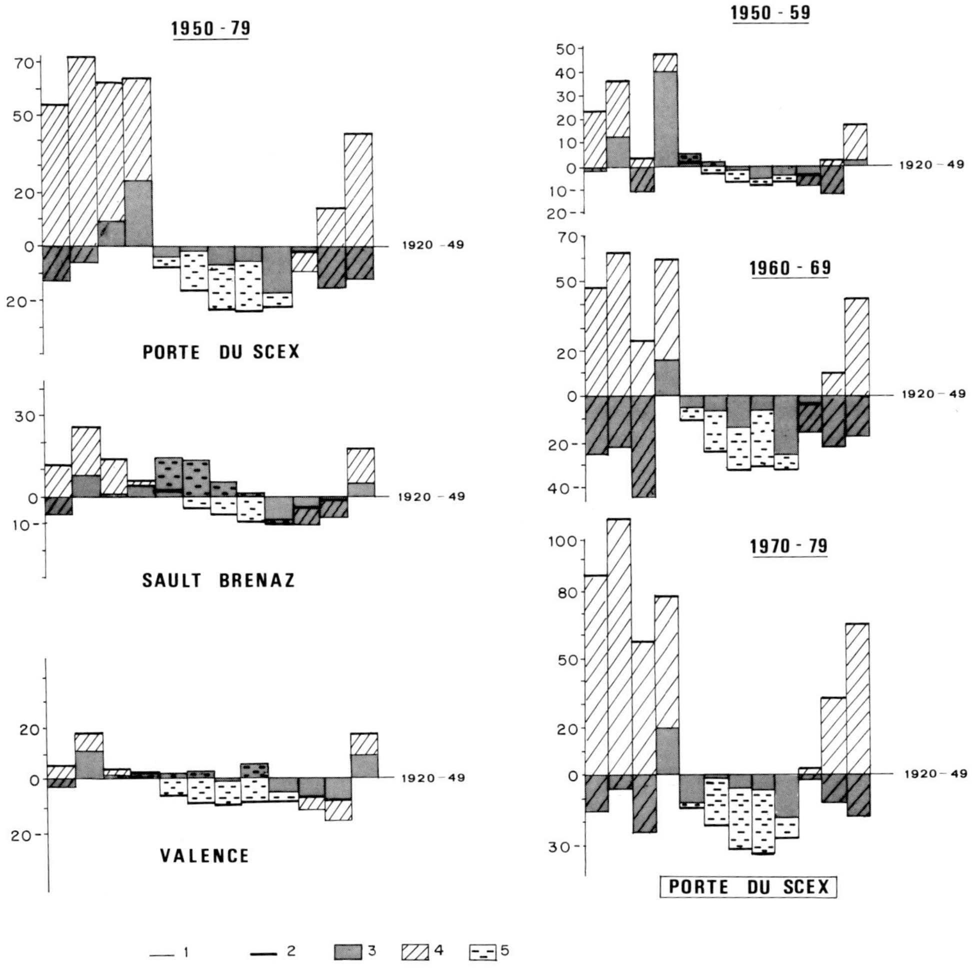

FIGURE 6

induced fluctuations of the river. The difference in relation to the 1920–1949 discharges (expressed in percentage of the reference discharge) quantifies these fluctuations. The actual discharges reflect both the natural and anthropological influences. These are also defined as a percentage of the reference discharge and by comparing these two percentages, marked with a plus (+) or minus (−) depending on the direction of change, it becomes easy to deduce the relative effects of anthropological transformations (Table 3, Figure 6). The heavily developed monthly percentages for the two upstream stations differ sharply from the downstream ones, where the importance of the natural flow (especially the Saône) is to attenuate the anthropological influence. At Porte du Scex this percentage reaches +76 per cent, at Valence the maximum value is only −11.6 per cent.

In Table 3 it is interesting to note the similarity between the discharge pattern at the three gauging stations: everywhere during the recent period there has been an increase in

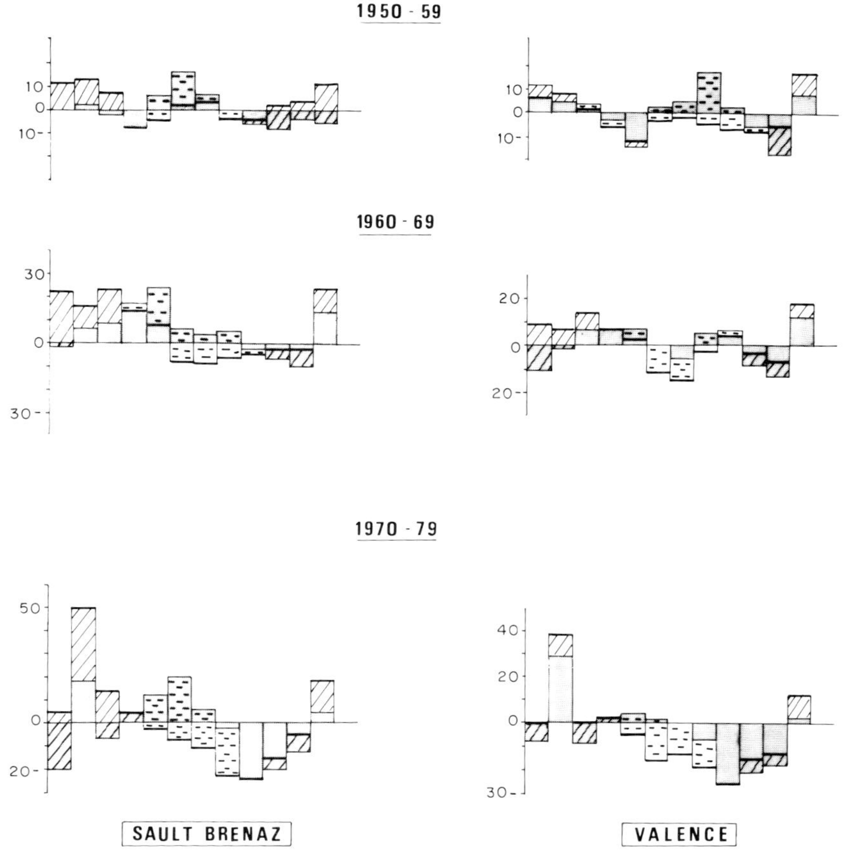

FIGURE 6. Natural and anthropological influence (in percentage of the reference discharge rates 1920–1949).
Key: 1. Natural discharge or reconstituted natural discharge rates expressed as a percentage of the reference values; 2. Actual discharge; 3. Natural influence; 4. Positive anthropological influence; 5. Negative anthropological influence

the monthly flow from October or November to March or April, and a decrease from May to September. At Porte du Scex and Sault Brenaz the winter increase affects the low discharge rates, and the summer decrease the high discharge rates. At Valence the increase accentuates the highest yearly discharges while the decrease reduces the low summer discharge rates still further. The first two gauging stations record the effect of water storage which operates parallel to the natural regime of the alpine rivers—a deduction made from the summer discharge. At Valence this regime resembles the Rhône's alpine tributaries (especially the Isere river) which have, at this point on its course, a pluvial regime with a winter maximum and summer minimum.

The three months with maximum positive percentages are December, January and February. The average percentages are: Porte du Scex, +66 per cent; Sault Brenaz, +17 per cent; and Valence, +7 per cent. The maximum negative percentages are in

June, July and August. The average decrease for these same stations are respectively: − 16 per cent, − 13 per cent, and − 12 per cent. A comparison with percentages representing the natural influence demonstrates that natural change plays a part in the fluctuations observed in the measured discharge rates.

At Porte du Scex, during the winter, natural influences raised the March and (especially) the April discharge rates by + 9 per cent and + 25 per cent respectively (whilst the artificial increase is + 51 per cent and + 37 per cent). During the summer, the reduction in glacial melt and the decrease in precipitation reduced the natural flow by − 7 per cent, − 6 per cent, − 17 per cent in July, August and September respectively (the artificial decrease being − 17 per cent, − 18 per cent and − 5 per cent for these months). The decrease recorded in September and the increase of April appears to be a natural variation. The marked increases for December, January and February however, are caused only by anthropological actions, since the natural discharges at this time have decreased.

At Sault Brenaz the natural influence contributes + 5 per cent and + 8 per cent to the increase recorded from December to February (artificial increase from + 13 per cent to + 19 per cent). During the summer, however, the anthropological influence has a great significance in reducing flows (− 18 per cent in June; − 11 per cent in July). It should be remembered that the natural summer increase for the 1950–1979 period as compared with the 1920–1949 series, was due to the occurrence of the well-defined dry decade, 1940–1949. This was especially noticeable during the summer, with an effect that was clearly more important than the drought of 1976.

At Valence, once again natural causes explain more than 50 per cent of the increase in the actual discharge values for February and December (+ 11 per cent out of + 18 per cent, and + 8 per cent out of + 15 per cent). The reductions in September and October (− 8 per cent) also appears to be mostly natural. Along the middle Rhône, the summer reduction is perhaps not entirely due to the effect of a decrease in rainfall and of water storage in the alpine reservoirs. The agricultural requirements of the C.N.R. necessitate a further water reduction of 175 $m^3 s^{-1}$ to ensure all supplies to irrigation. It is for this reason that the low water of the middle Rhône hydrosystem (canals and water table) must be effectively managed when agricultural needs are great. The situation at Beaucaire seems similar; however, the discharge values are not corrected, and so the previous type of comparison is impossible. It should be recalled that the use of the Durance waters does not affect the Rhône, since this tributary represents an insignificant proportion of the total flow throughout the year. The additional reduction of low waters at Beaucaire (Figure 3)—made evident by a comparison of the average discharges at the beginning of the century and from 1951–1981—might be due to low precipitation but the abstraction of flow for 40 000 ha of irrigated land in the valley will have an important influence. The January-February-March increases are partly caused by water releases from the hydropower plants, but even more so by the precipitation increases recorded throughout the drainage basin since 1950.

Sequential changes of the flow regime

That the identified fluctuations, and the importance of anthropological actions for certain months, have come about progressively and in an increasing order, may be shown by analysing the discharge data for each decade. Data analysis and the calculation of the relative

TABLE 4. Chronological evolution, 1950 to 1979, of the natural and anthropological influences (percentage)

Station		J	F	M	A	M	Jn	J	A	S	O	N	D
Porte du Scex	1	+2	+12	−11	+40	+5	+1	−1	−5	−5	−9	−12	+1
1950–59		+21	+25	+14	+7	+4	−4	−6	−5	−2	+5	+13	+18
	2	−25	−22	−43	+16	−5	−7	−13	−5	−25	−16	−21	−18
1960–69		+74	+85	+65	+44	−5	−17	−19	−25	−6	+13	+31	+61
	3	−16	−7	−23	+20	−11	−0.6	−6	−7	−19	−2	−12	−19
1970–79		+103	+119	+81	+59	−3	−21	−25	−25	−8	+4	+45	+86
Sault Brenaz	1	0	+2	−1	−5	+7	+16	+6	0	−5	−5	−3	−3.5
1950–59		+11	+10	+8	+0.2	−12	−15	−5	−2	+0.5	+10	+6	+15
	2	−1	+5	+9	+17	+24	+6	+4	+5	−2	−5	−10	+13
1960–69		+23	+11	+14	−2	−15	−12	−12	−9	−2	+3	+9	+10
	3	−20	+18	−6	0	+12	+20	+6	−2	−23	+20	−12	+15
1970–79		+24	+32	+21	+4	−15	−27	−17	−20	−0.5	+6	+8	+14
Valence	1	+12	+5	+3	−3	−13	+3	+5	+18	+3	−6	−18	+9
1950–59		−4	+3	−2	+1	−0.5	−7	−7	−22	−9	−1	+12	+8
	2	−10	−1	+8	+9	+9	0	−6	+5	+5	−9	−13	+12
1960–69		+20	+7	+5	−0.6	−8	−11	−8	−7	−1	+6	+5	+6
	3	−7	+29	−9	+0.4	+5	+2	−0.2	−6	−25	−20	−17	+2
1970–79		+7	+10	+9	+1	−11	−17	−12	−13	+0.5	+6	+4	+9

Note: +2 (natural influence)
+21 (anthropological influence)

importance of both influences was carried out for the periods: 1950–1959, 1960–1969 and 1970–1979; while still using the 1920–1949 period as reference. Table 4 summarizes the percentages attributed to the natural and anthropological influences, while Figure 6 shows the progressive changes undergone by the combination of these two flow parameters. For the three stations studied, the increase in the anthropological influence is quite clear for the key months of the regime: December–January–February–March on the one hand, and May–June–July–August on the other. Nonetheless, a slight difference should be noted depending upon the decade and the gauging station concerned: the artificial control of the discharge rates diminishes significantly downstream.

At Porte du Scex on the Valais Rhône, the anthropological influence is present especially during the winter months as early as 1950, and increases regularly and strongly until 1979 with an increase of more than 100 per cent for the December to March months which should otherwise have reduced naturally. Man's influence is not so great, in absolute terms, during the summer months; but there is clear progression (the percentages quadruple between 1950 and 1979). During May to September natural and anthropogenic influences contribute to the decrease in discharge, but the latter are dominant.

At Sault Brenaz the progression of anthropological change from 1950 to 1979 remains apparent, the winter average flow varying from +10 per cent to +20 per cent and the summer monthly average flow from −7 per cent to −16 per cent. However, the increases do not exceed +32 per cent (February, 1970–1979) and decreases are less than −27 per cent (June, 1970–1979). Since 1950, the detail of the three series shows that January, February and March are the months most disturbed by Man. The percentage contributions triple or quadruple between the first and third decades. Furthermore, a natural increase of +18 per cent for February during the last decade should be noted. Natural causes also seem important in explaining the recent decrease in the autumn discharges (−20 per cent to −23 per cent).

At Valence the importance of anthropological change is noticeable especially during the summer months (with a maximum of −12 per cent to −17 per cent). This is due to the impact of all the upstream hydroelectric power schemes to which must be added several deductions for irrigation. From September to November the discharge reduction is due to natural causes, mainly the rainfall deficit (−17 per cent to −25 per cent). The anthropological influence of supplementing flow by reservoir releases far from compensates for this reduction. Natural changes also widely affect the February discharge for the 1970–1979 series which increased by 29 per cent (natural change) and 10 per cent (man-made change). The influence of various water works during the winter season persists from December to March but only account for less than 10 per cent (a percentage which may only represent the error factor in the data).

CONCLUSION

After examining the discharge data and reflecting upon the importance of the changes in the annual and monthly discharge rates of the Rhône throughout the century, and on the respective part played by natural and anthropogenic influences, four conclusions may be drawn. First, the discharge values (measured at the gauging stations) and the chronological series indicate that little change occurred until 1950, thereafter anthropological influences are evident. Secondly, the changes to the Rhône's hydrological regime decrease rapidly downstream, and upstream from Lyon the changes are less apparent than in the Valais. Thirdly, this downstream reduction in the magnitude of change may be explained by a switch from a simple regime to a compensated regime, the compensation takes place due to both hydroelectric and climatic influences. And, finally, for certain months, there are important changes due to either a combined natural and anthropological influence (February–March–June–July–August), or else the pre-eminent action of one of these two influences.

REFERENCES

Bezinge, A. 1983. Aménagements hydroélectriques du Valais: leurs influences sur le Rhône. *Rencontres solaires suisses, Sierre, Octobre 1983*, 12 pp.

Compagnie Nationale du Rhône 1981. *Développement Programme—Achievement*, Campagnie Nationale du Rhône; Lyon.

Edouard, J. L. and Vivian, H. 1982. La nouvelle hydrologie d'une vallée alpine: La Maurienne. *Revue de Géographie Alpine*, **3**, 258–278.

Edouard, J. L. and Vivian, H. 1984. Une hydrologie naturelle dans les Alpes du Nord? Les nouveaux paramètres de l'hydrologie alpine: les aménagements hydroélectriques. *Revue de Géographie Alpine*, **2/3**, 165–188.

Juramy, S. and Montfort, I. 1987. *Evolution des Lits Fluviaux—l'Exemple d'une Rivière Amenagée: la Durance*, Doctorat 3ème cycle, Université d'Aix Marseille II, Institut de Géographie, 2 vol., 555 pages.

Kasser, P. 1969. *Influence of changes in the glaciated area on summer runoff in the Porte du Scex drainage basin of the Rhône*, Paper presented at the Symposium of the Hydrology of glaciers—Cambridge.

Martin S., Mounier, J. and Marchand, J. P. 1981. Variations pluviométriques durant la dernière période séculaire en Europe Occidentale. *La Houille Blanche*, **7/8**.

Parde, M. 1942. Quelques nouveautés sur le régime du Rhône. *Mémoires et Documents de l'Institut des Etudes Rhodaniennes de l'Université de Lyon*, 172 pp., 15 annexes.

Vivian, R. 1975. *Les Glaciers des Alpes Occidentales*. Edition Ophrys, Gap, 514 pp.

Vivian, H. 1983. Les fluctuations de l'abondance annuelle, mensuelle et des rythmes journaliers du Rhône supérieur à Sault Brenaz. *Revue de Géographie Alpine*, **4**, 311–329.

Vivian, H. 1984. Changements observés dans l'hydrologie nord-alpine: essai de détermination de la part des fluctuations des précipitations et des actions anthropiques. *23è Congrès International de Météorologie Alpine*, OPATIJA, Yougoslavie Septembre 84, Zbornik meteoroloskih radova—Beograd 319–324.

Vivian, H. 1986a. Le Rhône supérieur franco suisse: fluctuations naturelles et artificielles du régime fluvial. *Revue de Géographie Alpine*, **1/2**, 154–164, Mélanges J. Loup.

Vivian, H. 1986b. Un exemple de régime influencé: hydrologie et hydroélectricité dans les cours d'eau des Alpes du Nord. *Hydrologie continentale*, **1**, 1, 63–78.

Historical Change of Large Alluvial Rivers: Western Europe
Edited by G. E. Petts
©1989 John Wiley & Sons Ltd

5

Use of Cartographic Sources for Analysing River Channel Change with Examples from Britain

J. M. HOOKE and C. E. REDMOND

Department of Geography, Portsmouth Polytechnic, Portsmouth, Hampshire, UK

INTRODUCTION

Historical maps have the potential to provide much information on large rivers including considerable detail of river channel changes. A general background on the location and timescales of changes of channel morphology, particularly of channel bars, is important for understanding the dynamics of lotic ecosystems. A knowledge of the history of the river is of particular value in interpretations of floodplain development, and will help understanding of its composition and prediction of the location of certain types of deposit and ecological resource patches. Most studies are made of individual rivers and particular locations. Historical evidence may be needed to establish whether a section of channel has been stable or not over a certain time period, to delimit stable and mobile reaches, and to date old channels or floodplain segments. Furthermore, such evidence may help to date earlier river engineering works and channelization schemes and, in conjunction with other sources, can indicate the extent and nature of channel management and interference in the past. It can also show the timing and nature of morphological adjustment to such work. In practical terms, a knowledge of the history of a river can be important in dealing with the undermining of structures and design of future structures; in understanding the distribution, scale and effects of bank erosion; in managing problems associated with land drainage; and in designing maintenance programmes.

Historical studies are often used to complement process studies of erosion, sediment transport or deposition to assess the impact of the process and its degree of continuity. The purpose of study may be prediction of future changes and behaviour of the channel, and this is much more likely to be valid if an historical perspective is taken rather than relying solely on contemporary observations. The degree of detail required may vary from general prediction of instability to forecasts of the direction and style of change of each bend of a river. Historical data on modes of change can be derived from maps and, if the sources are sufficiently accurate and large-scale, then actual rates of movement can be measured. Variations in rate and river behaviour over time may also provide some clue as to possible contributory causes of change, though other historical documents may be necessary to establish changes in conditions such as land use which have taken place. Overall an historical perspective shows the persistence of a characteristic, whether there has been a trend or fluctuations over time, and the rate at which changes have taken place.

MAP INFORMATION AND ITS LIMITATIONS

Since maps show rivers in plan, it is only information on characteristics of patterns or networks that can realistically be derived, though occasionally engineers' plans will also show cross-sections. However, maps can be supplemented by descriptions of channels and of flood events to increase understanding and interpretation of changes. Ideally, maps at a sufficiently large scale and of sufficient accuracy, and number of editions, will be available to enable changes in morphology to be measured and analysed in detail. The general categories of information on channel pattern which can be derived from maps are given in Table 1. A number of problems arise in using such sources, some of which are associated with the historical nature of the sources, some have to do with the style and accuracy of the maps themselves, and some are related to the methods of analysis.

It is inherent in the use of historical sources that the user has little control over sampling design but has simply to accept the map evidence available, which is usually much scantier in terms of revision dates than is desirable. The maps only provide a picture of the situation at the time of mapping, so changes between the dates have to be inferred from other sources and from field evidence, as well as from river behaviour itself. This leads to particular problems in assessing the role of large events and determining the exact timing of changes. However, on meandering rivers, if the changes have been consistent in direction over time on several different maps and field evidence of old channels supports this, then it is usually safe to assume that the river has not gone back and forth between map dates. No such assumptions can be made for braided rivers which are much more irregular in their style of change.

In terms of the sources themselves, checks must be made initially on the known accuracy and provenance of the type of source, though often the background of major historical

TABLE 1. Information that can be derived from historical maps

Data	Map compilation	Analysis technique
Location and distribution of instability and old channels	Superimposed or sequence	Visual inspection, map abstract
Length of change Sinuosity, degree of braiding	Separate or superimposed	Direct measurement Comparison for amount of change
Distances and rates of movement Areas of movement	Superimposed Superimposed, shaded	Direct measurement Frame-grabber and image analysis
Direction of movement	Superimposed	Direct measurement, vector plots
Type of change	Sequence or superimposed	Classification
Change in characteristics	Digitized courses	Bend or reach definition Calculation
Floodplain ages	Superimposed	Mapping and zoning

sources has been investigated by historians and historical geographers. The scale of map to be used depends on the size of the river and purposes of analysis but in general the river should be shown as a channel several millimetres wide on the map for quantitative information to be at all reliable. Once a source has been identified as suitable and its coverage of a particular area established, then checks of the specific accuracy of that map sheet must be made. These are discussed fully in Hooke and Kain (1982). Major problems with maps of river courses are the conventions and methods of survey and of depiction used. These may have varied over time and spatially with different surveyors and cartographers. They relate especially to the level at which the water is mapped, which affects the pattern and width portrayed. Rivers with many bars are particularly problematical in this respect.

The greatest safeguard against misinterpretation of map information is to use it in close association with fieldwork and understanding of the river. Very often the old channels on the maps can be matched with features on the floodplain and, even if channels are not distinct, zones of channel movement of relative age can usually be distinguished by topography and vegetation. The map evidence should always be corroborated as far as possible both by other documentary evidence and by field evidence. Types of errors and limitations of maps, together with checks to be made on the reliability of information are provided in Table 2.

CARTOGRAPHIC SOURCES IN BRITAIN

Maps which are of potential use in assessing channel changes along British rivers can be divided into four main groups: early manuscript maps, 'county' maps, Ordnance Survey (O.S.) maps, and deposited plans. The details of availability, coverage, features, accuracy and background information are provided in Table 3. In other countries pilot charts have proved to be valuable for large navigable rivers but in Britain such charts are applicable only to the major estuaries.

Early manuscript maps are of two main types: those drawn for individual purposes and usually covering a limited area, such as estate maps, and those which were part of a larger

TABLE 2. Errors and limitations of historical maps

Limitations	Checks
1. Nature of historical sources	
Only particular dates documented so no sampling control	Field checking
Difficult to infer nature and timing of changes between map dates	Consistency, documentation of floods
Use only in repository	
2. Nature of maps of rivers	
Methods of survey	Surveyor, instruments,
Cartographic representation	cartographer, conventions
Perception of change	originality, purpose
3. Individual maps	
Shrinkage and stretching	Measurements
Planimetric accuracy	Measurements
Content accuracy	Documentary and field corroboration

TABLE 3. Details of historical maps available for Britain

Map source	Date	Scale	Form	Location	Coverage	Accuracy	References
Estate	Some late C16th Majority 1700→	Usually 1:2376	MS	PRO Scottish Record Office Scotland British Library Bodleian, Oxford National Library of Wales LROs	Good for England and Over 20 000 available for consultation	Variable	Harley (1972)
Enclosure	C18th and C19th		MS	PRO Libraries CROs	England and Wales	Variable	Tate (1978)
Tithe	1840s	Usually 1:2376 or 1:4752	MS	CROs PRO	3/4 of England and Wales	Variable	Kain and Prince (1985)
Deposited plans	Post 1794	Min. scale 1:15 840 after 1807	MS	CROs House of Lords Records Office	Britain	Variable	Harley (1972)
County	C18th and C19th	1:63 360 1:31 680	MS and published	CROs Museums	England and Wales	Variable	Skelton (1970) Rodger (1972)
Military Survey of Scotland (Roy)	1747–1755	1:36 000	MS		Scotland	Reasonably accurate	Moir (1973)
Ordnance Survey Old Series 1st ed.	1795–1873	1″ = 1 mile 1:63 360	Printed	British Library Museums	England and Wales	Poor	Harley (1965)
New Series 2nd ed.	1840–1893	1:63 360 1″ = 1 mile	Printed	British Library Museums	England and Wales	Moderate	Harley (1965)
3rd ed.	1901–1913	1″ = 1 mile	Printed	British Library Museums	England and Wales	Moderate	
4th ed.	1913–1926	1″ = 1 mile	Printed	British Library Museums	England and Wales	Moderate	
County Series	1840–1893	1:10 560	Printed	Museums Libraries	England and Wales	Good	Harley (1965)
1st ed.	1853–1893	1:25 000		British Library CROs			
1st revision 2nd revision	1891–1914 1904–1923						
National Grid Series	1948–1973	1:10 560 6″ = 1 mile (1:10 000 since 1969) 1:25 000	Printed	Libraries British Library	Great Britain	Good bank lines ±1 m	Harley (1975)

movement or instruments of national legislation, namely tithe and enclosure maps. Many estate maps date from the late eighteenth century or early nineteenth and were surveyed for particular landowners to show the extent and tenancy of their lands. It is often a matter of luck whether such maps are available for a particular stretch of river. They vary considerably in scale and standard of survey and cartography, and also in the amount of detail. Often only restricted areas or fragmented parcels of land are depicted but others show extensive estates and, since rivers are often boundaries, these are carefully delineated. Accurate estate maps can prove extremely valuable since they are often of earlier date than other reliable maps as McEwan (1985) has found.

The oldest widespread and large-scale maps generally available are the tithe maps which cover 3/4 of England and Wales. These are detailed maps which were produced on a parish basis in compliance with the Tithe Commutation Act of 1836 (Kain and Prince, 1985). These maps are mostly at scales of three, four or six chains to the inch (1/2376, 1/3168 and 1/4752 respectively) and surveyed between 1838 and 1845. The maps show details of field boundaries and buildings and include river courses. The original maps are now in the Public Record Office at their Kew repository but usually another copy is available in the appropriate County Record Office. The tithe maps vary enormously in standard of survey, cartography and preservation so tests of accuracy should always be made. Hooke and Perry (1976) compared English tithe maps with 1/2500 Ordnance Survey County Series maps and overall found planimetric errors averaging 3–4 per cent but with large variations both within and between maps. It appears that the tithe maps are less accurate planimetrically for upland areas.

Enclosure maps are available for many areas of England and Wales but the coverage tends to complement that of tithe maps (Harley, 1967). They usually only show fragmented plots of land and although they may incidentally show a river, particularly where it forms a boundary, they are of much less value than tithe maps.

The second category, county maps, encompass a variety of manuscript and published maps. They date from Elizabethan times to the nineteenth century. As sources of data these maps are very variable. The larger scale and later maps tend to be the more reliable and useful, of course, and they have been used for the study of coastal changes (e.g. Carr, 1969). The Scottish equivalent of the county maps in terms of scale and content and a precursor of the Ordnance Survey was Roy's Military Survey of Scotland (1745–1755) in which maps were produced at a scale of one inch to 1000 yards (1/36 000). Generally the survey is considered fairly accurate and has been used in relation to river channels by Ferguson and Werritty (1983) and McEwan (1985).

The first maps to be produced by the Ordnance Survey were at a scale of one inch to one mile (1/63 360) and were published between 1805 and 1873. These are of dubious accuracy (Carr, 1962; Harley, 1968) so that their use is not recommended for the study of change in topographic features (Hooke and Kain, 1982). From the 1870s onwards large-scale maps at 1/10 560 and 1/2500 were produced by the Ordnance Survey. There were several editions of these County Series maps (Table 3), later editions incorporating revisions and with the number of editions varying over the country. After 1939 these maps were superceded by the National Grid Series maps; the number of editions published generally depends on the amount of development and construction in an area. With regard to accuracy and reliability, the Ordnance Survey make their own tests but 1/2500 maps are considered to have river bank lines shown to an accuracy of ± 1 m. County Series maps were found to be of comparable accuracy to the National Grid Series maps. A detailed study of O.S. maps is provided by Harley (1975).

Deposited plans form a large category of documents in many County Record Offices; they comprise plans and maps on which proposed courses of canals, roads, railways and utilities were indicated. Some simply show the route superimposed on an O.S. first edition 1/63 360 map but others are at a larger scale and specifically drawn. Many only show detail adjacent to the route and so only depict rivers where they are crossed. Bridge plans are available for some counties, for example in Lancashire there are books of plans dating from about 1803 of all road bridges upkept by the county.

ANALYSING RIVERS ON MAPS

Methods of compilation

The choice of method of compilation depends on the form of map availability and the compatability of map scales as well as on the information required. If only qualitative information or isolated pieces of data are needed then it may be sufficient to assess these factors direct from the maps. This does not make full use of the map potential but may be an advantageous approach where maps are in the restricted custody of museums or cannot be handled. This approach does not allow analysis of the whole channel behaviour and has the disadvantage that it is difficult to go back for extra information.

If details of planform alteration including measurements of distances and rates of movement and of change in form are required then it is necessary to compare the map courses at a common scale. Again, because of the restricted availability of these maps the easiest strategy is often to trace each map individually. Care must be taken that a stable tracing material is used and that all necessary information is transferred including enough fixed points and ground detail for matching up maps of different date. If all the maps being used are of common scale then it is possible to make a base map of one map edition, the latest usually being recommended, and to trace courses of other dates directly onto this at the map repository (using different colours or symbols to distinguish the map dates). Other forms of copying or reproduction are likely to incorporate distortion which makes planform analysis impossible. Photography is expensive and may not always be feasible in map repositories. Once traces of individual maps have been obtained of the channels from the original maps these need to be compiled at common scale to elucidate the changes. If the maps are of differing scales then this can be done using a plan-variograph or similar machine. Care should be taken to match only small sections at once to avoid distortion and sufficient ground detail should be matched to ensure accuracy. This is often a problem in floodplain areas since from their very nature they lack structures or boundaries. Photographic prints can of course be produced at whatever scale is needed but a means may have to be found to superimpose the courses.

As digitizers and computers become more portable then it may be feasible to take these instruments to the map repository and digitize the courses of the river at source. Adjustments of scale can then be made and plots produced for any section and at any size from the computer. Some of the questions related to digitizing are discussed below.

Abstraction of data

The maps of the channel can be used directly to demonstrate and analyse changes. Very many studies using historical sources simply show the maps either in sequence or superimposed. There are some problems of cartographic design to show multiple courses on a superimposed map clearly especially in black and white but the clearest depictions are probably those using solid lines for both banks and shading for different dates or with some courses having pecked lines if multiple dates and complex changes, as shown in Figure 1. Some authors show overlap, others do not. Braided channels are best examined as a sequence otherwise the patterns become too confused, unless colour can be used.

A recent technological advance which may facilitate analysis direct from maps or map traces is to use a video frame-grabber. This allows the image of a shaded map or photograph to be stored then analysed according to the areas of different density on the image. A suitable density slicing can be selected and pixel counts to give area can then be made of each density class. There are problems, however, such as with lighting of the image, obtaining enough density contrasts, and with the size of map which can be analysed.

If quantitative analysis of the channel changes is required, then digitizing of the courses may be advantageous. Choice of digitizing method, whether constant spacing or timing, may have to be made and the interval has to be selected. This must be compatible with the scale of map and features to be analysed but too close an interval becomes time-consuming and expensive in computer usage. Equal spacing of 0.5–1.0 channel width is recommended. A problem in digitizing is that of line definition. If the wish is to produce the whole map or to analyse the movement of bank lines then both banks may need to be digitized. In many cases a single line to depict the river may be more suitable for subsequent analysis so the choice must be made between centre line, thalweg or one bank; centre line is usually the most suitable. Again technological developments may soon help

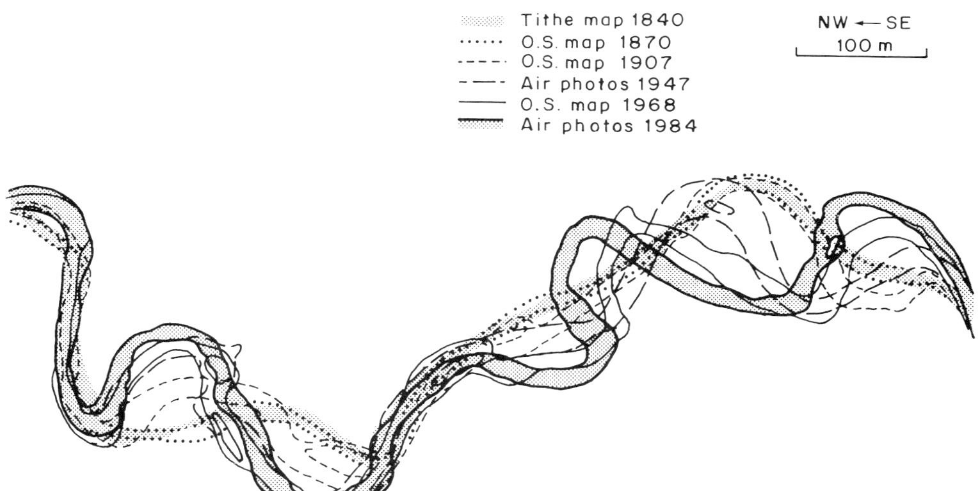

FIGURE 1. Section of the River Dane in Cheshire, N.W. England showing changes over *c*. 150 year period

here as equipment becomes cheaper and more widely available. One example is laser-scanning which can digitize up to 15 times faster than a manual digitizing procedure as well as produce more accurate results.

Methods of analysis

Many techniques are available for analysing cartographic evidence of the history of river channels and these may be grouped in various ways. In the following discussion the techniques have been categorized according to the form of the source material, i.e. whether

TABLE 4. Classifications or typologies of change

Author	Model/classification
Kondrat'yev (1968)	Graphical model of change from simple bend through to cutoff.
Daniel (1971)	Fitted a sine-generated curve to meanders. Identified five types of movement involving expansion, rotation and transition.
Keller (1972)	Processes of change and relationship to bedforms in a five-stage model.
Brice (1974)	Classification of morphology defined by circle combinations and analysis of sequences of change.
Hickin (1974)	Pattern of meander change developed from analysis of erosion pathlines.
Hickin and Nanson (1975)	Predictive model for determining rates of channel bend migration.
Kellerhals, Church and Bray (1976)	Identified six types of lateral activity: —downstream progression —progression and cutoffs —mainly cutoffs —entrenched loop development —irregular lateral activity —avulsion
Hooke (1977)	Used Daniel's (1971) primary elements of movement in double and triple combinations to compose a suite of 70 models of movement. Primary elements: —extension —translation —rotation —enlargement —lateral movement —complex change
Hooke and Harvey (1983)	Seven categories of change defined: —simple migration —confined migration —growth (extension) —lobing and compound growth —retraction and cutoff —complex changes (islands, abandonment etc.) —stable bends, no change
Brice (1984)	Classified changes in meander from according to mode of bank erosion

the map itself (original or traced) or derived data (usually digitized) and to the type of information required, whether qualitative or quantitative.

Qualitative analysis of maps

1. Description of the location and nature of the channel changes: this is most easily done from superimposed maps of the courses but can be made using visual sequences. It can show the general extent of instability and the types of change, e.g. whether simple meander shift, chaotic cutoffs or increase in braiding. It may be particularly useful as a pilot survey using just two map dates at small-scale to investigate whether examination at larger-scale and with intervening map dates is worthwhile.
2. Classification of *channel planform morphology* at different dates: this technique is rather more structured than a completely free description. Change is analysed by comparing the classes obtained over time. Separate maps can easily be used for this. The main problem is that there is no single satisfactory and agreed classification. Several which identify major groups (such as braided, meandering and straight) either do not have exclusive classes or else have criteria which are hard to distinguish or apply. Few have enough subdivisions within the meandering category and many were developed with reference to sand-bed rivers in North America. All are subjective in application (e.g. Kellerhals *et al.*, 1976; Ferguson, 1981; Brice, 1984).

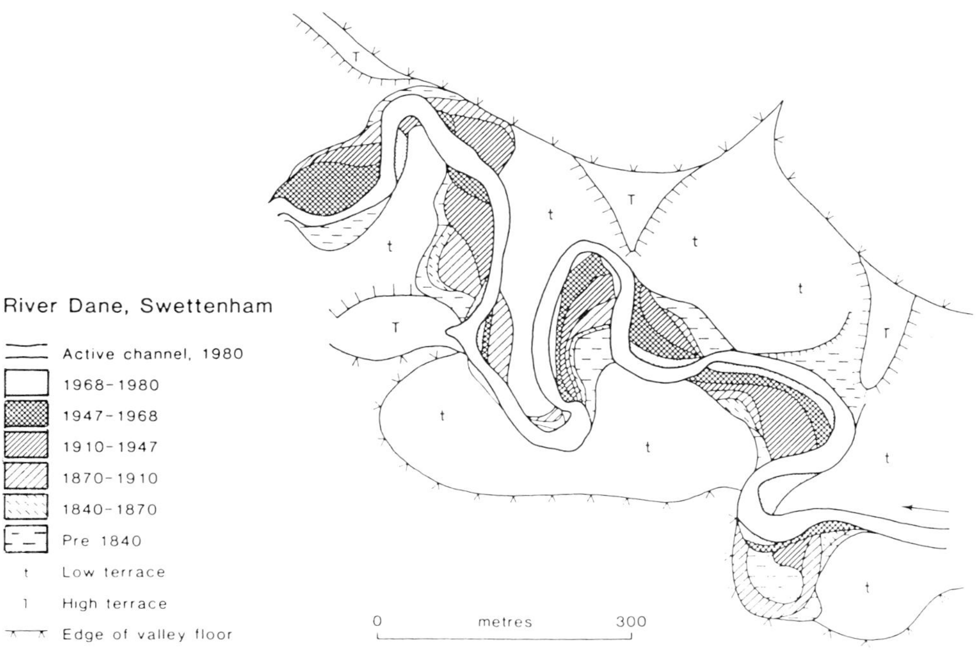

FIGURE 2. Section of the River Dane with floodplain zones dated from historical maps. Almost the whole floodplain can be dated, implying incision prior to this period

3. Classification of *type of change*: this ideally requires superimposed maps of the courses and entails comparing the type of change in each bend or section with an existing classification. Again there are a number of classifications to choose from as indicated in Table 4. Many of the classifications have been derived from studies of changes themselves then applied to further examples so again there is a lack of standardization. The process of classification itself is subjective to a large extent. It works best for meandering channels with distinct bends but very soon a problem of bend definition is encountered. Experiments have been done with sophisticated methods of coding but the possible numbers of combinations of change soon become so large that almost every bend is unique. Some general categories adopted, after such experiments, by Hooke and Harvey (1983) were quick and easy to use.
4. Dating and zoning of floodplain areas: if the purpose is to investigate the history of the floodplain then further maps can be derived from superimposed maps of the courses by shading the dates of movement in each area of change. This will demonstrate the ages of different parts of the floodplain, the extent of reworking, and distinguish the stable and active zones as can be seen in Figure 2. Zones of sedimentation can be identified and continuity of processes assessed.

Quantitative analysis from maps

1. Measurement of channel reach characteristics: this includes direct measurement from the maps of length of channel and sinuosity at different dates to assess change. The proportion of the course which has changed can also be calculated. Distances of movement at various points can be measured and rates of erosion calculated. Alternatively, areas of change between map dates can be measured and mean rates of erosion and area of land reworked can be derived. Channel width can also be measured but this may be of dubious accuracy unless the river is large and the map accuracy and convention of mapping width established. The number and length of bars can be measured and braiding indices calculated though care must be taken in their interpretation.
2. Measurement of meander loop characteristics: the conventional parameters of meander size and shape such as wavelength, amplitude and ratio of radius of curvature to width, can be measured directly from maps. However, there are several problems including those of subjectivity of position of measurement, variations in definition of the parameters and the fact that single measures do not portray complex forms adequately. Sampling problems also arise, for example whether all bends in a reach are measured, only those on one side, or only those at intervals. Bend definition soon becomes a severe problem again. Vectors of change in position of like points may facilitate analysis of direction and type of change.
3. Use of frame-grabber: it may be possible to do image-processing using maps directly under a frame-grabber. If channels of different date or areas of change are made to show up as different colour or density, then density slicing can be used and numbers of pixels, and therefore areas of different density, can be measured.

Quantitative analysis from digitized data

1. Automated bend division and bend parameter calculation: if curvature (change in direction) is calculated between equally spaced points then points of inflexion can

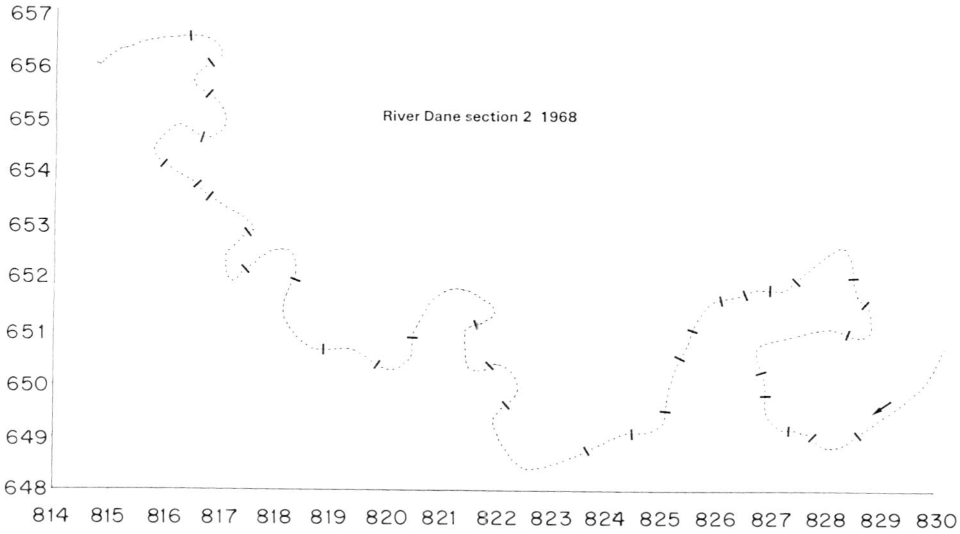

FIGURE 3. Digitized points and bends defined by points of inflexion for section of River Dane. Numbers on axes are national grid coordinates

be located (where curvature changes direction) and these can be used to define bends (Figure 3). However, even this is not straightforward as certain criteria of minimum bend size must be established. Hooke (1984) has used a combination of minimum curvature and length to eliminate small fluctuations. O'Neill and Abrahams (1986) have suggested a more elaborate method but based on similar principles. Parameters such as wavelength can then be calculated for each loop between points of inflexion (e.g. Table 5). There is still some problem of comparability of bends if change is very large.

2. Distribution of curvature: the sequential patterns of curvature in each bend may be analysed and types of distribution may be classified according to the position and number of peaks in the distribution.
3. Classification of type of change (see the section *Qualitative analysis of maps*): method 3 may be applied more easily once the bends have been defined and delimited by points of inflexion.

TABLE 5. Characteristics of meanders in a reach of the River Dane at different dates derived from digitized data of map courses

Date	Number of bends	Rate of movement (m yr^{-1})	Channel length (m)	Wavelength (m)	Total curvature (radians)	Maximum curvature (radians)
1840	10	0.23	174.3	116.4	2.08	0.39
1870	10	0.41	158	115.9	2.28	0.49
1907	10	0.69	167.8	109.3	2.73	0.55
1947	10	1.19	146.8	102.3	2.16	0.47
1968	10	0.97	172.7	105.1	2.42	0.62

4. Spectral analysis: the pattern can be decomposed into constituent cyclic components using this technique and it should show the relative contributions of the different wavelengths. It has been found useful to distinguish the general type of pattern but the peak definitions are not sensitive enough for the technique to be used as a measure of wavelength of meanders.
5. Mathematical curve-fitting: curves can be fitted to digitized data but this is largely of theoretical value and not of practical value except perhaps in channel design and navigation considerations. It does not facilitate analysis of change.

APPLICATIONS

Simple map compilations can be used to indicate where channels have changed their courses. The identification of zones of instability can take place at a variety of scales at which different types of explanation may also emerge. Four scales may usefully be defined. At the regional scale, an analysis of changes in channel pattern in northwest England as revealed by a comparison of O.S. 1/10 560 scale maps for two dates, *c.* 1870 and *c.* 1955, shows that channel changes have taken place on several of these rivers in the last 100 years and that most of the changes are in the piedmont areas where the streams come onto the lower land from the uplands. Commonly, however, comparison of individual rivers reveals a diversity of channel changes. Even neighbouring rivers with similar catchments can show contrasting behaviour, as in the case of the Rivers Bollin and Dane, located south of Manchester: the former exhibits a considerable number of cutoffs and the latter significant lengthening and increase of sinuosity, implying that they have experienced different conditions. Historical maps and other documentary sources may be used to establish possible causes of such changes.

Thus, in Britain, historical sources have been widely used to study channel changes. Lewin has examined many different aspects of mobile gravel-bed rivers in Wales (e.g. Lewin, 1976, 1983); Mosley (1975) and Thompson (1987) have both examined active rivers in Northern England, identifying the types and sequences of changes and inferring causes from the exhibited behaviour and timing; and Werritty and Ferguson (1980) have applied similar methods of historical analysis to braided river sections on Scottish rivers to elucidate the mechanisms and timescales of change. Hooke (1977, 1984, 1987) has provided information on spatial patterns of channel change and on rates of change, and has examined quantitative relations of meander morphology and change. Floodplain history and sequences of sedimentation have received much attention for example from Lewin (1978) whilst changes in river networks have been examined through historical sources by Ovenden and Gregory (1980). Most of these studies encompass time periods of the past 100–150 years, the period for which sufficiently reliable information is widely available in Britain.

At the scale of a single river, it is found that there are reaches of stability and instability. Often the reaches are quite distinct and the transition between each very rapid. This implies certain thresholds for change and it should be possible by comparison of conditions in stable and unstable reaches to establish these thresholds.

Finally, at the scale of an individual reach, comparison of channels of different date can indicate the nature and type of change, the distances and rates of alteration of the course (e.g. Table 5). Analysis at this scale has shown, for example, that some meandering rivers do migrate steadily downstream, but that many others exhibit much more complex change with growth in amplitude and development of compound forms being common (as can be seen in Figure 1). Also at this scale, cartographic data sources may be used to establish floodplain history by dating zones (e.g. Figure 2) to understand and predict the floodplain sedimentary composition, and to identify areas vulnerable to reworking.

For many studies, the stability and timescale of development of channel bars is of interest but such information is often difficult to derive from historical maps because the detail of mapping varies especially with water level and also bars, as transient features, are liable to change several times between map dates. The relative degree of braiding can, however, be derived as Lewin and Weir (1977) have shown and if the changes are slow or the map coverage very good, then details may be ascertained. Historical evidence may extend the time perspective of studies based on short-term repeat surveys (e.g. Thompson, 1987).

Study of channel adjustment after engineering works or drainage basin alterations can show the extent to which the adjustment is taken up by channel pattern, the mode of adjustment, and the rate of change. For example, redevelopment of meanders after channel straightening has been examined by Lewin (1976) and Hooke (1977). For a full understanding of mechanisms and causes of channel change the changes need to be related to change in river discharge regime and sediment load, though it is often difficult to derive this information historically. The changes in morphology themselves are often used as evidence that such changes in processes have taken place.

CONCLUSIONS

The potential of historical maps for providing information on river channel changes has been reviewed and the stages in data abstraction and analysis have been discussed. A variety of aspects of historical behaviour of rivers may be studied through maps and an enormous amount of data can be derived. Most of the methods of use are fairly simple but users should always be aware of the limitations of historical maps and corroborate the evidence. Although the details given apply to Britain, the principles apply elsewhere and historical map coverage is comparable and often better in many other European countries. The importance of an historical perspective to any process study or practical application of rivers cannot be overestimated.

REFERENCES

Ackers, P. and Charlton, F. G. 1970. Meander geometry arising from varying flows. *Journal of Hydrology*, **11**, 230–252.

Antropovsky, V. I. 1972. Equations of regression between channel changes and flow for incomplete meandering and channel and floodplain braiding. *T.G.G.I.*, **195**, 4–19.

Brice, J. C. 1974. Evolution of meander loops. *Bulletin of the Geological Society of America*, **85**, 581–586.

Brice, J. C. 1984. Planform properties of meandering rivers. In Elliott, C. M. (Ed.), *River Meandering*, American Society of Civil Engineers, New York, 1–15.

Carr, A. P. 1962. Cartographic record and historical accuracy. *Geography*, **47**, 135–145.

Carr, A. P. 1969. The growth of Orford Spit: cartographic and historical evidence from the sixteenth century. *Geographical Journal*, **135**, 28–39.

Chitale, S. V. 1970. River channel patterns. *Journal of Hydraulics Division, Proc. Am. Soc. Civil Engineers*, **HY1**, 96, 201–220.

Chitale, S. V. 1973. Theories and relationships of river channel patterns. *Journal of Hydrology*, **19**, 285–308.

Culbertson, D. M., Young, L. E. and Brice, J. C. 1976. Scour and fill in alluvial channels. *U.S. Geol. Survey Open-File Report*, 58 pp.
Daniel, J. F. 1971. Channel movement of meandering Indiana streams. *U.S.G.S. Prof. Paper*, **732–A**.
Dury, G. H. 1969. Relation of morphometry to runoff frequency. In Chorley, R. J. (Ed.), *Water, Earth and Man*, Methuen, London, 419–430.
Ferguson, R. I. 1981. Channel forms and channel changes. In Lewin, J. (Ed.), *British Rivers*, George Allen and Unwin, London, 90–125.
Ferguson, R. I. and Werritty, A. 1983. Bar development and channel changes in the gravelly River Feshie, Scotland. In Collinson, J. and Lewin, J. (Eds), *Modern and Ancient Fluvial Systems. Spec. Publs. Int. Assoc. Sediment.*, **6**, 181–193.
Harley, J. B. 1965. The re-mapping of England, 1750–1800. *Imago Mundi*, **19**, 56–67.
Harley, J. B. 1967. Enclosure and tithe maps. *Amateur Historian*, **7**, 265–275.
Harley, J. B. 1968. Error and revision in early Ordnance Survey maps. *Cartographic Journal*, **5**, 115–124.
Harley, J. B. 1972. *Maps for the Local Historian, a Guide to the British Sources*, London, National Council of Social Service.
Harley, J. B. 1975. *Ordnance Survey Maps, a Descriptive Manual*, Southampton, Ordnance Survey.
Hickin, E. J. 1974. The development of meanders in natural river channels. *Am. J. Sci.*, **274**, 414–442.
Hickin, E. J. and Nanson, G. 1975. The character of channel migration on the Beatton River, north east British, Columbia, Canada. *Bulletin of the Geological Society of America*, **86**, 487–494.
Hooke, J. M. 1977. The distribution and nature of changes in river channel pattern. In Gregory, K. J. (Ed.), *River Channel Changes*, Wiley, Chichester. 265–280.
Hooke, J. M. 1984. Changes in river meanders: a review of techniques and results of analysis. *Progress in Physical Geography*, **8**, 473–508.
Hooke, J. M. 1987. Changes in meander morphology. In Gardiner, V. (Ed.), *International Geomorphology*, Wiley, Chichester, 591–609.
Hooke, J. M. and Harvey, A. M. 1983. Meander changes in relation to bend morphology and secondary flows. In Collinson, J. D. and Lewin, J. (Eds), *Modern and Ancient Fluvial Systems*, Oxford, Blackwell, 121–132.
Hooke, J. M. and Kain, R. J. P. 1982. *Historical Change in the Physical Environment: A Guide to Sources and Techniques*, Butterworth, Sevenoaks.
Hooke, J. M. and Perry, R. A. 1976. The planimetric accuracy of tithe maps. *Cartographic Journal*, **13**, 177–183.
Kain, R. J. P. and Prince, H. C. 1985. *The Tithe Surveys of England and Wales*, Cambridge University Press, Cambridge.
Keller, E. A. 1972. Development of alluvial stream channels: a five-stage model, *Bull. Geol. Soc. Am.*, **83**, 1531–1540.
Kellerhals, R. M., Church, M. and Bray, D. I. 1976. Classification and analysis of river processes. *Proceedings of the American Society of Civil Engineers, Journal of the Hydraulics Division*, **102**, 813–829.
Kondrat'yev, N. Ye. 1968. Hydromorphic principles of computations of free meandering, 1. Signs and indexes of free meandering. *Soviet Hydrology*, **4**, 309–335.
Leopold, L. B. and Wolman, M. G. 1957. River channel patterns—braided, meandering and straight, *U.S.G.S. Prof. Paper*, **282–B**, 39–85.
Lewin, J. 1976. Initiation of bed forms and meanders in coarse-grained sediment. *Bulletin of the Geological Society of America*, **87**, No. 2, 281–285.
Lewin, J. 1978. Meander development and floodplain sedimentation: a case study from mid-Wales. *Geological Journal*, **13**, 25–36.
Lewin, J. 1983. Changes of channel patterns and floodplains. In Gregory, K. J. (Ed.), *Background to Paloeohydrology*, Wiley, Chichester, 303–319.
Lewin, J. and Hughes, D. 1976. Assessing channel changes on Welsh rivers. *Cambria*, **3**, 1–10.
Lewin, J. and Weir, M. J. C. 1977. Morphology and recent history of the Lower Spey. *Scottish Geographical Magazine*, **93**(1), 45–51.
Moir, D. G. (Ed.) 1973. *The Early Maps of Scotland to 1850*, Royal Scottish Geographical Society.

Mosley, M. P. 1975. Channel changes on the River Bollin, Cheshire, 1872–1973. *East Midland Geographer*, **6**, 185–199.

McEwan, L. 1985. *River channel planform changes in upland Scotland, with specific reference to climatic fluctuations and land use changes over the last 250 years*, Unpublished PhD thesis, University of St. Andrews.

O'Neill, M. P. and Abrahams, A. D. 1986. Objective identification of meanders and bends. *Journal of Hydrology*, **83**, 337–353.

Ovenden, J. C. and Gregory, K. J. 1980. The permanence of stream networks in Britain. *Earth Surface Processes*, **5**, 47–60.

Rodger, E. M. 1972. *The Large Scale County Maps of the British Isles, 1596–1850: A Union List*, 2nd edition, Oxford, Bodleian Library.

Schumm, 1963. Sinuosity of alluvial rivers on the Great Plains. *Bulletin of the Geological Society of America*, **74**, 1089–1100.

Skelton, R. A. 1970. *County Atlases of the British Isles 1579–1850*, London, Carta Press.

Tate, W. E. 1978. *A Domesday of English Enclosure Acts and Awards*, edited with an introduction by M. E. Turner, Reading, The University of Reading.

Thompson, A. 1987. Channel response to flood events in a divided upland stream. In Gardiner, V. (Ed.), *International Geomorphology*, Wiley, Chichester 691–709.

Thorne, C. R. and Lewin, J. 1979. Bank processes, bed material movement and planform development in a meandering river. In Rhodes, D. D. and Williams, G. P. (Eds), *Adjustments of the Fluvial System*, London, George Allen and Unwin, 117–137.

Werritty, A. and Ferguson, R. I. 1980. Pattern changes in a Scottish braided river over 130, and 200 years. In Cullingford, R. A., Davidson, D. A. and Lewin, J. (Eds), *Timescales in Geomorphology*, Wiley, Chichester, 53–68.

Historical Change of Large Alluvial Rivers: Western Europe
Edited by G. E. Petts
©1989 John Wiley & Sons Ltd

6

Cartography of Rivers in France

J.-P. BRAVARD

Université Jean-Moulin Lyon 3 et URA 260 du CNRS, 74 rue Pasteur, 69007 Lyon, France

and

J. BETHEMONT

Université de Saint-Etienne et URA 260 du CNRS, 74 rue Pasteur, 69007 Lyon, France

INTRODUCTION

If one considers only the public services' archives, the French documentation on rivers is surprisingly abundant. Maps, and among them, river maps, have been collected by the 'Archives Nationales' in Paris and the 'Archives départementales'; by public libraries, such as the Bibliothèque Nationale in Paris and the libraries of many cities; by the Army; by different ministries such as Ponts et Chaussées, Agriculture; and finally by towns and cities (Archives municipales). These referenced collections exclude the thousands of maps, drawings and plans contained in the Archives and often still unused ('Séries' dealing with the 17–18th century, with the French Revolution, with factories, routeways and rivers, forests etc.) Furthermore, very little is known about the private archives belonging to noble families, for instance. One can also use a wide range of documents dating from the 18th century, or explore foreign libraries because wars often resulted in the transfer of maps from one country or state to another. From the end of the 19th century onwards, photographs and postcards are of particular interest because they often show rivers. Postcards display a large range of topics: bucolic landscapes, picturesque scenes of gorges or torrents, bridges and boats, or symbols of an industrializing era. Unfortunately, the quest for this type of documentation is time consuming and costly. Of particular interest is the *Atlas photographique du Rhône*, published in 1931 for the section Geneve–Lyon (A. Cholley and F. Seive); it begins the rich series of air photographs taken since the end of the 1930s by the Institut Geographique National. This paper reviews the cartographic data sources available for historical studies of rivers in France. Examples cited in this chapter are located on Figure 1.

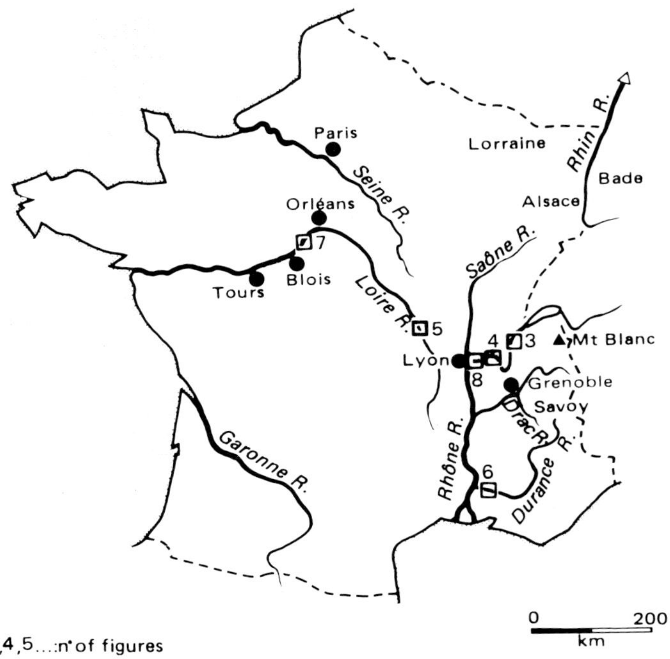

FIGURE 1. Map locating sites referenced in text and examples given in subsequent figures

THE MAPPING OF FRENCH RIVERS

A typology of river mapping

River mapping in France has been motivated by a range of circumstances because the interest in streams has changed over time. The mapping of France was rather late compared with Italy and Flanders. The first map of the French territory was published in 1482 at Florence, while the first to be published in France was issued at Paris in 1525. Progress was slowed by the Wars of Religion in the second half of the 16th century and by the fact that the residences of the Kings and Princes were the castles of the Loire valley until 1594. At this time, the maps of some interest for studies of rivers were located mainly in the provinces of the central part of France and the regions close to Lyon, which is where printing began in France. Maps were political and administrative tools for royal governments. Louis XIV created the 'Corps des Ingénieurs Géographes du Roy' to produce this type of map. In 1747, Louis XV entrusted Cassini with the systematic mapping of the kingdom at the scale of 1/86 000° but this programme was not yet completed in 1815 and was changed into the survey of the so-called 'Carte d'Etat-Major' (1818–1868) for military purposes.

The most interesting maps were drawn to support war campaigns, for reconnaissance surveys of foreign countries, or for the construction of forts along the frontiers. For instance, a map was drawn in 1536 and given to King Francois I by the Governor of Grenoble, an important stronghold of the time facing the foreign country of Savoy. This map justified a request for subsidies in order to control the floods of the Drac river which threatened the town. 'Portraits' of the same river were drawn for the same purpose in 1541 and 1599 (Bouchayer, 1925).

In 1604, M. de Sully, a minister of King Henry IV, created the so-called 'Corps des Ingénieurs du Roy' in charge of fortifying the towns of the kingdom. They studied their art at Venice and in the United Provinces and began their work at the beginning of the 18th century (Broc, 1986; Archives Nationales, 1987). In 1688, Louis XIV created the Dépôt de la Guerre (War Ordnance) in order to collect all types of military maps. Again the 'Corps des Ingénieurs du Roy' was in charge of drawing military maps. For instance, a map of the Rhine river boundary was drawn in 1722 at the scale of 1/21 660° by the Engineer Broutin working for the Duke of Lorraine. In fact this map surveys the fortresses, villages, routeways, forests and the banks of the river rather than the river itself (Levy, 1986). One particularly interesting map, drawn at the scale of 1/33 000° to delineate the states of France and Piémont–Sardaigne along a 50 km section of the upper Rhône river, provides an accurate image of channel braiding, land use and dykes in 1760.

Conflicts about land possession concerned also the communities of opposite banks in the same country. Hundreds of maps and sketches were drawn of the Rhône river during the 18th and the beginning of the 19th centuries when the braided channel used to erode fields and build new bars and islands. The finest example is probably the map of Grandvoinet (1781–1785) drawn at the scale of 1/3 500° and showing a 140 km section downstream of Lyon. The latter is, in fact, a fluvial cadastre. This type of document was surveyed at the scale of 1/2500° as early as 1728 in Savoy where it was called '*mappe sarde*'. The French equivalent was surveyed between 1804 and the 1840s and replaced the old '*terriers*' (land use plans) drawn since the Middle Ages; at the beginning of the 19th century each French commune is so represented at the scale of 1/2000° (sectional map) and 1/10 000° (assemblage map), but the large rivers are often roughly drawn as the channels were unstable and braided.

Many maps or sketches were surveyed on the occasion of large floods and give, usefully, the spatial extent of low frequency events, such as the exceptional flood of 1856. These allow the delimitation of inundated areas along several French valleys. In the Loire valley, two maps of 1707 and 1709, drawn at a scale of 1/5000°, represent the breaks of the 'levees' (earthen dykes) after large floods (Dion, 1961).

Most of the fluvial maps were drawn to aid development schemes, such as embankments against the risk of flooding or submersible embankments built for improving steam navigation. The Rhine river, along the Alsatian and Badish section was first surveyed in 1807–1808, and then in 1828, 1845, 1863 and 1869, at the scale of 1/20 000° for assessing the natural conditions and the changes after Tulla's works. The Loire river was surveyed at the same scale in 1848; and the Rhône river partly in the year 1840, and totally between 1857 and 1866 (scale: 1/10 000°). After 1830, the revolution in transport changed many small French rivers. The channelization of rivers such as the Garonne river, the Loire river, and the Saône river, mainly for suppressing shallow riffles, completed the network of canals built since the 17th century. A vast amount of information concerning these schemes is scattered in archives and in engineering literature.

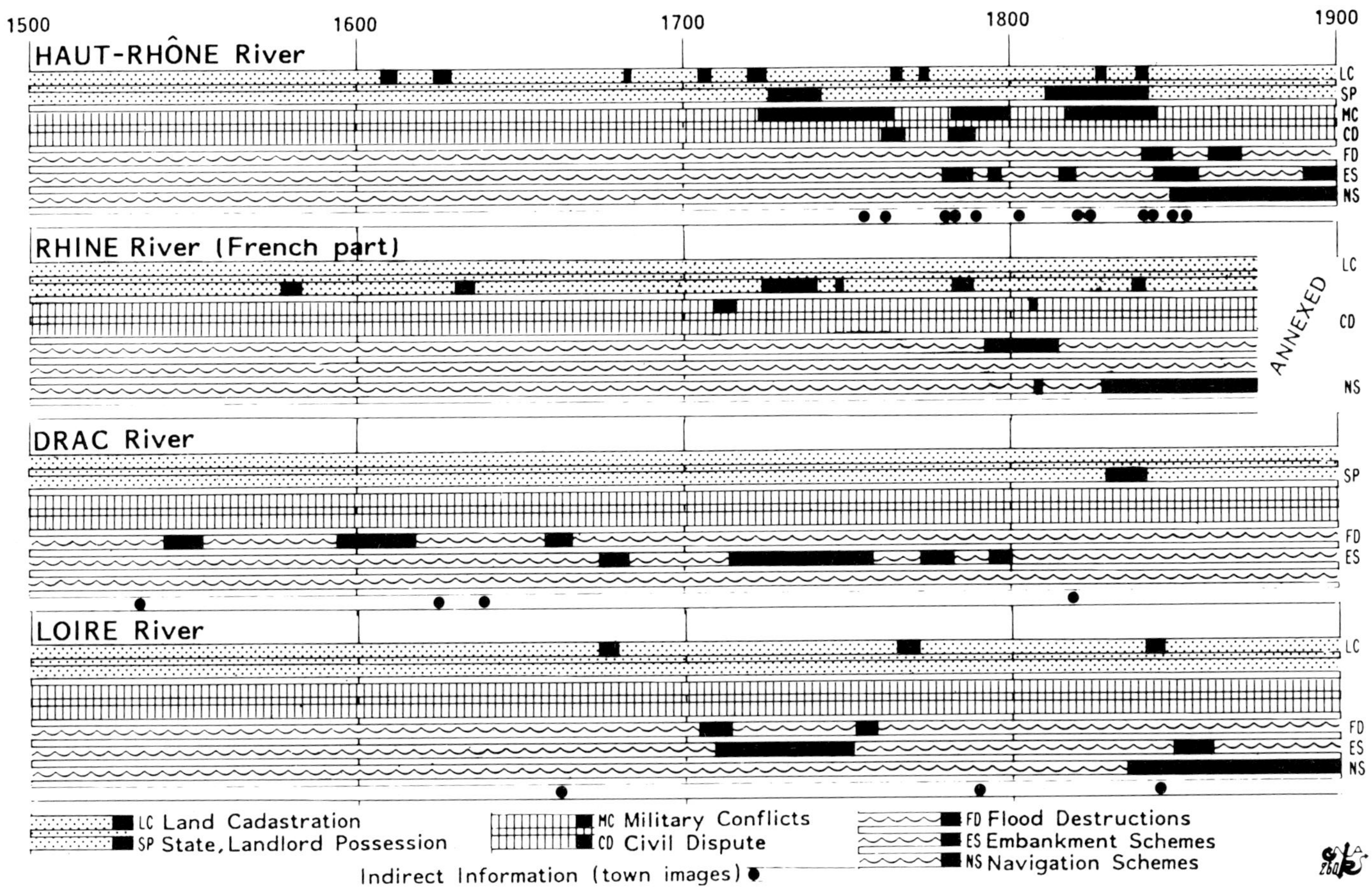

FIGURE 2. Cartographic sources available for French rivers

A particular kind of map, most of them drawn on behalf of King Henry IV, concerns the reclamation of great marshes which were royal properties in the northern part of France (Somme marshes), on the Atlantic shore (*marais poitevin*) and near the Rhône delta called Camargue (H. de Dienne, 1891; G. Boutonnet, 1963). Some of these maps were drawn by Dutch engineers. The salt marshes were also of great interest as fiscal resources and were mapped in the Camargue as early as the 16th century.

Finally, there are maps registered in town archives that show the use of water for urban needs: transport, mills, and horticulture. They are often very accurate and display a former urban riverscape destroyed in the second half of the 19th century (Guillerme, 1983). The most interesting of these maps include those concerning Amiens, Avignon, Beauvais, Strasbourg or Toulouse (Fortuné, 1988).

Iconographic models (Figure 2)

The scientific value of river maps depends on the number available, the age of former documents and the accuracy of these sources which relate to the goals of the surveyors. For each river, it is then necessary to check the list of maps as completely as possible; and the origin of each one must be assessed. Archives may provide the historical background explaining the topics of the survey, the techniques adopted, the limitations of the maps etc. Many sketches are of interest only because of the text they illustrate and should not be studied separately.

A collection of maps cannot be said to be complete because many sources are still unknown. The existing French documentation displays a typology of river mapping which depends on the history and geographic location of the river throughout the country. Four representative examples have been chosen:

—The Drac river is associated with two kinds of maps: the oldest representing the flood devastated valley, the second set representing the dykes after 1676, resulting from the initiatives of Colbert, a minister of Louis XIV, who provided funds for the protection of Grenoble (Bouchayer, 1925).

—Three generations of maps were produced concerning the Rhine river, illustrating first military conflicts, then protection of the banks against river erosion and, finally, navigation. Additional maps may well exist and this river would probably display a very complete cartography, as the Rhône river. Problems of braiding the frontier changes are the two major reasons for these early maps.

—Compared with the above, the cartography of the Loire river refers to an inner model with the epoch of levee construction (18th century) and the epoch of navigation improvement (after 1835).

—River cartography is not homogeneous and different kinds of maps may occur on the same river with different purposes on different sections. On the Rhône, for instance, the map showing the flood of 1856 covers the whole area between Switzerland and the delta, maps produced for border conflicts show only the Upper Rhône which was the border between France and the Kingdom of Piémont-Sardaigne until 1860 (Table 1), and downstream of this sector the only purpose for mapping was conflict about land possession.

TABLE 1. Chronological list of maps available for studying a 20 km section of the French Upper Rhône river (Chantagne). Sources are indicated in brackets

1706	Carte des Etats de Savoye et de Piémont, le Dauphiné, La Bresse. . . H. Jaillot, Paris—Petite échelle (Grenoble, Museé dauphinois).
1724	Tabula generalis Sabaudiae—BLAUE, Amsterdam—1/200 000e (Chambéry, A.D.: Archives dèpartementales).
1728–1732	Mappes du cadastre sarde—1/2500e (Chambéry, A.D.).
1741	Extrait d'un ancien plan des iles d'Anglefort. . . (Anglefort, archives municipales).
1744	Carte de France—Planches 117–118—CASSINI de THURY—1/86 000e.
1760	Carte Géométrique d'une partie du cours du Rhône depuis Genève jusqu'au confluent du Guyers pour servir à la nouvelle limitation des Etats de France et de Savoye—VILLARET, Ingénieur cartographe, 1/33 000e (Chambéry, A.D., C 625).
1772	Plan ou fragment de carte du cours du Rhône depuis Seyssel jusqu'à Rochefort (Paris, Archives Nationales, H 165).
1772	Carta corographica degli stati di S.M. il Re de Sardegna, data in luce dell' ing: Borgognio nel 1683, corretta e accrescinta nell' anno 1772 (Chambéry A.D.).
1820	Dessin manuscrit représentant les digues du Rhône (Chambéry, A.D., Fonds sarde, 2424–2425).
1829–1955	Série de plans partiels ou généraux au 1/10 000e levés par l'Administration des Ponts et Chaussées (Lyon)—en particulier: 1815, 1863, 1867, 1874, 1875, 1876, 1883, 1890, 1893, 1896, 1902, 1922, 1924, 1928, 1931, 1936, 1937, 1945, 1955.
1844	Carte d'état-major—Feuille Chambéry. 1/80 000e.
1855	Quadro d'unione della gran carta degli stati sardi—Feuille Ruffieux. 1/50 000e.
1857–1866	Carte topographique du cours du Rhône entre le Parc et le pont de Donzère—1/10 000e (Lyon, Ponts et Chausseés).
1880–1885	Tableau d'assemblage—1/10 000e—cadastre (Chambéry, Bourg et communes).
Since 1935:	Cartes topographiques de l'Institut Géographique National—1/20 000e puis 1/25 000e; 1/50 000e.

THE DOCUMENTATION OF RIVER CHANGE

The reconstitution of the pre-industrial conditions in French alluvial plains

One fascinating theme for research is the reconstitution of historic patterns which were prevalent, for example, during the 18th century before the harnessing of rivers by embankments. This study is specially worthy of interest when considering large braided rivers (e.g. Figure 3).

The Loire river is a sandy braided river with some islands that have been stable since the end of the 18th century because it made navigation difficult along some river sections (Babonaux, 1970; Bomer, 1976).

The maps of the braided Rhône river were drawn at low flow but nothing is known about the precise water stage, which is a constraint upon further interpretations (Hooke and Kain, 1982; Werrity and Ferguson, 1980). However, using this kind of map dating from 1780 (Figure 4), for the section upstream of Lyon it is possible to evaluate the braiding index and to approximate the fluvial rejuvenation defined as the growth rate of gravel bars within the braided belt. Some maps dating from 1850 allow the description of large scale geomorphic features, including the types of bars and the cross-sections of channels when water depth is mentioned or when cross-sections have been surveyed (Bravard, 1986).

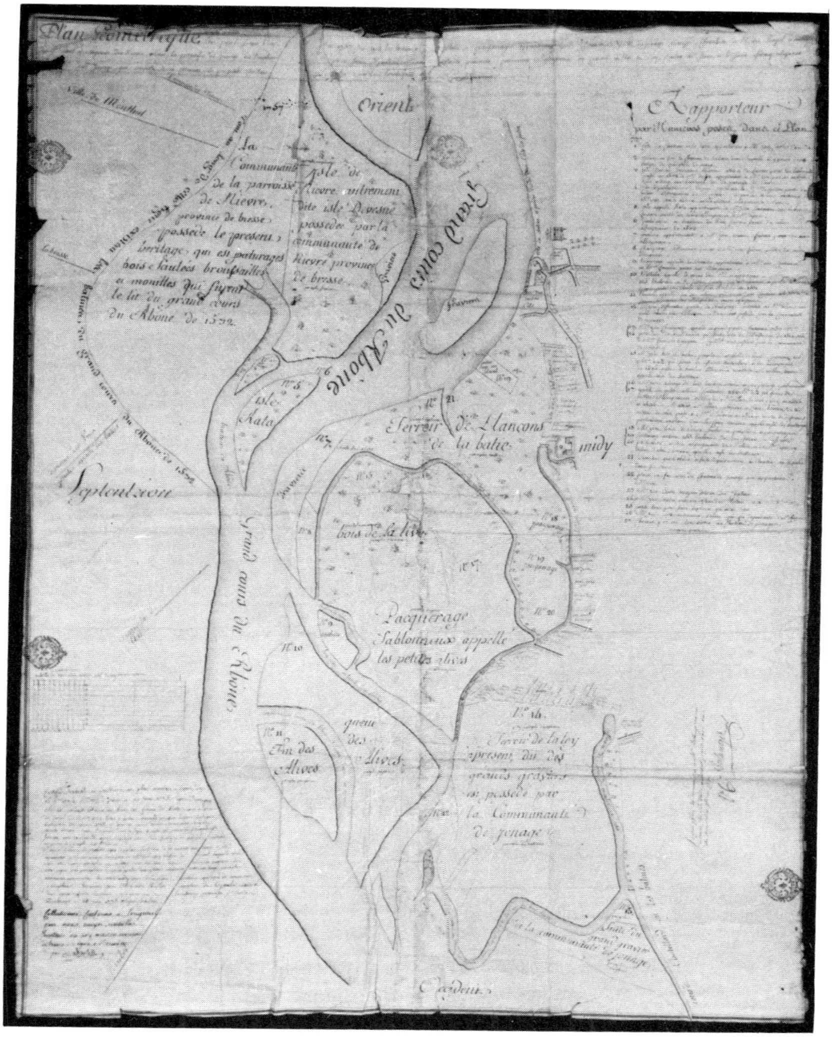

FIGURE 3. A map of the Rhône river drawn in 1757, showing the gravel bars and the grazed alluvial forest in a braided section. This survey was used for pleading after a contestation of property between two landlords: . . . ; on the left part the course of the main channel of 1532 is drawn . . . (Arch. Dép. Ain, E. 472), scale: 1:6000

Along some sections of the Rhône river, an integrated study of maps and written archives allows a reconstruction of the fluvial dynamics. This method was successfully used by Sundborg (1956) and Le Roy-Ladurie (1967) in other environments. For instance, in a small region called Chautagne (see Roux *et al.*, this volume), the riverine communities have recorded the sequence of erosion and overbank sand alluviation (Figure 5) on the occasion of large floods. The cadastres give interesting information about land occupation in active

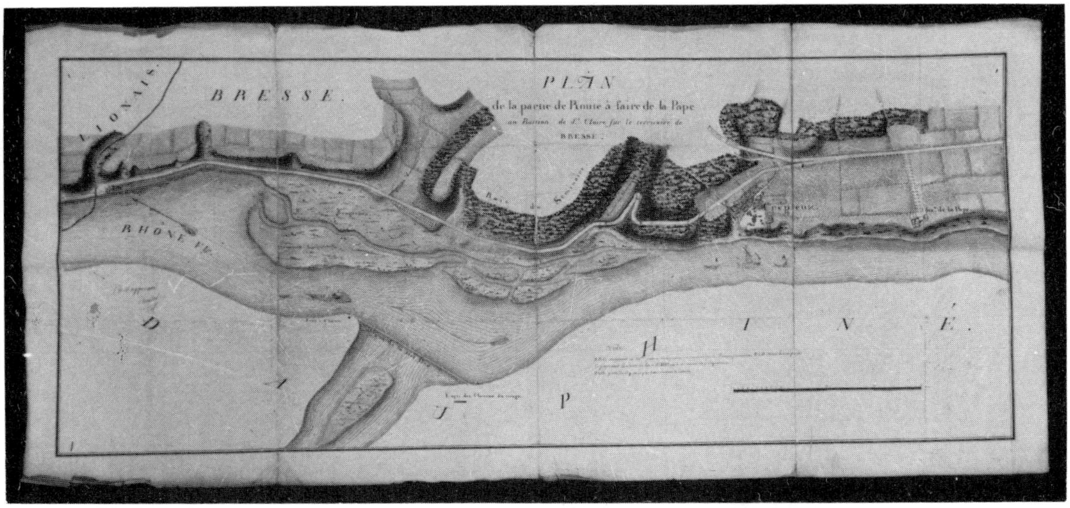

FIGURE 4. The Rhône river in the vicinity of Lyon in 1780, showing the grazed and wooded islands (called 'Brotteaux'), different kinds of boats for trade and fishing, the lodge for hawling horses, the vineyards and the woods on the valley sides and the road from Lyon to Geneva. (Arch. Dép. Ain, *c*. 1048) scale: about 1:35 000

braided belts, such as wheat cultivation on side gravel bars, thickets for domestic fuel on swept bars, vegetables in the silted downstream-opened former channels, and hay in the isolated former channels even with values of productivity. This type of information was also available concerning some backswamps of the upper Rhône river (Bravard, 1981).

Dating former channels

Abandoned channels are conspicuous features of alluvial plains which may be dated using maps and written archives. These dead arms relate to the continuum of fluvial patterns (i.e. braiding, anastomosis, meandering, which explain their specific geomorphic characters, the nature of the ecological successions herein, and the speed of evolution to the terrestrial environments by progressive siltation and organic sedimentation (Amoros *et al*., 1986). Maps dating from the pre-industrial era are often available for European alluvial plains to determine the geomorphic pattern of reference. Furthermore, when supplemented by absolute dating, these maps enable predictive scenarios relating to human impacts, to be established. For instance, the dead arms of an alluvial plain on the Rhône river were investigated to identify which relict aquatic environments persist the longest (oxbow-lakes and former anastomosed arms, and which most quickly disappear. In this case the development of dead arms and ecological successions was documented from the middle of the 19th century and the impact of submersible embankments, and of dyke breaching in 1917, were assessed throughout the 20th century (Bravard *et al*., 1986a).

The richness of the cartography of the Rhône river enables the dating of dead arms and talwegs over about five centuries in some areas. The Ain and Rhône rivers confluence has

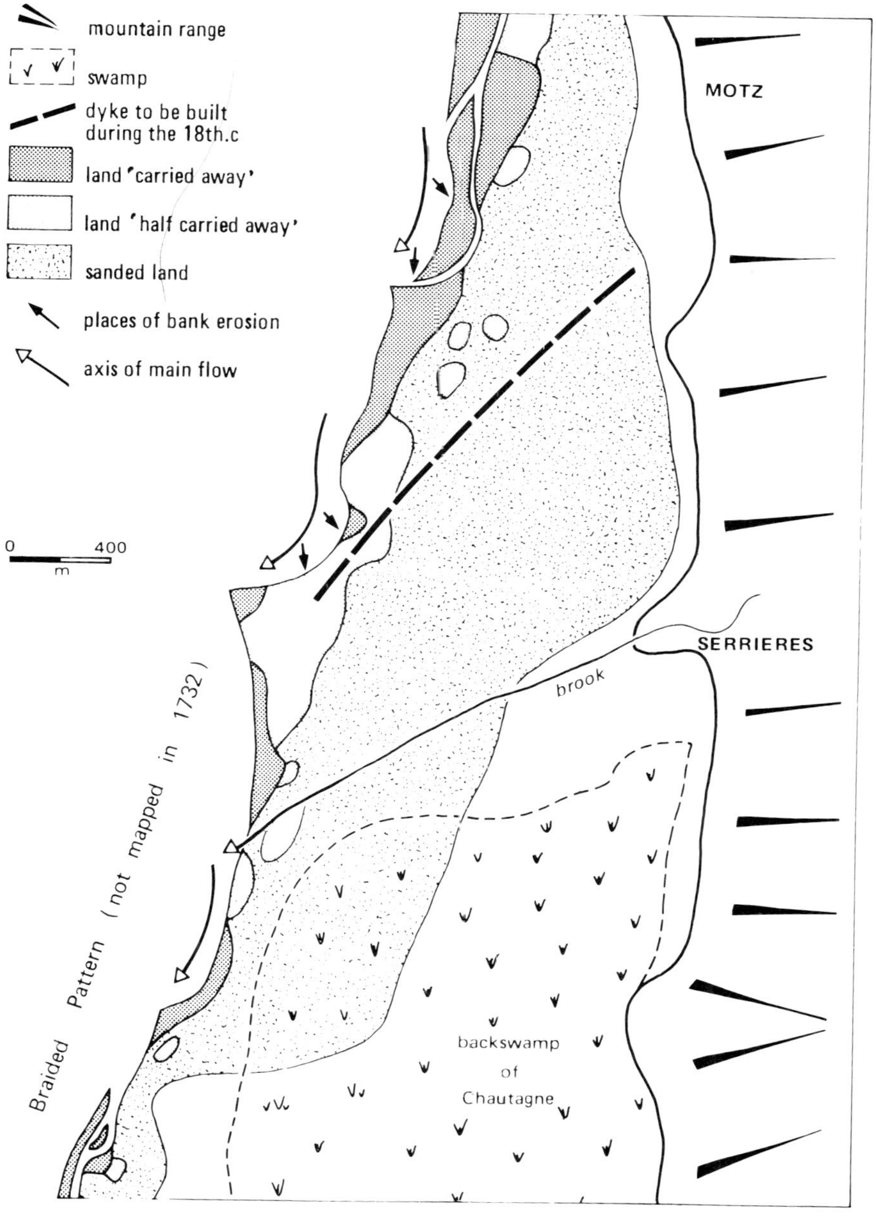

FIGURE 5. Areal extent of changes after the 1732s flood in the Chautagne floodplain (after Bravard, 1981)

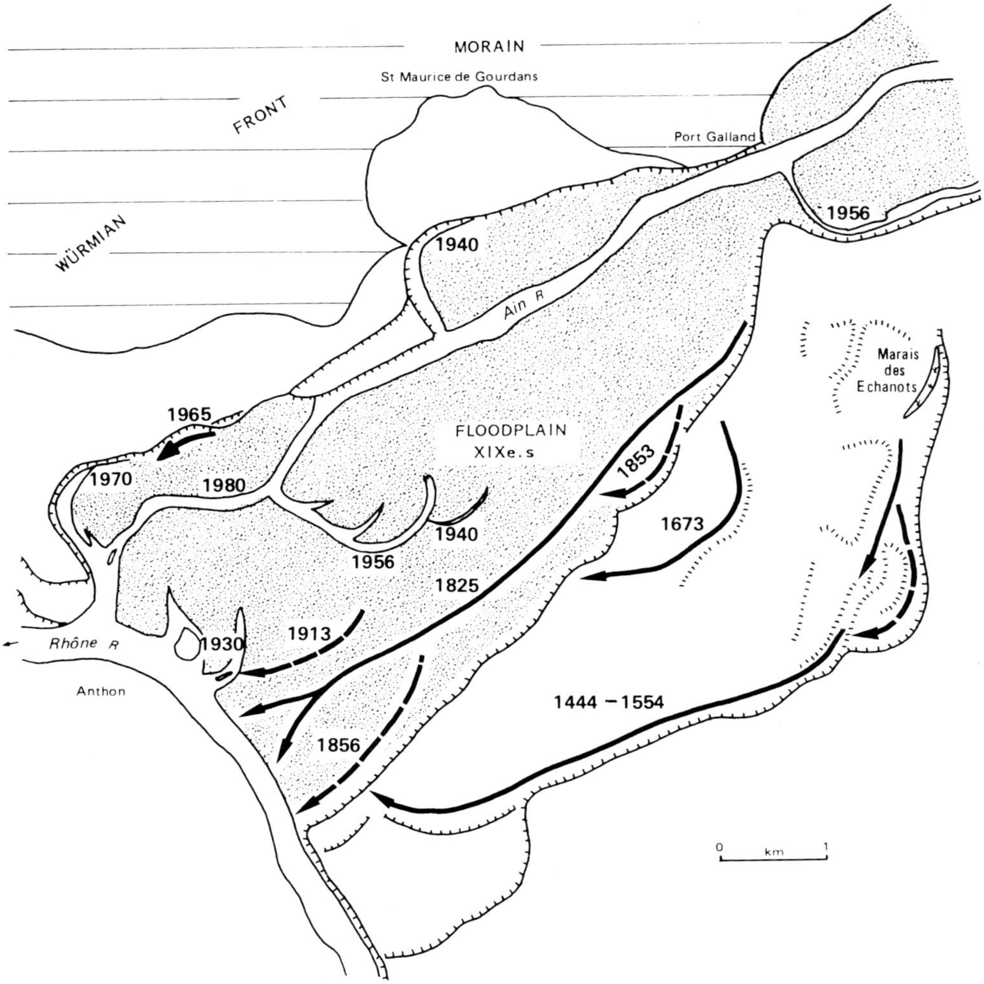

FIGURE 6. The former courses of the Ain river since the 15th century (after Bravard, 1986)

shifted westward and the two rivers have been degrading, during the Holocene. Many dried and unsiltated channels are still visible on low terraces. The Ain river course dating from the 14th and 15th centuries was redrawn in 1783, on the occasion of a law-suit between the landlord and the community, after medieval land surveys (Figure 6). Maps were used till 1930 and were then replaced by vertical air photographs (Bravard, 1986b). Other examples of the ecological implications of historical geomorphology is for phytoecological diagnosis of dead arms (Castella and Amoros, 1984), or for understanding the spatial organization and evolution of the terrestrial vegetation on floodplains (Amoros *et al.*, 1986).

Maps provide a point–time image of a river. It is often necessary to approximate the dating of meander cut-offs: this is often a compromise between the date of the flowing channel and the first date of the isolated channel. The question is not so easy in braided patterns because the isolation and siltation of this type of channel tends to be progressive, and because floods are able to regenerate the channels. It is possible to date areas of abandoned braided belts or former channels for instance, when the date of the abandonment is certain. Clearly, the more complete the sequence of maps and documents, the more accurate is the exercise. In France, the main streams were classified by the Administration as 'navigable' or 'floatable'; when one of their channels was cut off, it was withdrawn from the main stem and could belong to the riverine landowner, provided that a small boat could no longer access it from the main channel. Law-suits between Administration and farmers were responsible for the drawing of many large scale and precise, but very local, drafts which complement the official river maps. Indeed, the previous case studies stress the successions or sequences of maps rather than the accurate dating of channel cutoffs.

River pattern changes

It is well established that changes of channel pattern are geomorphic responses to changed conditions in a watershed (e.g. Schumm, 1977). Discussion of the 'external factors' responsible for the metamorphosis of french rivers is found elsewhere (see Roux *et al.*, this volume); herein we shall simply present some cases demonstrated by map evidence.

In the Alps, using cartographic data it is possible to approach the question of long-term changes of the braided pattern of mountain streams. For instance, on the Drac river, by combining historical documents and iconography, Bouchayer (1925) suggested that the Drac river has changed from a single channel before the 14th century to a multichannel river at the end of the 14th century. Mills were destroyed by floods, as were the walls of the city of Grenoble, the braided belt became wider and aggraded with cobble bars and log jams. Deforestation has been identified as the cause of these severe changes but the climatic change known as the Little Ice Age probably played a pre-eminent role as exemplified by the following example.

On the Arve river, which is a proglacial tributary of the upper River Rhône, upstream from Geneva, a synchronic survey of the *Mappe Sarde* (1730) and use of semiquantitative parameters (width of the active tract, sinuosity of the main channel) demonstrated downstream changes in the braided pattern related to a decrease of slope and to hydrological change prior to early direct anthropogenic impacts on the fluvial dynamics (Peiry, 1988). On the other hand, a diachronic reconstitution of the river pattern using a series of old maps in the middle Arve river has shown an increase of the braiding activity at the beginning of the 18th century. The Little Ice Age progression in the Massif of Mont Blanc was responsible for heavy sediment inputs in the fluvial system during the 17th century (Grove, 1966). Sediment was transported progressively downstream throughout the following century, building alluvial megaforms over the valley flat (Peiry, 1988).

A different type of change has occurred along the Loire river. Bomer (1972) describes the spatial development of vegetated islands and the contraction of channels (Figure 7). The proposed explanation is of hydrological change since the beginning of the 20th century, namely a decrease of peak floods. Maps of 1830 and subsequent surveys at a scale of 1/20 000° or 1/25 000° have been compiled for the section between Amboise and Tours. The width of the channel for mean discharges has been reduced by 17 per cent, from 330 to 275 m (Bomer *et al.*, 1979).

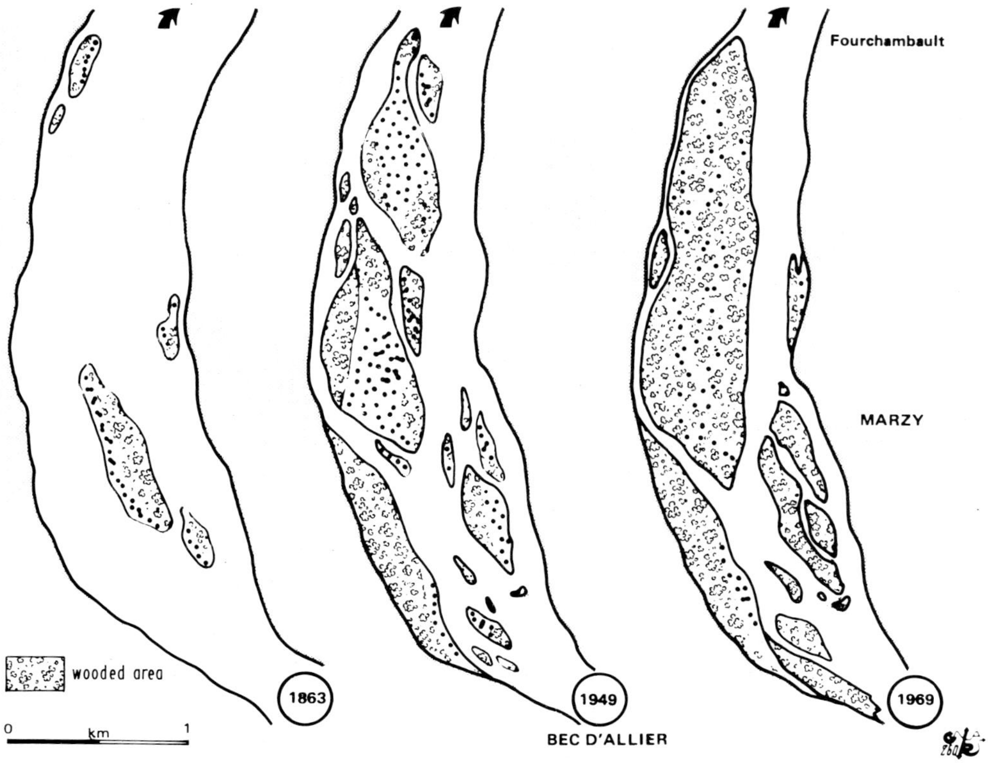

FIGURE 7. Evolution of sand bars and islands along a section of the Loire River, between 1863 and 1969 (after Bomer, 1972)

The main changes are man induced. The downstream course of the Ain river, a tributary of the Upper Rhône river, is rather poorly mapped, but the evidence shows a braided river until 1930 and a subsequent metamorphosis to a meandering pattern, probably due to the combination of several factors: a natural rarefaction of large floods since the end of the 1950s, a decrease of cobble bedload and a lowering of peak discharges after completion of upstream reservoirs, the river constriction by bridges which reduced lateral migration, and the growth of vegetation on river banks and the reforestation of the floodplain following the dereliction of grazing and woodcutting. Meandering and accelerated bed degradation (2 m during the last century) have changed the ecological conditions in the floodplain, which exhibits a set of fossil and present ecological units (Amoros *et al.*, 1986).

The maps of the lower Durance (Figure 8)—the only important tributary of the Rhône in Provence—display a sample of impacts related to the dam of Serre-Poncon, erected in 1960, which impounds a water body of more than a cubic kilometre. Before damming, the braided channel was more or less restrained by earthen dykes but its instability prevented the formation of riverine forests. After dam closure, the river bed experienced a drastic decrease in the intensity of fluvial dynamics and a fast growth of the riverine forest. Today, the left bank is occupied by a highway and the unique channel must be cleared artificially (Pieri, 1984).

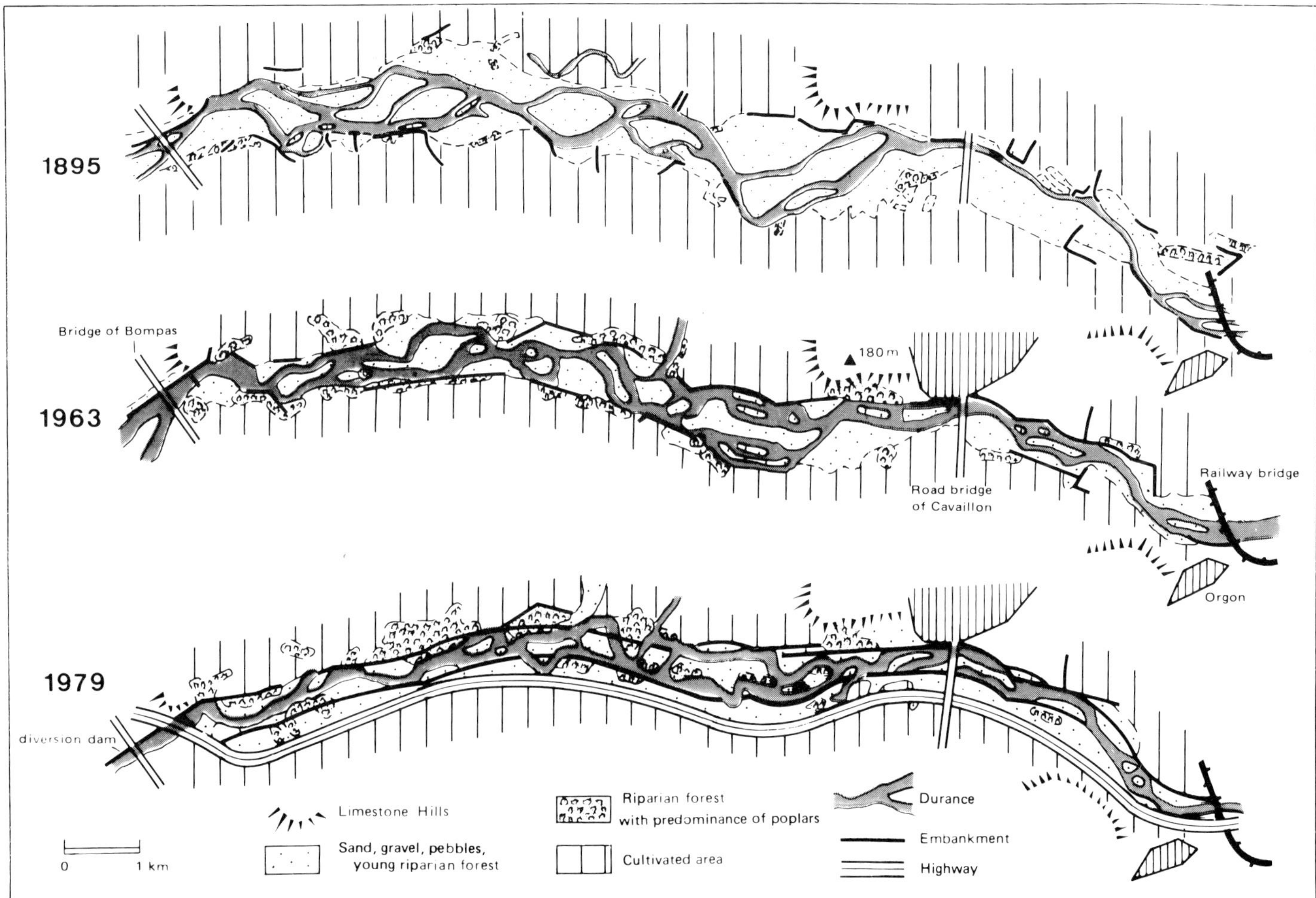

FIGURE 8. Evolution of the lower course of the Durance River since the end of the 19th century (after Pieri, 1981)

Most of large French rivers have experienced changes of pattern, exemplified by the above cases. The dominant change has been the channelization of former braided patterns. This evolution proceeds directly from civil engineering works which selected for embankment the main stem of the multichannel network and proceeds from delayed impacts, namely the progressive siltation of upstream cut-off arms. It is usually possible to document these changes because complementary embankments, or repairs, were frequently undertaken from the end of the 19th century and during the 20th century. Some sections of the Rhône river have so been mapped every ten years at a scale of 1/10 000°; nevertheless this type of document must be carefully criticized because some are merely copies of former ones, concern unimportant areas, or focus on peculiar bank sections.

Changes in the long profile

Interpretation of channel planform changes may be supplemented by observations on river flows and channel profiles. Mobile sand or gravel river beds may aggrade or degrade in response to changed conditions of sediment load or hydrology. Gauging stations provide long records for rivers such as the Loire and the Rhône. The level of the lowest yearly discharge has been recorded since 1830 at Orleans and since 1861 at Blois; the degradation of the bed approximates 1.50 m and its rate has been recently increasing. This tendency was initiated by the construction of embankments but was enhanced by sand extractions (Figure 9). On the Rhône river the records date from 1840 and the data show that the historical tendency for aggradation continued until 1950 at Seyssel but was then reversed by dam closure upstream and by gravel extraction.

Archives may provide reliable data about vertical changes and one of the world's oldest records is probably that for the Drac river at Grenoble. In 1787, an anonymous memoir

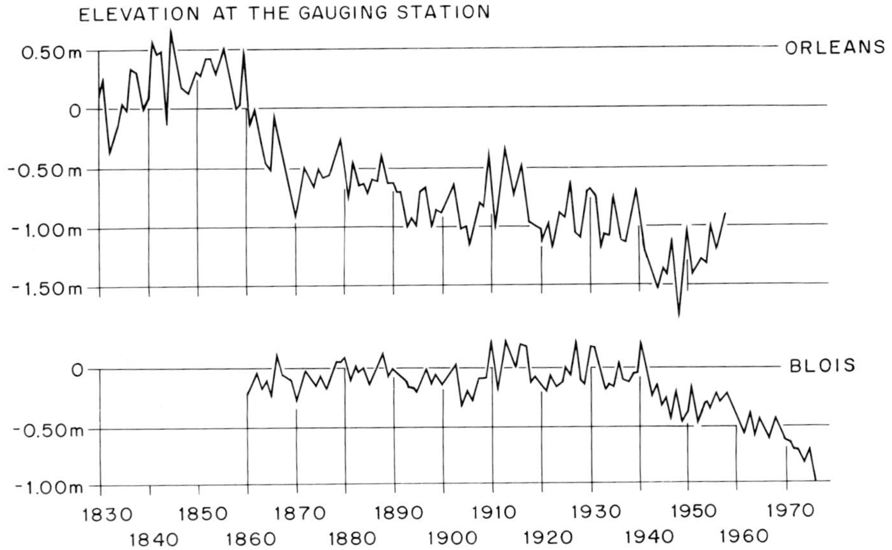

FIGURE 9. Elevation of the Loire river at low flow showing the bed degradation since 1830 at Orleans (after Lorain, 1981)

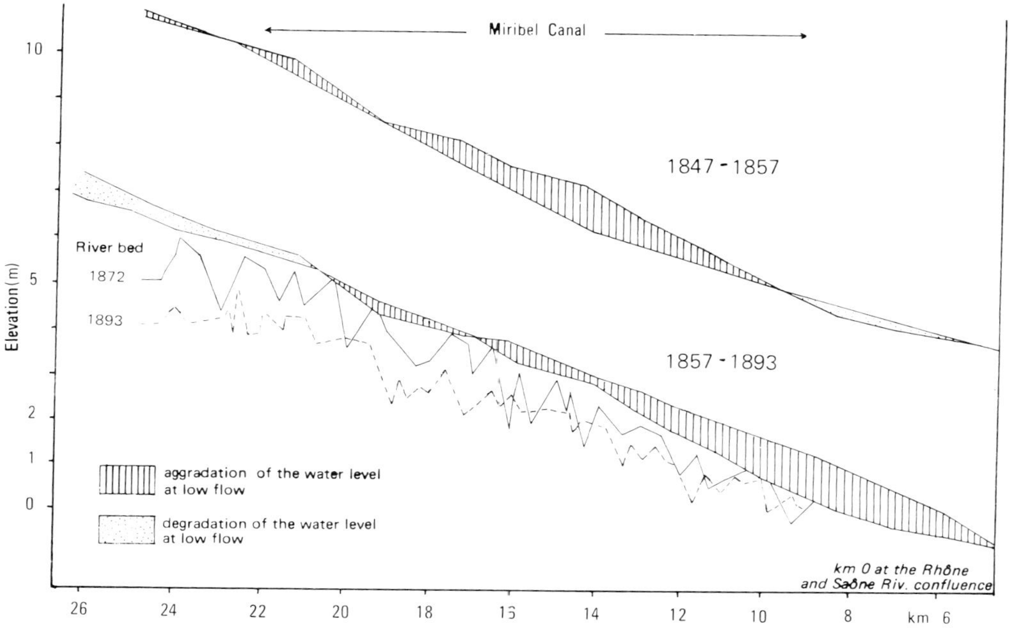

FIGURE 10. The Canal of Miribel (upper Rhône River): the tilting of the long profile since 1857 (after Gemaehling and Chabert, 1965)

describes the process of aggradation at the confluence of the Isere and Drac River; the slope of the Drac was between 0.0067 and 0.009, while the slope of the Isere was between 0.0028 and 0.0034. It was then explained that 'when two torrents or rivers join, if one of them is steeper than the other, of necessity the steepest will aggrade the common bed at the junction'. Between 1514 and 1787 the total aggradation was up to 7.20–7.50 m on the downstream course of the Drac river (in Bouchayer, 1925). Aggradation was a major problem during the 19th century because floods worsened at this time. Along the Drac river, cross-sections were surveyed from 1854, and these demonstrate a yearly aggradation of 3 cm between 1855 and 1878 at one critical section (Bouchayer, 1925).

As early as 1848, the Rhône river provides interesting examples of bathymetric maps and cross-sectional surveys in braided sections where embankments were planned. Long profile surveys undertaken during the construction of the Miribel Canal, upstream of Lyon (1848–1858), demonstrate a tilting of the bed along a 20 km reach (Figure 10). The confinement of high discharges between two straight dykes increased the energy of the river in a steep section and degradation of up to 5 m occurred upstream of the canal between 1857 and *c.* 1930 (Winghart and Chabert, 1965; Bravard, 1986).

CONCLUSION

Maps of rivers in France are manifold and French geographers used them as early as the beginning of the century. However, this field of research has not been investigated for many years. The result is that inventories of available material do not exist in many cases and one need is for these to be compiled prior to a systematic research on the interaction between man and river in the modern times. Nevertheless it must be pointed out that maps cover, at best, four centuries. The importance of maps is in providing a framework for other work, such as archaeology, palinology, etc. In this perspective, their interest is widely related to their accuracy. This one must control with the help of various methods ranging from critical examination of texts to remote sensing analysis. In fact field analysis as well as laboratory research is often made easier and safer with the help of ancient maps. But the authors must confess that the main interest of maps is related to Man: how did he live along rivers which were both a danger and a resource of importance; how did he try to impound them; did he change them intentionally, or not, and what was the response of the rivers? Cartographic analyses are clearly a fundamental tool for historical studies of change in fluvial hydrosystems but in many areas their potential is still to be exploited.

ACKNOWLEDGEMENTS

We are grateful to C. Daniere (URA 260) for drawing the figures.

REFERENCES

Amoros, C., Bravard, J. P., Castella, E., Girel, J., Jacquet, C., Pautou, G., Reygrobellet, J. L., Richardot-Coulet, M., Richoux, P., Roux, A. L., Roux, C. 1986. Recherches interdisciplinaires sur les écosystèmes de la basse plaine de l'Ain, France: potentialités évolutives et gestion. *Doc. Carto. Ecol.*, **XXIX**, 166 pp.

Archives Nationales 1987. *Espace français. Vision et aménagement, XVI°–XIX° siècles*. Catalogue d'exposition, Paris, Ministère de la Culture et de la Communication, 260 pp.

Babonaux, Y. 1970. Le lit de la Loire. Etude d'hydrodynamique fluviale. *Paris, Comité des travaux Hist. et Scientif. Mém. Sect. Géo.*, **5**, 252 pp.

Bomer, B. 1972. Les îles de la Loire évolution ou stabilité. *Etudes Ligériennes*, 69–80.

Bomer, B. 1976. *Aspects et évolution du lit mineur de la Loire entre les confluents de la Vienne et de la Maine*, Actes du 99° Congrès National des Sociétés Savantes, CTHS, Paris, 15–21.

Bomer, B., Busnel, L., Duault, C., Trousselet, Y. 1979. Milieux naturels et interventions humaines en Touraine. Résultats de recherches récentes. *Bull. Assoc. Geogr. Fr.*, **460**, 139–143.

Bouchayer, A. 1925. 1. Le Drac dans la plaine de Grenoble, de 1280 à 1651. 2. Le Drac et ses affluents. *Revue Géogr. Alpine*, **12**, 115–172 and 287–357.

Boutonnet, G. 1963. Histoire du dessèchement des marais du Bas-Languedoc, Nîmes, *Revue Bas-Rhône–Languedoc*, **14–17**, 53 pp.

Bravard, J. P. 1981. La Chautagne, dynamique de l'environnement d'un pays savoyard. *Lyon, Instit. Etudes Rhodaniennes, Mémoires et Docs*, **16**, 182 pp.

Bravard, J. P. 1986a. *Le Rhône, du Léman à Lyon*, Lyon, La Manufacture, 450 pp.

Bravard, J. P. 1986b. La basse vallée de l'Ain, dynamique fluviale appliquée à l'écologie. *Doc. Carto. Ecol.*, **XXIX**, 17–43.

Bravard, J. P., Amoros, C. and Pautou, G. 1986. Impact of civil engineering works on the succession of communities in a fluvial system. A methodological and predictive approach to a section of the upper Rhône river, France. *Oikos*, **47**, 92–111.

Broc, N. 1986. *La Géographie de la Renaissance*, Paris, Editions du CTHS, 260 pp.

Castella, C., Amoros, C. 1984. Répartition des Characées dans les bras morts du Haut-Rhône et de l'Ain et signification écologique. *Cryptogamie, Algologie*, **V** (2–3), 127–139.

Cholley, A. and Seive, F. 1931. *Atlas photographique du Rhône*, Lyon, J. Desvignes ed.

Dienne, H. de 1891. *Histoire du dessèchement des lacs et marais en France avant 1789*, Paris, Librairie Champion, 570 pp.

Dion, R. 1961. *Histoire des levées du Val de Loire*, Paris, 312 pp.

Fortuné, M. 1988. *Usages parsés et écologie de la Garonne*. Unpublished Thesis Inst. Nat polytéchnique de Toulouse, 162 pp.
Guillerme, A. 1983. *Les temps de l'eau, la cité, l'eau et les techniques*, Seyssel, Eds. Champvallon, 263 pp.
Grove, J. M. 1966. The little ice age in the Massif of Mont-Blanc. *Trans. of the Instit. of British Geographers*, **40**, 129–143.
Hooke, J. and Kain, 1982. *Historical Change in the Physical Environment*, Butterworth, London, 236 pp.
Le Roy-Ladurie, E. 1967. *Histoire du climat depuis l'an Mil*, Paris, Flammarion, 367 pp.
Levy, M. 1986. Géomorphologie et archéologie. In: *Buttenheim. Une motte castrale en Alsace*. Ann. Soc. Hist. sundgaurienne, pp. 89–110.
Pieri, M. 1983. *Analyse des facteurs dynamiques pourant modifier le lit d'une rivière et la nappe subordonnée. Exemple: la Basse-Durance occidentale*. Unpublished Thesis Aix-Marseille III, Fac. Sc. Luminy, 446 pp.
Peiry, J. L. in press. *Dynamique Fluviale Historique dans le Bassin de Cluses*, Colloque CTHS, Lyon, avril 1987.
Peiry, J. L. 1988. *Approche géographique de la dynamique spatio-temporelle des sédiments dans un cours d'eau intra-montagnard: l'exemple de la plaine alluviale de l'Arve*, Unpublished Thesis. Department of Geography, University of Lyon 3, 350 pp.
Schumm, S. A. 1977. *The Fluvial System*, Wiley & Sons, New York, 338 pp.
Sundborg, A. 1956. The river Klarälven, a study of fluvial processes. *Geogr. Annaler*, **38**, 127–316.
Werritty, A. and Ferguson, R. I. 1980. Pattern changes in a Scottish braided river over 1, 30 and 200 years. In Cullingford, R. A., Davidson, D. A. and Levin, J. (Eds), *Timescales in Geomorphology*, Wiley, Chichester, 53–68.
Winghart, J. and Chabert J. 1965. Le Haut-Rhône à l'amont de Lyon, étude hydraulique de l'île de Miribel-Jonage. *La Houille Blanche*, **7**, 1–20.

Historical Change of Large Alluvial Rivers: Western Europe
Edited by G. E. Petts
©1989 John Wiley & Sons Ltd

7

Evolution of the Po River: an Example of the Application of Historic Maps

G. BRAGA

Dipartimento di Ingegneria del Territorio, University of Pavia, Italy

and

S. GERVASONI

Dipartimento di Scienze della Terra, University of Pavia, Italy

INTRODUCTION

In order to investigate the geomorphological evolution of the River Po over the last few centuries, a reach between the confluences of the Rivers Sesia and Adda has been examined. Changes of the river channel in response to natural and artificial influences have been analysed by using three sources of data. First, aerial photographs of the present-day river and alluvial plain have been used to map the major geomorphological elements, such as terrace scarps, abandoned channels and bars. Secondly, by using historical data sources—maps and documents—the abandoned channels have been dated. Finally, analyses of the natural events that have influenced morphological evolution and the spatial pattern of landforms allow consideration of the effects of geological and neotectonic activity to be evaluated.

THE PO RIVER BASIN AND ALLUVIAL PLAIN

The Po river is the main river in Italy. Its 70 091 km^2 catchment area corresponds to about a quarter of the whole of the Italian territory. The river starts from 'Pian del Re', at 2 022 m a.s.l., on the northern side of the Monviso mountain, and flows for 673 km eastwards, until it reaches the Adriatic Sea. The river is characterized by considerable loads of suspended and bed material. The climatic zone in which the river flows is called the 'temperato mesaxerica' (Tomaselli *et al.*, 1973) with annual mean temperatures ranging from 0° to 10°, rainfalls of 1400 mm yr^{-1}, and rainy seasons in spring, summer and autumn, the river lacks a dry season. In fact, two different hydrologic regimes, referred to as the Alpine and the Apennine, combine to give a discharge of 47×10^9 m^3 yr^{-1}. The

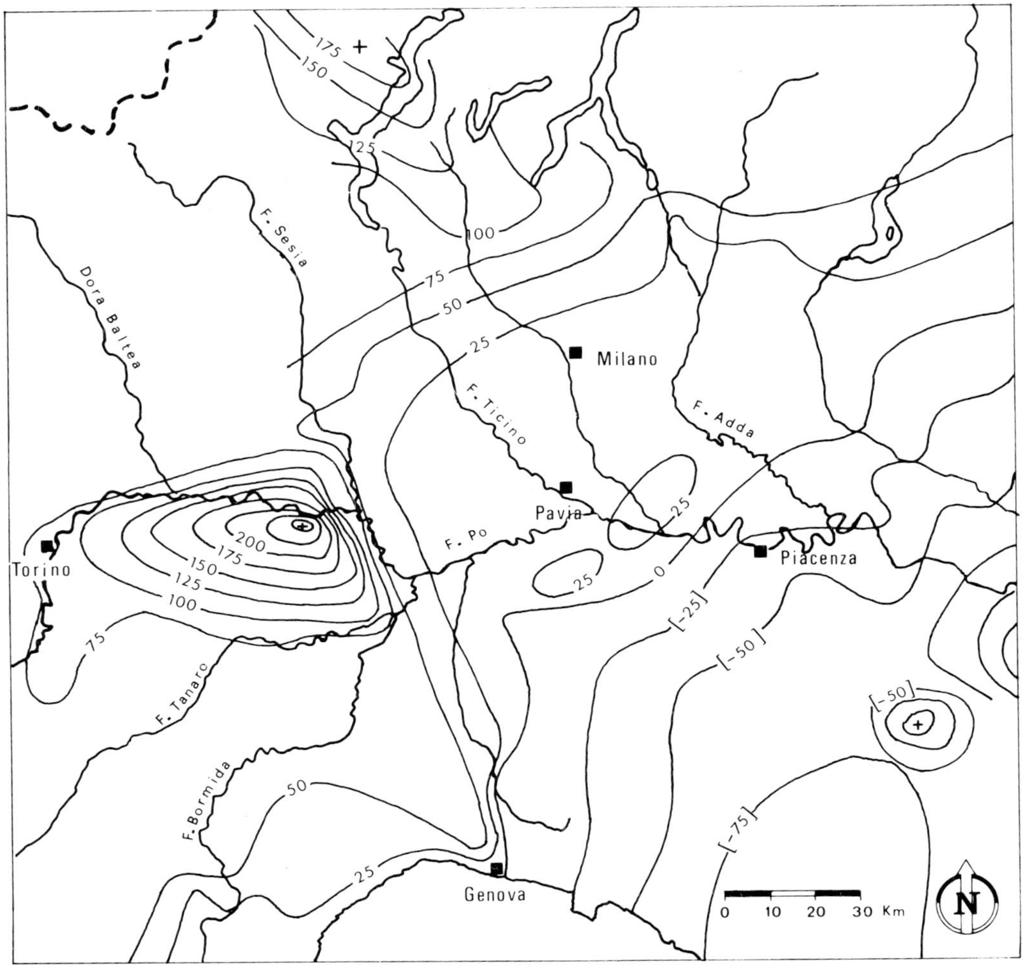

FIGURE 1. Map of recent tectonic movements within the central area of the Po basin showing vertical movements (mm) from 1897 to 1957 (modified after Arca and Beretta, 1985)

Alps, which are mainly on the north side of the Po river valley, contribute snow-melt runoff in summer. In contrast, the Apennine drainage basins, located on the southern side, have a pluvial regime with a marked low flow season in summer and two high flow seasons in spring and autumn.

The alluvial plain of the Po river has evolved since the Pliocene when an eastwards open gulf, connected with the Adriatic Sea, stretched between the Alpine and the Apennine chains. The River Po and its tributaries have filled this basin, and Pleistocene fluvioglacial and Holocene alluvial deposits have a thickness ranging from tens to several hundred metres. The bottom of the basin on which these deposits rest is not flat but has been deformed by a number of buried tectonic structures (folds and faults) mainly elongated parallel to the Apennine direction (NW–SE). The top of the anticlines form occasional outcrops within the alluvial plain, as happens in the Monferrato high structure and in the S. Colombano Hill. Movements are continuing even at the present time (Figure 1).

The good climatic and soil characteristics have led to the alluvial plain being intensively cultivated because of the high fertility. The need for flood control and progressive floodplain reclamation, to create new agricultural land, has resulted in the development of an intricate flow control system that has confined the river within progressively narrower artificial limits over the past three centuries.

METHODOLOGY

Any investigation of the evolutionary development of a fluvial channel requires the consultation of historical data sources, but the first step must be the analysis of the present

TABLE 1. Chronological list of consulted maps with references to the location and catalogue numbers

17th century	*Corso del Ticino e del naviglio Grande fino a Boffalora* (Archivio di Stato di Milano—MM piane 6D—Acque p.a. cart. 1237).
1588	*Nova, vera et integra descriptio totius fluminis Padi, incipiendo a Castro Arenae, usque ad Castru Novu Bucae Abduae, et cum toto territorio placentino, et eius confinibus ultra Padum* by P. Bolzoni (Archivio di Stato di Parma, tav. 3, vol. 46).
1700	*Carto topografica del corso della Sesia dai monti superiori di Masserano, sino all'imboccatura del Po tra Casale e Valenza; colle città e luoghi adiacenti, ed osservazioni militari* by La Marche (Archivio di Stato di Torino—catal. 126—Sesia no. 2).
1731	*Corso del Fiume Po superiormente ed inferiormente alla città di Valenza, con indice ed osservazioni circa un taglio progrettato per introdurre l'acqua dello stesso fiume a percorrere contro la ripa al di sotto della città—7 ottobre 1731* by Bona (Archivio di Stato di Torino-Valenza 21 A VII).
1745	*Plan du passage du Tanaro effectué par l'armée gallispane—27 novembre 1745* (Historical and Geographical Department—University of Pavia).
1762	*Continuazione del corso presentaneo del Po da Piacenza, sino allo sbocco d'Adda, nello stesso fiume, rilevato da me sottoscritto in misura di passi andanti nel mese di luglio 1762, con tutti gli Stati di Giurisdizione di S.A.R.D. Filippo di Borbone, infante di Spagna e Duca di Parma, Piacenza, Guastalla, ecc. durante questo corso*—M. Morelli (by Archivio di Stato di Parma—tav. 42—volume 50).
1772	*Carta corografica degli stati di S.M. il Re di Sardegna, data in luce dall'ingegnere Borgonio nel 1683, corretta, ed accresciuta nel 1772 da Stagnoni*—incisione, divisa in 16 fogli (Archivio di Stato di Torino—catalogo 126—Piemonto no. 23).
1773	Carta topografica del Pavese Principato, la Lomellina e l'Oltrepo—di Siro Severino capsoni dé Predicatori (Geography Department Pavia).
1820	*Topografia della Provincia di Cremona, tratta dalla celebre carta dell'ex Stato di Milano e dedicata al sig. Cavaliere Antonio Persichelli dal ragioniere Giuseppe Legnani—approvata dall'I.R. Governo* (Geography Department Pavia).
1833	*Carta topografica del Regno Lombardo Veneto, costruita sopra misure astronomico-' trigonometriche ed incisa a Milano nell'Istituto Geografico Militare dell'I.R. Stato Maggiore Generale Austriaco*—Vienna—scala 1:86 400 (Geography Department Pavia).
1852	*Gran carta degli Stati Sardi in terraferma* Pubbl. dal Corpo Reale di Stato Maggiore nell'anno 1852, scale 1:50 000—divisa in 91 fogli—no. 47 Casale, no. 48 Mortara, no. 55 Alessandria (Archivio di Stato di Torino).
1889	Tavolette I.G.M.I. (Italian Military Geographyc Institut)—scale 1:25 000 (Geography Department—Pavia).
1980	Aerial photos scale 1:20 000 (Regione Lombardia, Milano).

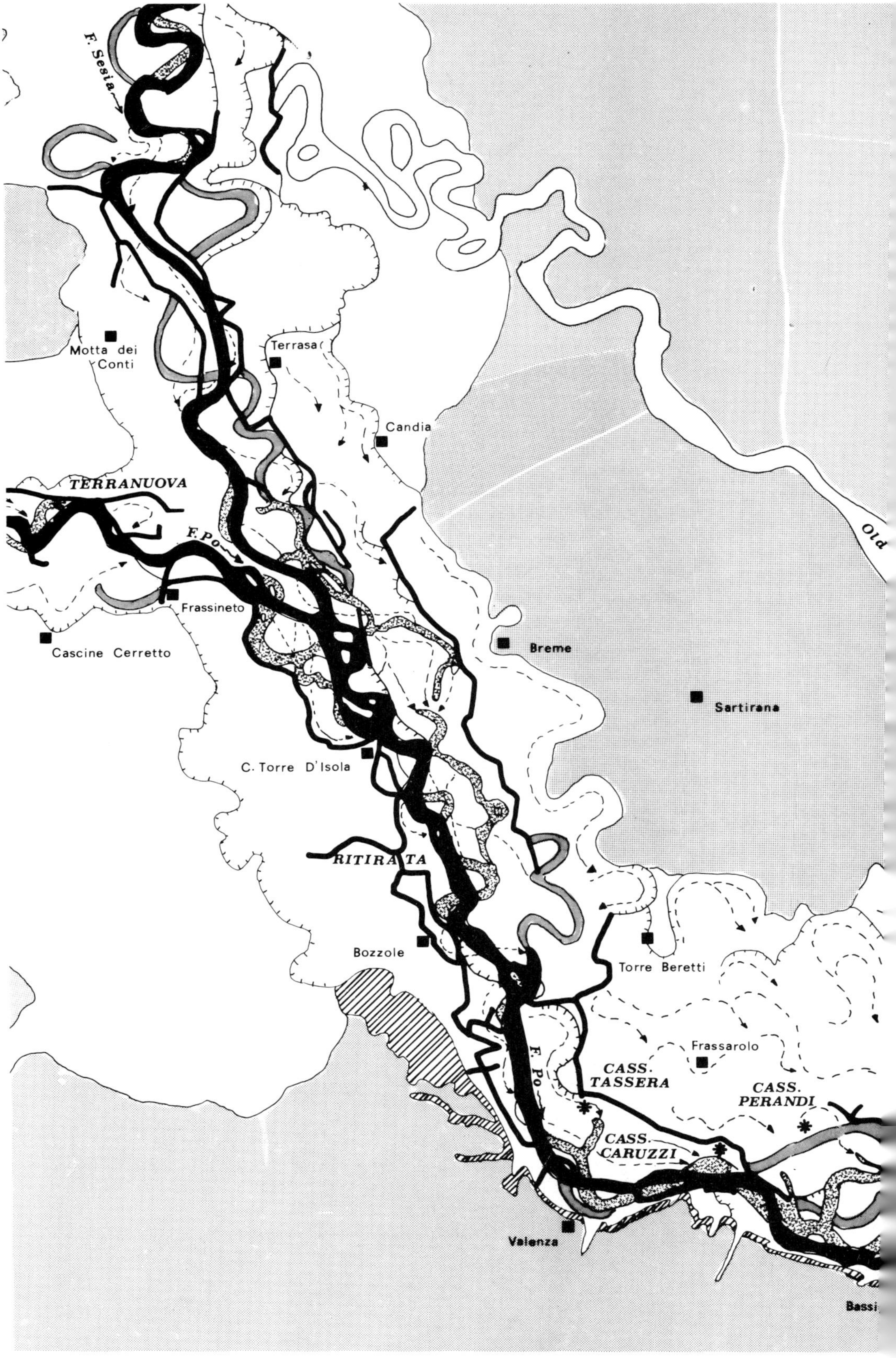

F. Sesia
Motta dei Conti
Terrasa
Candia
TERRANUOVA
F. Po
Frassineto
Cascine Cerretto
Breme
Sartirana
Old
C. Torre D'Isola
RITIRATA
Bozzole
Torre Beretti
Frassarolo
F. Po
CASS. TASSERA
CASS. PERANDI
CASS. CARUZZI
Valenza

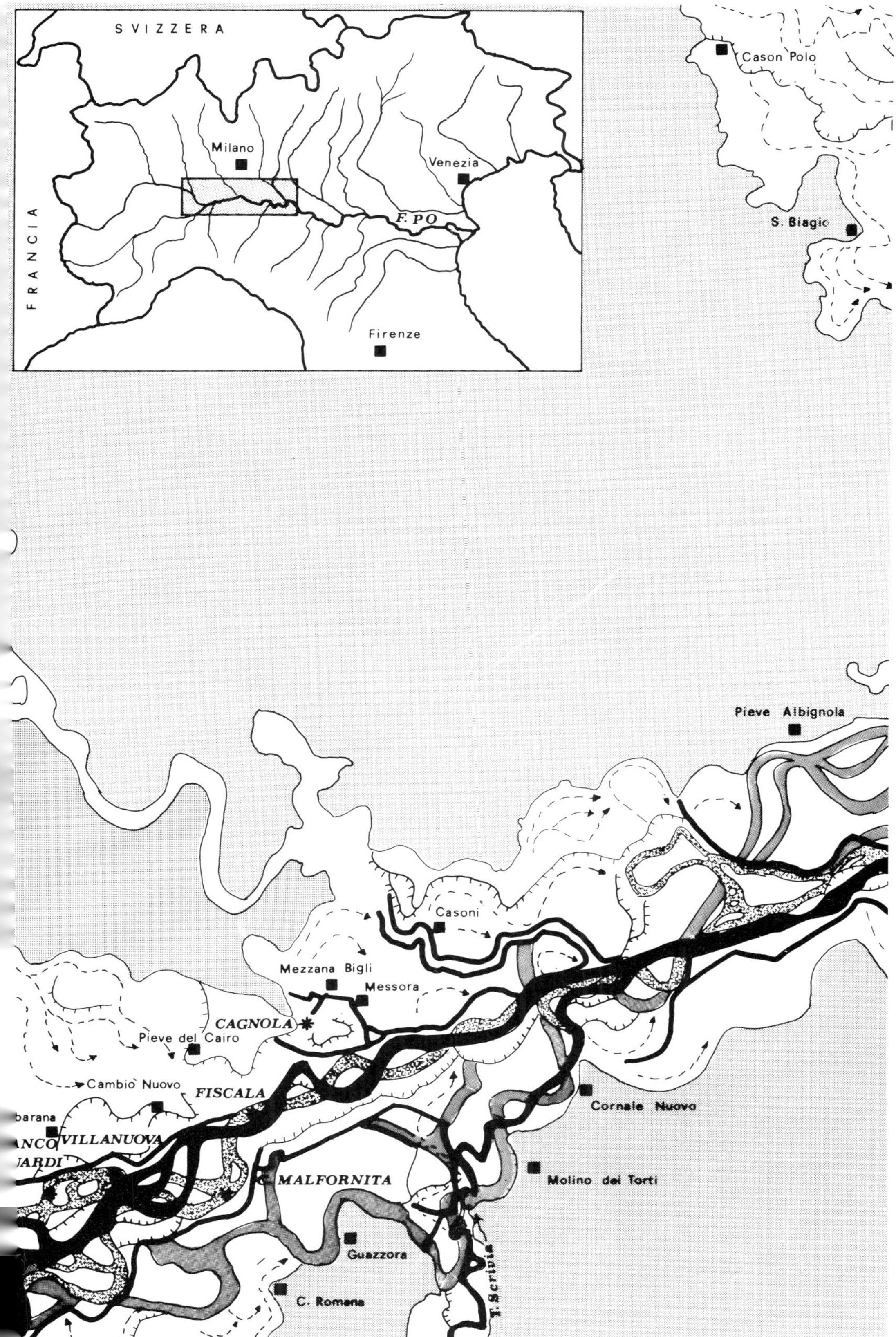

SVIZZERA
FRANCIA
Milano
Venezia
F. PO
Firenze
Cason Polo
S. Biagio
Pieve Albignola
Casoni
Mezzana Bigli
Messora
CAGNOLA
Pieve del Cairo
Cambio Nuovo
FISCALA
VILLANUOVA
C. MALFORNITA
Cornale Nuovo
Molino dei Torti
Guazzora
C. Romana
T. Scrivia

F. Ticino
Bereguardo
Parasacco
Torre D'Isola
PAVIA
Sabbione
Carbonara Ticino
Mezzano
Zinasco Vecchio
Sommo
Rea
CAMPAGNA
Pancarana
Cervesina
RIVER BED IN THE 20th CENTURY (1980)
RIVER BED IN THE 19th CENTURY (1889)
RIVER BED IN THE 18th CENTURY (1745 – 1773)
RIVER BED IN THE 16th CENTURY (1588)
RIVER BED IN THE 15th CENTURY (1466)
RIVER BED IN THE 13th CENTURY (1230)
UNDATED ABANDONED CHANNEL

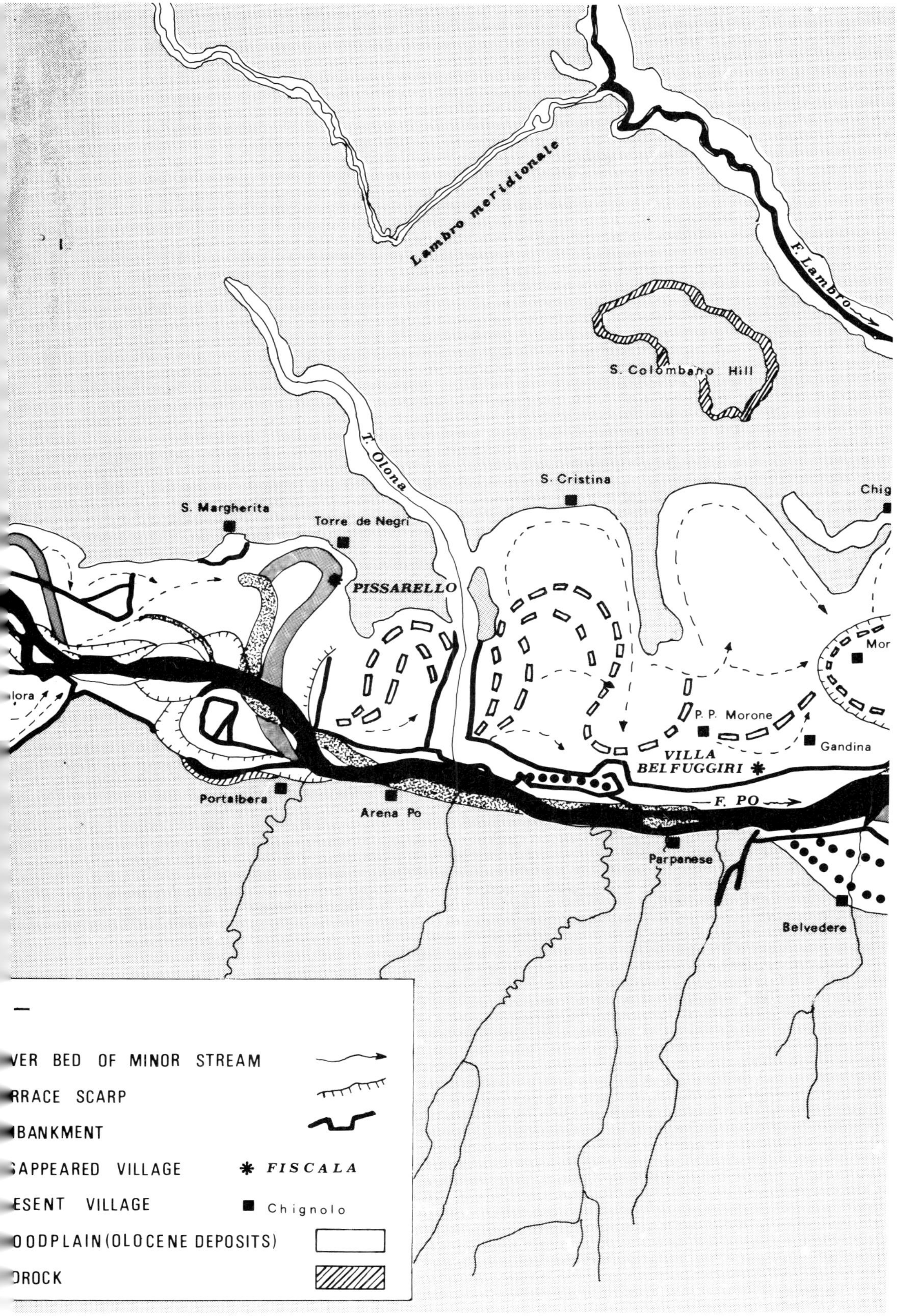
Lambro meridionale
F. Lambro
S. Colombano Hill
T. Olona
S. Cristina
Chig
S. Margherita
Torre de Negri
PISSARELLO
Mor
lora
P. P. Morone
Gandina
VILLA
BELFUGGIRI
F. PO
Portalbera
Arena Po
Parpanese
Belvedere
VER BED OF MINOR STREAM
RRACE SCARP
IBANKMENT
SAPPEARED VILLAGE
FISCALA
ESENT VILLAGE
Chignolo
OODPLAIN (OLOCENE DEPOSITS)
DROCK

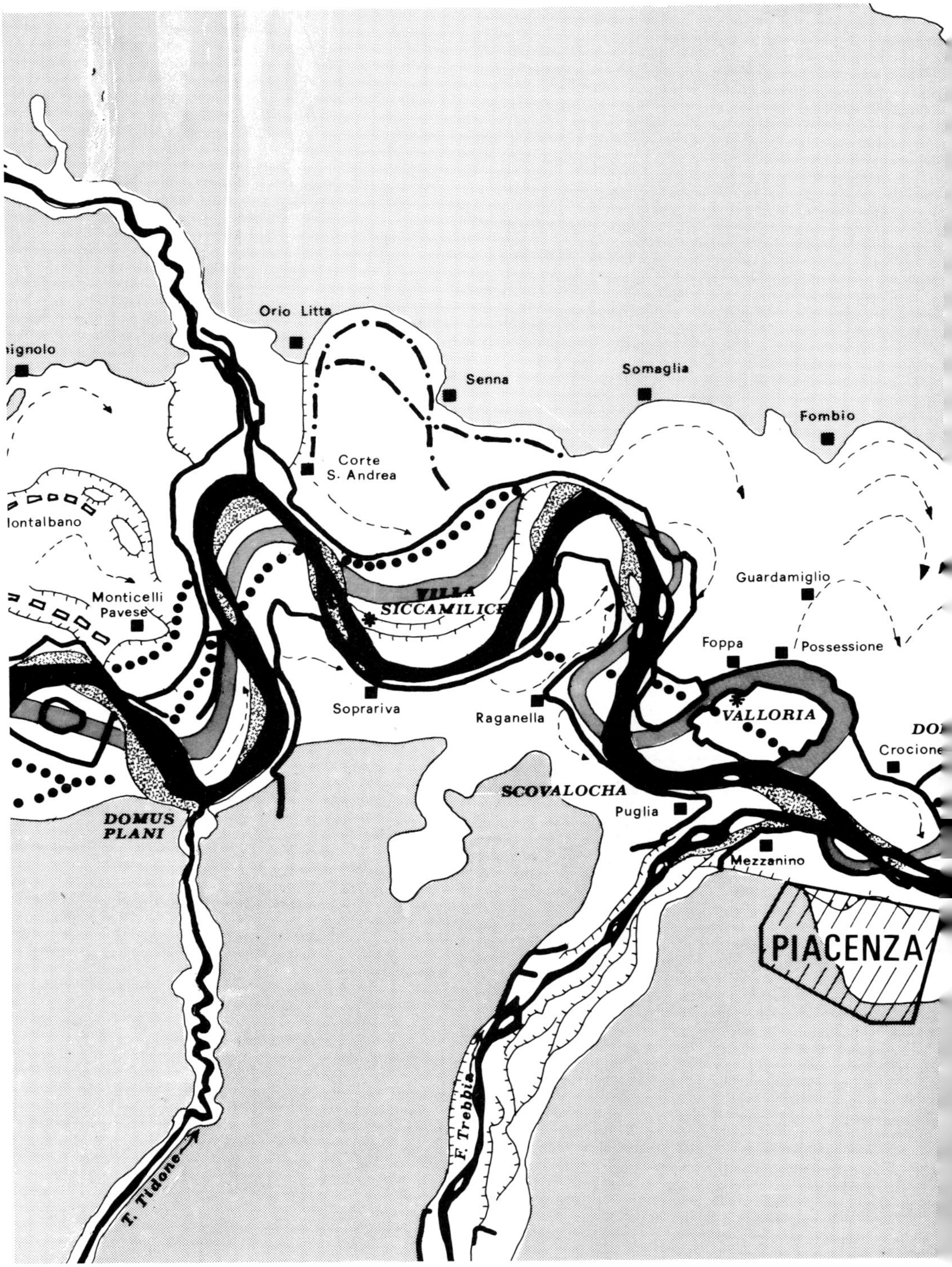

Orio Litta
ignolo
Senna
Somaglia
Fombio
Corte
S. Andrea
Montalbano
Monticelli
Pavese
VILLA
SICCAMILICE
Guardamiglio
Foppa
Possessione
Soprariva
Raganella
VALLORIA
Crocione
SCOVALOCHA
DOMUS
PLANI
Puglia
Mezzanino
PIACENZA
F. Trebbia
T. Tidone

FIGURE 2. The historical evolution of the Po River from the 12th century to the present day. Some possibly earlier forms are also shown

situation. Abandoned channels form elongated and/or curvilinear shaped elements, such as terrace scarps, shallow depressions and humid areas. Most of them are remnants of meander bends, which now form an intricate pattern, with the older ones occurring on the outer parts of the plain, most remote from the present river channel. This sequence allows a *relative dating* of the landforms and identification of the major *directions* of channel change.

It is at this point of the research that the use of old maps becomes of real interest, allowing the development of an *absolute chronology* of the fluvial geomorphic events. The history of modern cartography begins with Italian surveyors as early as the 16th century and mapmaking in Italy provides a valuable data source. However, there is a major problem concerning the reliability of data derived from the different survey methods, instruments, accuracies in representation, scales, and the migration of landmarks such as villages over time, due to catastrophic streambank erosion. Maps drawn in the 18th and 19th centuries generally provide valuable information, but the interpretation of maps from the 16th and 17th centuries is usually problematic.

Historical data sources

In spite of losses suffered during the First and Second World Wars it is still possible to find many historical sources at the State Archives and at the University Departments in Italy (Table 1). These include maps starting from the 16th century, the surveys of the grid of Roman agricultural boundaries (centuriations–cadaster) and other official documents. When using cartographic sources it must be remembered that the very morphological dynamics that form the focus of research also introduce problems. For example, it is easy to compare the 1980 and the 1889 situations, by superimposition of the two maps with the same scale (1:25 000), but to compare them with older maps of the river is more difficult. This limitation is due not only to the problem regarding the reliability of older maps, but also to the difficulty in finding maps reproducing all the study-area at an appropriate scale, such as 1:25 000 or 1:50 000, or to the difficulty in working with maps giving the following problems:

—Accuracy intrinsically connected with the different scales;
—Reproducing only small zones of the study area;
—Two different dates, one for the survey and one for the publication.

For the above reasons, it has been decided to reproduce in Figure 2 only three clear and complete situations with an interval of about 100 years (regarding the 20th, 19th and 18th centuries). Data for the preceding centuries is restricted to a few discontinuous tracts, related to the 16th, 15th and 13th centuries. The intermediate maps of 1833 and 1852 have been used only to verify the evolutionary trend.

NATURAL AND ARTIFICIAL CHANGES

The evolution of channel morphology over an historic timescale reflects both natural and artificial influences. For a river like the Po, with a hanging bed in the eastern part, surrounded everywhere by agricultural and residential areas, most changes from the 18th century until today have been caused by anthropogenic works such as embankments, cut-offs and bank defences. In order to emphasize and analyse the natural trend due to neotectonic activity, it is essential first to recognize and exclude the following artificial ones.

Embankments

A flood defence system made of earth embankments, both on the left and right bank, borders the Po river channel for 2290 km. The whole embankment system has generally an irregular character and was built relatively recently upstream of Piacenza, but is older and more continuous eastward from this town. This change is of direct concern to the study area (Figure 2). Downstream of Piacenza the 18th century river bed was completely constrained by old embankments. Upstream of this point instead, in the 18th century there were many long stretches still able to migrate freely. Some examples of artificial changes include the following. First, an embankment on the left side from Portalbera to the Tidone confluence was carried out in the 14th century during Galeazzo Visconti's rule. This embankment which is still effective today, made the river bed straight for 15 km, removing the major meanders south of S. Cristina. It is of real interest to point out that no embankment has ever been built on the right side of this same reach. This is due to the natural protection of the river bank provided by outcrops of bedrock, structurally connected with the buried part of the external margin of the Apennines. This geological control reflects the neotectonic activity that still occurs. Secondly, during the decades from 1466 to 1476, thanks to the Duke Galeazzo Visconti, the meander bend running next to Chignolo was abandoned by cutting a new artificial river channel passing to the south of the village of Monticelli Piacentino, that from this moment on has been called Monticelli Pavese. However, earlier embankments date back to the 13th and 14th centuries. These were set back from the river and all evidence of disappeared villages, abandoned as a result of river erosion, remain inside this artificial boundary.

Natural changes

The most obvious evidence of natural geomorphological changes from the 18th century until the present day, has been the 2 km northward shift of the Po river channel downstream of the Tanaro confluence. On the other hand, the sharp change in the flow direction of the River Sesia occurred before the 16th century. A well-preserved abandoned channel, is clear evidence of a sudden event which diverted this stream from the old direction parallel to that of the Ticino river, to the present N–S direction. Four other channel changes are worthy of note. First, around Valenza it is possible to recognize a braided pattern that indicates a major phase of aggradation. Secondly, the Ticino river confluence has moved downstream about 2000 m in 200 years. Thirdly, the natural migration of the river bend from near S. Margherita (eastern of Pavia) in the 18th century to the present situation, has been continuous and regular. Such a strange evolution of a meander bend that regresses instead of enlarging may be explained by looking at its position upstream of the major channelization works from Portalbera to the Tidone confluence, carried out in the 14th century. Because of the increased slope, flow velocity increased and the stream probably tried to cut-off the bend. However, the neck of the meander bend was still too wide, so the meander bend began to regress to the present straight channel. Fourthly, historical sources show that the change in the river bed that occurred in 1230 close to Orio Litta relates to a natural event, and this caused the village of Corte S. Andrea to pass from the right to the left side of the present Po river.

DISCUSSION

The above analysis of channel patterns based upon historic maps has revealed two general types of change. First, the sudden and dramatic change in the course of the old Sesia river, from one parallel to the Ticino river (NW–SE) to the present southward direction, occurred long before the 16th century. Such an event may be related to neotectonic movement that is still active on the northeastern side of the Monferrato chain of hills where there is a buried overthrust. Contemporary (1897–1957) vertical movements of the ground are shown in Figure 1. Secondly, the more recent effects of the neotectonic movements are causing

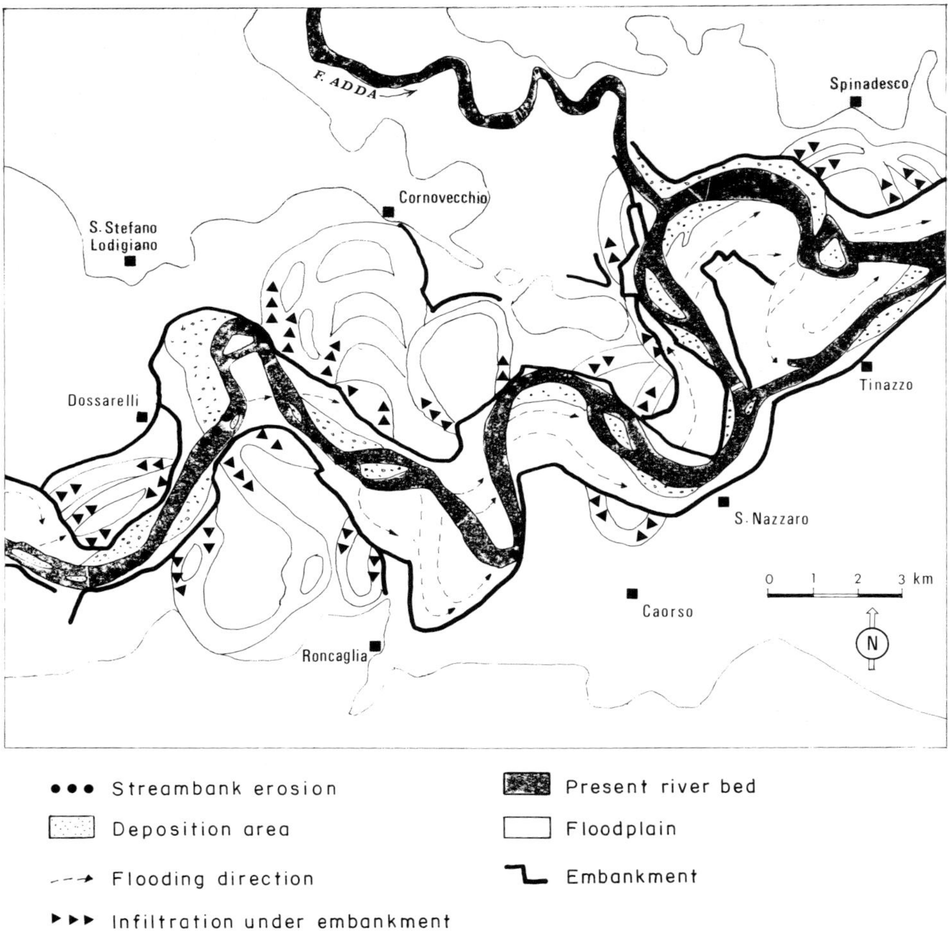

FIGURE 3. Hydrological hazard map of a section of the River Po based on knowledge of channel and floodplain evolution

the raising of the external margin of the Apennine chain, and the northward migration of the Po river bed. Evidence for this movement includes:

— The presence of bedrock outcrops along the right bank from Valenza to Bassiniana and near Portalbera;
— The northward shifting of the old bed of the Po river, downstream of the Tanaro confluence (18th century);
— The lack of embankments on the south side downstream from Portalbera to the Tidone river;
— The braided pattern of the Po and Sesia rivers upstream of Valenza during the 18th and 19th centuries.

The synthesis of this kind of historical and geomorphological survey, allows consideration of the hydrogeomorphology of the studied reach of the Po river channel, in order to assess future trends.

Knowledge of historical changes is important in order to develop models to predict future changes, changes that may reflect natural evolutionary pathways, climatic change, and/or human influences. However, such data also have a number of applications of immediate concern. In particular, it is possible to define some of the potential hazards and resources. Some examples of these are identified on Figure 3. Streambank erosion due to shifting meanders within the embankments can be extensive: for instance, along the bank north of Soprariva (east of Piacenza town) some 50 ha were removed between 1889 and the present day. A number of villages shown on old maps have subsequently disappeared. Foci of deposition may also be identified by comparison of the 1889 and 1980 maps, the latter showing many new meander bars, channel bars and islands. These changes are viewed as a resource, rather than as a hazard providing new land and raw materials for quarry activities. Abandoned channels, may also be problematic. Due to their genesis, channel fill deposits are impermeable at the surface but highly permeable at depth. These characteristics, allow subsurface flow under the embankments during floods, causing flooding or water logging of the soils over parts of the alluvial plain.

REFERENCES

Arca, S. and Beretta, G. P. 1985. Prima sintesi geodetico geologica sui movimenti verticali del suolo nell'Italia settentrionale (1897–1957). *Bollettino di Geodesia e Scienze Affini*, **2**, anno XLIV.

Braga, G. and Gervasoni, S. 1983. *Evoluzione storica dell'alveo del fiume Po nel territorio Lodigiano–Piacentino: rischi idrogeologici connessi*, Estratto da: Atti del Convegno Nazionale 'Il suolo come risorsa', Piacenza.

Castellarin, A. *et al.* 1985. Analisi strutturale del fronte Appenninico Padano. *Giornale di Geologia, serie 3ª*, **47**/1–2, Bologna.

Chevallier, R. 1983. *La romanisation de la Celtique du Pô*, Ecole française de Rome, Palais Farnese, Roma.

Cati, L. 1981. *Idrografia e idrologia del Po*, Pubbl. no. 19 dell'Ufficio Idrografico del Po, Istituto Poligrafico e Zecca di Stato, Roma.

Ministero dei Lavori Pubblici — Magistrato per il Po 1984. *Per il Po e i suoi affluenti*, 2° Convegno di Idraulica Padana, Parma.

Tomaselli, R. *et al*. 1973. *Carta bioclimatica d'Italia*, scala 1:2 000 000, Istituto di Botanica, Università di Pavia.
Tozzi, P. Z. 1987. *Memoria della terra. Storia dell'uomo*, Pubbl. Fac. Lettere e Filosofia Università di Pavia, **41**, Dip. Scienze Antichità, La Nuova Italia Editrice, Firenze.

Historical Change of Large Alluvial Rivers: Western Europe
Edited by G. E. Petts
©1989 John Wiley & Sons Ltd

8

Evidence for Historical Heavy Metal Pollution in Floodplain Soils: the Meuse

M. C. RANG and C. J. SCHOUTEN

Department of Physical Geography, State University of Utrecht, Box 80.115, 3508 TC Utrecht, The Netherlands

INTRODUCTION

For centuries the large European rivers, including the Meuse, have been used as public sewers into which domestic and industrial waste could freely be discharged. Large quantities of domestic, industrial and mining wastes were discharged directly into the rivers, particularly in the 19th century, when little was known about the possible deleterious effects of certain waste substances. The pollution was not restricted to the water phase. Through the import of fresh silt and through the production of organic material (algae, weeds, etc.) the heavy metals are continuously absorbed on the solid material. In time the solid particles flocculate, then sink and become part of the river bed. During periods of peak flow and high current velocity the settled material can be picked up again and transported (Rang *et al.*, 1986). The sedimentation of this material has led to the severe pollution of large areas of the floodplains along the large rivers. In recent decades this pollution has reached an unacceptable level in The Netherlands (Salomons *et al.*, 1982). Along the Meuse system severe pollution occurred much earlier, because of the exploitation of reserves of lead, zinc and coal in the catchment.

This chapter illustrates the application of pedological and geochemical surveys of floodplain sediments (1) to evaluate the spatial effect of pollution within hydrosystems and (2) to determine the historical changes of pollution loading. Focussing on the heavy metal pollution of the River Meuse sediments, particular attention is paid to the relationship between soil development and frequency of flooding on the one hand and, on the other hand, the degree of pollution. This not only gives a good spatial picture of the extent of the pollution but also reveals the historical developments involved.

MAPPING SOIL POLLUTION IN FLUVIAL ENVIRONMENTS

One of the aims of this study was to ascertain the spatial variability of the heavy metal content of the alluvial soils of the Meuse River and to prepare a map of the soil pollution, for regional planning. For soil surveys, the optimal sampling density depends on the expected spatial variation and the scale of the final map. The pollution maps required a scale of

1:10 000 or 1:25 000. From the outset of the study it was clear that at these scales, interpolation from point observations selected by regular grid sampling or random sampling of homogeneous units was virtually impossible. Due to the large spatial variation of the heavy metal content of the topsoils, these methods of sampling would have required a vast number of samples, and therefore would have been too costly. For this reason it was decided to investigate the relationship between heavy metal pollution and sedimentological and pedological characteristics. If these relationships appeared to be unequivocal, these characteristics could be used when mapping the heavy metal load of the soil.

Sedimentological processes and the pollution of alluvial soils

Various researchers have investigated the heavy metal pollution of river systems. They have paid much attention to the nature and pollution of the suspended matter and to the interaction of this suspended matter with the fluid phase (Forstner and Wittmann, 1983; Salomons and de Groot, 1978). The reason for this attention is that most of the dispersion of heavy metals in river systems occurs via the transport of suspended material. In the aforementioned, largely geochemically oriented studies, little attention is paid to sedimentological aspects. However, when studying the pollution of bank soils, these aspects are very important, because the sedimentation of polluted silt is the most important cause of subsequent pollution. Because of this, spatial differences in the occurrence of processes of fluviatile erosion and sedimentation lead to differences in the degree of pollution. Lewin *et al.* (1977) describe the relationship that exists between different sedimentary environments and the level of pollution. They found that the concentrations of metals in sediments infilling old river cut-offs and in finer sediments deposited in low-energy environments are likely to differ from those in coarse-grained point bars. The reason for this is that metals are not homogeneously distributed over the various grain size fractions. Within the grain size spectrum the clay fraction shows relatively high metal contents.

Thus, when investigating the spatial distribution of heavy metals in alluvial soils, a distinction must be made between sediments in abandoned channels, in point bars and swales, and in the actual floodplain. Differences in the degree of pollution can also occur within these geomorphological units.

Given the temporal variation of the extent of the discharges of heavy metals in the river system, the degree of pollution of the sediments in abandoned channels will depend on the age of these sediments and thus on when cutoff occurred.

Differences in the heavy metal content may also be present within the floodplain. Sediments are deposited on the floodplain when it is inundated with floodwaters. The whole floodplain is not necessarily covered during each event. Irregular bank heights downstream ensure that the bankfull discharge varies along the channel and certain parts of the floodplain receive overbank deposits more frequently than others (Bradley and Cox, 1986). This may result in discrete zones of equal pollution rate, characterized by differences in frequency of inundation and rate of sedimentation.

Soil development and the pollution of alluvial soils

Not only the degree of soil pollution but also the degree of soil development is influenced by the frequency of flooding and the rate of sedimentation. Soil formation in the winter channel of the Meuse is closely related to the occurrence of fluvial erosion and sedimentation processes. In general, young soils are low-lying, are frequently inundated and consequently will be burdened more by heavy metals than the older, higher-lying soils. This difference is accentuated by the fact that the youngest soils are largely formed from recent, severely polluted sediments.

THE URBAN AND INDUSTRIAL DEVELOPMENT IN THE MEUSE CATCHMENT

The presence of coal and metal has largely determined the industrial development in the Meuse catchment. Coal was already being mined in the early Middle Ages, although large-scale

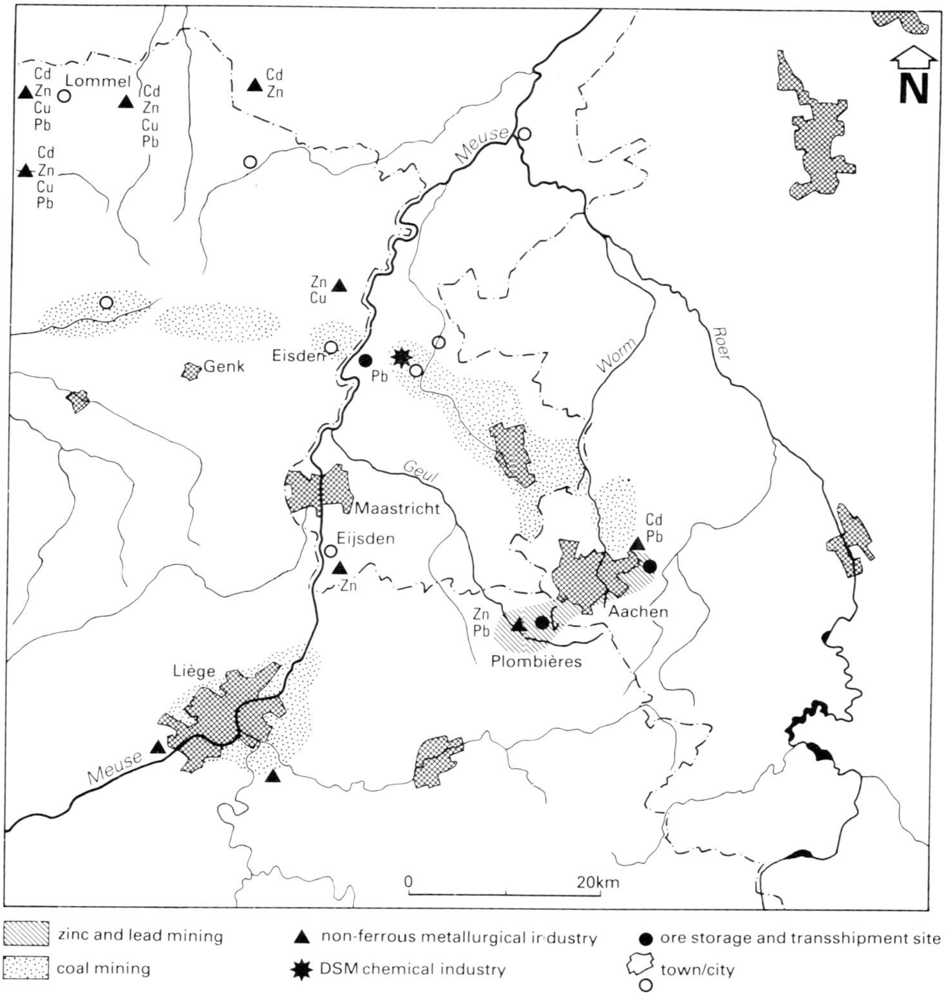

FIGURE 1. Some sources of heavy metals in South Limburg and adjacent areas

exploitation did not take place until the industrial development that followed the Second World War. In addition to coal, large amounts of matrix material, such as sandstone and clay stone were mined. The coal was separated from this matrix material by means of flotation techniques. The waste water, which still contained large amounts of material in suspension, was initially discharged untreated into the surface water system. As a result, until the beginning of this century, the concentration of coal particles in the Meuse was so high that during periods of low river discharge the water was black.

Important ore reserves were discovered near Aachen, in the catchment of the Inde, a tributary of the Roer, and near Plombieres, in the Belgian part of the Geul catchment (see Figure 1). The exploitation of zinc and lead ores began in the early Middle Ages and reached its peak between 1820 and 1880. The mines were exhausted in the 1930s and exploitation ceased. The extraction of lead from the huge waste tips on former mine sites continued into the 1950s.

The separation techniques used in the extraction of ores were based on the same principle as those used in coal production, i.e. on differences between the density of the ore and of the matrix material. The separation techniques were far from efficient, and therefore the waste water still contained high levels of metal-rich particles. This waste water was discharged into the river, thereby leading to an appreciable input of heavy metals into the river system (Rang and Schouten, 1987).

At present the Meuse catchment is a relatively densely populated and heavily industrialized area. This still has repercussions on the quality of the surface water. The French part of the catchment contains four towns of note, which together have a population of 130 000. The most important industries here are metallurgy and the foodstuff industry. Most of the urban and industrial waste water from this area is treated before being discharged into the river. The largest towns and conurbations in the Meuse catchment are in Belgium and Germany. Charleroi, Namen and the Liege conurbation together have *c.* 800 000 inhabitants. The most important industries are iron, steel and non-ferrous metallurgy and the textile and foodstuff industries.

Coal mining and the heavy steel industry have become less important in recent decades. In the region around Liege and Namen still only part of the waste water is treated; most is discharged untreated into the surface water.

In the German part of the catchment, sugar refining and textiles are the most important industries. The largest towns and conurbations here are Aachen and Munchen-Gladbach, together accounting for *c.* 1 600 000 inhabitants. Streams in the German part of the catchment flow into the Meuse mainly via the Roer and Niers. Approximately 90 per cent of the waste water discharged into the Roer is purified (Department of Public Works, 1981). In spite of this, the water in this river is of exceptionally low quality.

The most important towns in the Dutch part of the catchment are Maastricht, Eindhoven, Den bosch, Tilburg and Breda, with a total population of *c.* 1 700 000. Metals and chemicals are the most important industries here. In 1985 approximately 80 per cent of the waste water was treated.

LEVELS OF HEAVY METALS IN THE MEUSE WATER

It is difficult to ascertain levels of heavy metals, especially in the Meuse, which has a complicated drainage network. The total load of heavy metals is largely determined

TABLE 1. Percentage share of metals bonded to the suspended matter (period 1977–1988). (Source: State Department of Public Works)

	Eijsden	Lith	Keizersveer
Copper	58	59	50
Nickel	41	29	28
Zinc	57	65	57
Lead	93	91	78
Cadmium	81	86	69
Chrome	72	82	68

TABLE 2. Overview of mean levels of heavy metals in Meuse water in 1981. (Source: State Department of Public Works)

	Nameche	Liege	Eijsden	Heusden	Keizersveer
Susp. matter (mg l^{-1})	39.7	27.3	42.7	28.6	18.4
Zn. total ($\mu g\ l^{-1}$)	83	225	274	210	118
Zn. dissolved ($\mu g\ l^{-1}$)	—	109	88	—	66
Cd. total ($\mu g\ l^{-1}$)	5.3	4.5	1.7	1.5	0.9
Cd. dissolved ($\mu g\ l^{-1}$)	—	1.3	0.4	0.5	0.2
Cu. total ($\mu g\ l^{-1}$)	18	19	17	14	5
Cu. dissolved ($\mu g\ l^{-1}$)	—	37	13	7	2
Pb. total ($\mu g\ l^{-1}$)	31	21	11	13	15
Pb. dissolved ($\mu g\ l^{-1}$)	—	7	2	4	1
Co. total ($\mu g\ l^{-1}$)	0.0	0.0	1.4	1.4	0.6
Co. dissolved ($\mu g\ l^{-1}$)	—	—	0.7	0.3	0.1
Ni. total ($\mu g\ l^{-1}$)	0.0	0.0	7.7	7.7	4.9
Ni. dissolved ($\mu g\ l^{-1}$)	—	—	5.2	7.1	4.0
Cr. total ($\mu g\ l^{-1}$)	0.0	2	5	6	5
Cr. dissolved ($\mu g\ l^{-1}$)	—	0.0	2	2	1

by flow rate and the content of suspended silt. Table 1 gives an overview of the percentage of heavy metals attached to silt.

This suspended silt can settle if river discharge or current velocity are low. Thus, between Eijsden and Keizersveer the amount of suspended matter decreases, and therefore a smaller load of heavy metal is recorded. This apparent improvement in quality merely means that these metals have indeed disappeared from the aquatic environment. The material that settled during low current velocities is resuspended again when current velocities increase, and a larger load of metals is then recorded. This means that large fluctuations can be measured in the heavy metal content of the river, even though the emissions remain constant.

Table 2 shows the fall-off in the heavy metal levels downstream. The decline in heavy metal levels downstream from Eijsden is striking. This is partly attributable to a fall-off in suspended matter in this direction. Given that the levels of heavy metals that are present in dissolved form also decline, it can be concluded that the heavy metal load on the aquatic system decreases downstream and that the most important sources of heavy metals are upstream of Eijsden.

SOILS AND SOIL DEVELOPMENT IN THE MEUSE VALLEY

Pedogenesis in the Meuse valley depends greatly on the hydrological behaviour of the Meuse. In South Limburg the Meuse is only occasionally dyked, and therefore until recently the river was able to change its channel regularly here. As a result, soil development has locally been disturbed by the erosion of soil material, or by fresh sediment being deposited on this material. Hence the soils in the Meuse valley vary greatly in age, as does the degree to which the original fluvial deposits have been changed by soil-forming processes. In South Limburg the Meuse has continuously lowered its channel, and therefore the oldest soils are higher in relation to the river than the younger soils. And because low-lying deposits are inundated more frequently than high-lying deposits, the pedological difference between old and young soils has been accentuated. The following description of the pedogenesis makes use of the German system of soil classification, because this is strongly dependent on the genesis of the soil profile. Where it was felt to be relevant, the Dutch classification of the soil profile has been used.

Pedogenesis begins immediately after the deposition of coarse, gravel-rich material at low water level. When this material is colonized by a pioneer vegetation, the accumulation and biological decomposition of the plant material results in some enrichment with humus. During periods of high discharge more material will be deposited, and as a result over the years the levee is raised. The higher the levee, the finer-textured the deposits. After some years the levee is so high that is is only occasionally inundated. As a result, a greater accumulation of organic matter can take place in the top layer. Through the action of precipitation, vegetation and soil organisms the originally calcareous soil material will be decalcified, as a result of which the soil acidity will increase to a pH of about 6. Under such conditions of slight acidity internal weathering of the soil material will occur, resulting in iron being release from the soil minerals. This free iron, present in the soil as a hydroxide, is responsible for the characteristic brown colouring of these soils. In Germany, similar brown-coloured soils are called 'Braunerden' (Scheffer and Schachtsabel, 1979).

If the natural acidification of the soil continues, at a pH of approximately 5.5 clay eluviation can ultimately occur. The soil layer where the clay particles accumulate is called a 'brick layer', because of its reddish-brown colour and compactness. Soils in the Meuse valley having such a brick layer are classified as 'Radebrikgrond' in the Dutch classification (German equivalent: Parabraunerde).

The difference in the age of the soil is expressed not only in the degree of soil development but also in the properties of the parent material. Since the hinterland of the river was brought into cultivation, the fluvial system, and hence the composition of the deposits, has been strongly influenced by human activities.

The entire Meuse catchment was originally forest-covered. Therefore, only a small amount of sediment was washed into the river. During the deforestation and cultivation of the catchment, the sediment load in the river increased markedly because of the erosion of the loess, and the nature of the sediments also changed. Whereas the original sediments had a more clayey texture, the material deposited after the deforestation of the catchment is characterized by its extreme siltiness and clear brown colour. Soils that have developed in this preweathered soil material are not recognized as a separate soil type in The Netherlands, but in Germany are classified as 'Vega'. On the basis of the degree of decalcification a distinction is made between lime-rich Vegas (maximum depth of decalcification 40 cm),

TABLE 3. The soils in the investigated area classified according to the German, the Dutch and the USDA classification system

German system	Dutch system	USDA soil taxonomy
Parabraunerde	Radebrikgrond	Typic Hapludalf
Braunerde	Non-calcareous Ooivaaggrond	Fluventic umbric Dystochrept/fluventic Haplumbrept
Calcareous Vega	As above	As above
Lime-rich Vega	Calcareous Ooivaaggrond	Fluventic Eutrochrept/fluventic Haplumbrept
Young Auen soils	As above	Mollic Udifluvent/typic Udifluvent
Typische Gley	Non-calcareous Poldervaaggrond	Aeric Fluvaquent
Oxygley	As above	Aquic Udifluvent/mollic Udifluvent
Braunerde-Gley	As above	Fluvaquentic Dystochrept

calcareous Vegas (maximum depth of decalcification 80 cm) and lime-deficient Vegas (depth of decalcification more than 80 cm).

Iron translocation occurs locally at some distance from the river, under the influence of groundwater movements. This results in iron mottling in the soil profile. Where anaerobic groundwater wells up, the iron compounds in the soil can be reduced; this can be seen from the grey colours of the reduced iron. Groundwater-influenced phenomena of oxidation and reduction are indicated by the term 'gley'. The soils described above therefore belong to the class of gley soils (oxygleys and typic gleys, respectively). If, as well as gleying, brown iron mottling occurs, then the soil is classified as Braunerdegley or, if the brown mottling horizon extends to more than 80 cm, as Gleybraunerde.

The recent deposition of alluvium by the Meuse, i.e. the sediments that have been deposited since the industrial revolution, are characterized by anthropogenic pollution with slack and slag grit and a high calcium content. Locally this influence is so strong that one can speak of an anomalous parent material that is characterized by grey and black colours. We have termed the soils developed in this material 'young Auen soils' (after Rang and Schouten, 1987).

In Table 3 a summary of the soils in the study area and their classification according to the Dutch system of soil classification and the USDA Soil Taxonomy are given. The differentiating characteristics differ from system to system. Therefore, a comparison of soil classifications in the three systems has limited value.

FIELD METHODOLOGY

In order to obtain a good picture of the spatial variation in heavy metal concentrations in the Meuse floodplain, an extensive field study of the soils was carried out. To this end, soil was augered to a depth of 120 cm at some 500 locations in the Meuse valley. At each location the soil profile was sampled and described. After every augering a soil sample was taken from the top layer (A horizon). To avoid variations resulting from very localized differences in concentration, 15 subsamples were taken for every sample, over an area of 25 m^2. These subsamples were then mixed and a sample was taken from the mixture. Using atomic absorption spectrometry the soil samples were analysed for zinc, cadmium, lead, copper, cobalt, chrome and nickel.

TABLE 4. Distribution of sample points over the various classes of frequency of flooding

FF class	Recurrence (year)	Discharge ($m^3\ s^{-1}$)	No. of samples (n)	(%)
1	2	1500	74	14.8
2	5	1750	92	18.4
3	10	2000	108	21.6
4	50	2500	178	35.7
5	250	3000	47	9.4

The prime consideration when setting up the sampling scheme was a proportional distribution over the entire area. In addition, the sampling was primarily directed to sites where local variations in fluvial influence suggested the presence of large variations in levels of heavy metal. The degree of fluvial influence was determined from frequency of flooding and the local soil development. The frequency of flooding was estimated from the elevation of the site in question and data on the streamflow of the Meuse. The 1:25 000 maps of flooding frequencies prepared by the Department of Public Works were also consulted. The German system of soil classification was used when ascertaining the nature and degree of soil development, as described earlier. Table 4 gives an overview of the distribution of the sample points over the various frequency of flooding classes.

To ascertain the historical development of the heavy metal pollution of the Meuse, 26 locations were selected on the basis of the soil study, and the vertical distribution of heavy metals in their soil profiles was investigated. At each location soil samples were taken from two depths. In addition, the deposits in two silted-up former channels of the Meuse were sampled in detail to a depth of *c*. 3 m. Finally, five silt samples were taken from the present channel. The levels of zinc, cadmium, lead and copper in all the samples were determined.

RESULTS

The degree of pollution of the soil samples

When interpreting the levels of heavy metals in the soil it is important to know to what extent these deviate from the naturally occurring levels. The latter can vary greatly, depending on the mineralogical composition, the organic matter content and the texture of the parent material. Nevertheless, benchmark values were mostly used, regardless of the soil type.

The list of test values from the Dutch Guidelines for Soil Sanitation shown in Table 5 is an example of such values. From this list it can be generally inferred whether a soil is polluted and if so, to what extent. Three sorts of values can be distinguished. The A values are benchmark values. They are found in many places in The Netherlands, where soil pollution is not an issue, which is not to say that these values conform with natural background concentrations. The B values signal the need for further investigation. If the concentrations of certain metals exceed the B value, then further research is necessary, for example to ascertain their distribution. The C values signal the need for research directed towards soil sanitation. If the concentrations of certain materials exceed the C value, then

TABLE 5. Framework for assessing the levels of concentration of heavy metals (in mg kg^{-1}) in the river bottom and in Limburg loess soils

Metal	A-value	B-value	C-value	Loess soil
Zinc	200	500	3000	84
Cadmium	1	5	20	1.0
Copper	50	100	500	15
Lead	50	150	600	34
Cobalt	20	50	300	10
Nickel	50	100	500	16
Chrome	100	250	800	21

TABLE 6. Descriptive statistics for the heavy metal contents (in mg kg^{-1}) of approximately 500 samples from the soil in the Meuse valley

	Min.	Median	Mean	Max.	s.d.	>B-value (%)	>C-value (%)
Zn	42	402	622	4840	585	42.3	6.0
Cd	0.6	2.4	4.3	22.2	3.9	30.8	0.4
Cu	3	31	49	550	48	10.5	0.7
Pb	18	142	191	898	142	48.2	1.8
Co	4	16	16	104	6	0.6	0.0
Ni	13	27	28	72	7	0.0	0.0
Cr	14	28	33	140	15	0.0	0.0

sanitation research needs to be done at once. To enable comparisons to be made with soils lying outside the winter channel of the Meuse, Table 5 also shows the mean concentrations of the heavy metals as measured in research on the background concentrations of heavy metals in South Limburg (Rang, unpublished data).

Table 6 gives the statistics that describe the heavy metal contents of the samples from the top layer of the soil. To simlify interpretation, the percentages by which these values exceed the B- and C-values of the Guidelines for Soil Sanitation are also shown. It can be seen that levels of zinc, lead and cadmium are greatly elevated, and those of copper less so. Nickel, chrome and copper do not pose a problem in the research area: chrome and nickel never exceeded the B norm and cobalt only exceeded it once. Therefore, these three metals were not considered in further analysis of the data. In view of the above it must be realized that by following the directed sampling strategy it was impossible to extrapolate the frequency distributions directly to areas.

The pollution index

When mapping soil pollution we were confronted by the problem that it is virtually impossible to display the spatial variation of seven metals on one map. It therefore seemed desirable to express the degree of soil pollution in one variable, to simplify the making of a map of pollution. The variable used for this, the pollution index, was obtained by applying a method of data reduction to the results of the chemical analysis of the soil samples to be assessed. The different metals whose levels have been ascertained can only be mutually compared if the concentrations have been standardized. To do this, the measured

TABLE 7. Soil samples classified according to value of pollution index

Pollution index	Description	No. of samples (%)
0.0–0.5	Unpolluted to lightly polluted	39.5
0.5–1.0	Moderately polluted	25.4
1.0–2.0	Severely polluted	24.6
2.0–	Very severely polluted	10.5

concentration of a certain element can be divided by a norm concentration, for example, the one from the Guidelines for Soil Sanitation. The variable thus obtained, called the standard score, is thus a dimensionless quantity and enables the levels of concentration of different levels to be compared. A simple expression for the pollution index that can be used in practice is:

$$V \text{ index} = \sum_{j=1}^{n} c_j/(N_j * n) \quad (1)$$

where c_j = concentration of element j
N_j = norm value for the element j
n = number of elements analysed

In the present research the pollution index was calculated from the concentrations of the elements zinc, cadmium, lead and copper, and the norm values of the different elements were taken from the 'test value indicating further research' (the B value) as given in the Guidelines for Soil Sanitation (Rang *et al.*, 1985). On the basis of the value for the pollution index calculated using Equation (1), the soil samples were classified according to the degree of pollution (Table 7).

THE SPATIAL PATTERN OF POLLUTION RELATED TO FLOOD FREQUENCY

In order to test the relationship between the heavy metal content and the frequency of flooding, the samples were divided into five groups according to frequency of flooding. Table 8 shows that for all metals the median level increases with increasing frequency of flooding. From this it can be concluded that the fluvial influence is indeed the reason for

TABLE 8. The median (med.) and range (s.d.) of the levels of four heavy metals (in mg kg^{-1}) and the pollution index in the soil, grouped according to frequency of flooding

	Zinc		Cadmium		Lead		Copper		P index	
Recurrence interval	med.	s.d.	med.	s.d.	med.	s.d.	med.	s.d.	med.	s.d.
2 year	888	681	8.2	4.8	264	140	75	35	1.5	0.8
5 year	760	601	5.4	3.9	270	130	54	47	1.3	0.7
10 year	332	350	2.2	2.1	118	116	28	67	0.6	0.5
50 year	220	203	1.6	1.3	76	76	22	13	0.4	0.3
250 year	314	139	1.4	0.5	71	58	20	12	0.4	0.2

the strongly increased levels of concentration. This is confirmed by the correlation coefficient of −0.67 between the frequency of flooding and the pollution index. It can be postulated that more than 75 per cent of the soils that are inundated at least once every two years, and more than 50 per cent of the soils that are inundated at least once every five years exceed the B norm for zinc, cadmium and lead.

It is striking that the spread per class increases as the frequency of flooding increases. There are a number of reasons for this. First, there is a difference between the actual and calculated frequency of flooding. The frequency of flooding is calculated from data on the elevation and on the incidence of certain water levels. However, often these cannot be determined unequivocally, especially where the relief is very varied, as for example in abandoned channels. Moreover, the frequency of flooding changes over time. Sedimentation and engineering works can cause the frequency of flooding to decrease locally. Furthermore, there is also a seasonally related variation, caused by the differences between summer and winter vegetation. The presence of agricultural crops can cause local damming in the summer, with the result that different areas are inundated than in winter. Secondly, there is a difference in sedimentation. This too is not only related to frequency of flooding, but also to differences in current velocity and channel morphology. Thirdly, differences in the textural and mineralogical composition of the soil. Fourthly, there may be one or more non-fluvial sources of heavy metals. If another source of heavy metals is responsible for an appreciable input of heavy metals, the picture will become confused.

Using the ANOVA test, the percentage of the total variance in the pollution index that can be attributed to differences in frequency of flooding was found to be 46 per cent. After combining the two classes with the lowest frequency of flooding, which did not differ from each other significantly, the explained variance rose to 47 per cent.

THE SPATIAL PATTERN OF POLLUTION RELATED TO THE DEGREE OF SOIL DEVELOPMENT

In all, nine different soil types were distinguished in the area studied. Using the value of the pollution index, a Mann–Whitney test was done to ascertain the extent to which the pollution of these soil types differed. Soil types that were found to be polluted to a similar degree were then combined. This resulted in four soil groups that differed significantly from

TABLE 9. The median (med.) and range (s.d.) of the levels of four heavy metals (in mg kg^{-1}) and the pollution index in the soil, grouped by soil type

Soil type	Zinc med.	Zinc s.d.	Cadmium med.	Cadmium s.d.	Lead med.	Lead s.d.	Copper med.	Copper s.d.	P index med.	P index s.d.
A	840	638	7.8	4.9	244	125	75	35	1.5	0.8
B	700	598	4.8	3.7	254	130	49	44	1.1	0.7
C	326	244	1.8	1.4	96	99	23	52	0.5	0.4
D	204	141	1.6	1.3	68	47	22	28	0.4	0.2

A Young Auen soil
B Lime-rich Vega
C Calcareous Vega, Gleybraunerde and all gley soils
D Lime-deficient Vega, Braunerde, Parabraunerde

each other (see Table 9). As expected, the young Auen soils were the most polluted. These are the youngest soils, which are inundated most frequently. The uppermost horizons of the profile are developed in material that is visibly polluted with coal slack and fragments of slag. The numerous Vegas also have high levels of heavy metals in their upper layers, although less so than the young Auen soils. They too are young, frequently inundated soils, the upper material of which is, in many cases, polluted with coal slack.

The next group of soils includes all the Gley soils, including the Gley Braunerde and lime-deficient Vegas. In general, gley soils are found in depressions, i.e. appreciably low and at some distance from the river. As a result of this they are more severely polluted than the Braunerden and Parabraunerden, which appear to be the least polluted of all the soil types. An ANOVA test revealed that half of the difference in the degree of pollution can be attributed to differences in soil type.

THE HISTORICAL DEVELOPMENT OF THE DEGREE OF POLLUTION

Overbank deposits

At 26 sites in the floodplain a sample was taken at a depth of 100 cm as well as from the surface. Only the sample from 1 m can be expected to reflect the degree of pollution of Meuse silt at the time of the sedimentation. This is because only the soil material lower in the profile can be relatively dated; the upper layer may have been deposited during soil formation. Furthermore, at a shallow depth in the profile post-sedimentary transport of heavy metals may have occurred, e.g. from atmospheric deposition.

At greater depths the heavy metal content may also have changed since deposition, because of leaching. The mobility of heavy metals in the soil—and hence the chance of leaching—is determined by the organic matter content of the soil, the content of elutriable parts, the calcium content and the pH. Table 10 shows the relevant physical and chemical properties of the soil profiles studied. From the high levels of organic matter content and calcium and the high pH it can be inferred that no vertical translocation of heavy metals has occurred in the two youngest soils, i.e. the young Auen soil and the lime-rich Vega. The mobility of heavy metals is thought to be small in the two oldest soils too. Nevertheless, given the age of these soils, it cannot be excluded that there has been a redistribution of heavy metals in the soil profile.

TABLE 10. Physical characteristics of the soil profiles studied

Soil type	Depth (m)	<2 μm (%)	<16 μm (%)	Lime content (%)	Organic matter (%)	pH
Young Auen soil	0.1	15	35	4.6	14.6	6.8
	1.0	13	30	5.4	8.4	7.4
Lime-rich Vega	0.1	21	47	4.5	4.8	6.9
	1.0	21	44	5.3	1.2	7.3
Braunerde	0.1	22	44	0	2.8	6.5
	1.0	31	42	0	1.2	5.8
Parabraunerde	0.1	19	37	0	2.6	6.2
	1.0	28	45	0	0.8	5.2

TABLE 11. Mean heavy metal contents (in mg kg^{-1}) at a depth of 1 m in overbank deposits of different age (n = no. of samples)

Age (years)	Soil type	n	Zn	Cd	Pb	Cu
-1	Recent deposits	94	860	8.0	230	74
2–200	Young Auen soil	10	1680	12.0	700	150
200–1000	Lime-rich Vega	6	136	0.6	85	15
1000–2000	Braunerde	4	105	0.4	31	22
2000–	Parabraunerde	6	61	0.4	17	16

In spite of the above mentioned limitations, the heavy metal contents of the various soils at 1 m depth give a reasonable indication of the heavy metal contents of the Meuse silt at the moment of sedimentation. Using the degree of soil development and the position of the sampled soil profiles in the landscape an attempt was made to date the various deposits. The results, which are summarized in Table 11, suggest that there was an increase in heavy metal contents more than 1000 years ago. The differences in heavy metal content found between the two oldest soils could, however, also be caused by the translocation of heavy metals after deposition, under the influence of percolating soil water. It can safely be assumed that the increase in the heavy metal content of the Meuse silt certainly began more than 200 years ago.

Deposits in former channels of the Meuse

The oldest silted-up abandoned meander of the Meuse that can be dated from old maps and documents is at least 300 years old. The sampled profile is in a gravel pit near the hamlet of Grote Meers. Maps from before 1650 indicate the main channel of the Meuse here. A map from 1651 suggests that at that time the Meuse flowed more than 1 km further east. Old documents from the second half of the 16th century and the first half of the 17th century report rapid channel changes during this period (Van Mulken, 1949). A document from 1559 describes a disastrous flood that resulted in an appreciable shift of the Meuse channel.

The profile in the gravel pit in the infilled channel shows some extremely striking phenomena. The deepest part of this profile consists of a thick layer of gravel with clear zonation. There is scarcely any fine-grained matrix material in this layer. This gravel body contains layers with abundant rounded coal fragments, with diameters up to 50 mm. Animal bones and pieces of wood and charcoal occur, but remarkably few bricks. This deposit seems to be the result of one or two catastrophic floods in which the quays along the Meuse in Liege and Maastricht were washed away and many animals drowned. The gravel body contains a layer, more than 3 m thick, of clear brown calcareous loam. If this was indeed deposited after 1600, then the sedimentation rate was at least 1 cm per year. Table 12 gives the vertical distribution of the heavy metals in the loamy deposits. The oldest deposits are at least 300 to 350 years old. The levels of zinc and lead in these deposits are appreciably higher than natural levels. From this it can be inferred that the heavy metal pollution of the Meuse silt is not a recent phenomenon, but began at the end of the 16th century. This confirms the dating based on soil development.

TABLE 12. The vertical distribution of heavy metals (in mg kg^{-1}) in an infilled channel of the Meuse

Depth (cm)	Zn	Pb	Cd	Cu
0	234	89	1.4	21
60	118	53	1.0	17
120	136	57	0.8	20
180	101	58	0.8	18
240	176	80	1.2	27
270	154	84	1.0	27

The youngest deposits are clear brown in colour and just as polluted as the oldest. These deposits are not the same as the material from which the young Auen soils have developed. The profile in question should therefore be classified as a lime-rich Vega. The youngest deposits are therefore at least 200 years old. This means that the deposition of the 3 m thick body of loam must have occurred in an even shorter time period. Assuming that the oldest sediments are 350 years old, the accretion must have occurred in *c*. 150 years; this means a deposition rate of two metres per century!

A second silted-up Meuse channel was sampled near the confluence of the Geul and the Meuse. This channel is well exposed at the place where it has been cut through by the Geul. In the profile, layers of coal slack mixed with sand are found to a depth exceeding 3.5 m, with up to 10 cm thick layers of rounded coal cinders (Figure 2). By comparing historical and modern maps it is possible to generalize about the age of these deposits.

The Tranchot map of 1820 and the oldest map of the river, dating from 1849, show the banks of the Meuse lying some 200 m further south than today, i.e. at the point where the Oude Kanjel flows into the Geul. On the river map of 1892 the Meuse follows its present course. The latter map shows an embankment along the river, with the dates 1854–1862. The oldest deposits found in the profile are therefore at most 130 years old. By comparing the elevations shown on the 1892 river map with those on The Netherlands Topographic Service map of 1976 it can be inferred that approximately 1 m of silt was deposited in this period. The oldest deposits in the profile therefore date from before 1892 and after *c*. 1850.

Figure 2 shows that the heavy metal contents decrease with depth; this indicates that the older the deposits, the higher the heavy metal contents. Samples 154 and 156 are exceptions to this: they consist entirely of slag grit. From this increase it can be concluded that pollution used to be worse in the past than it is today. Given that the oldest deposits in the channel studied proved to be the most polluted it can therefore be inferred that the greatest pollution of Meuse silt occurred before the channel was silted up, i.e. at least 90–130 years ago.

Table 13 summarizes the heavy metal contents of the deposits whose age is known. From this table it can be seen that the increase in the heavy metal content of the Meuse silt began more than 350 years ago and peaked at least 90 years ago. The peak in pollution was most probably in the middle of the last century, coinciding with the flourishing of the lead and zinc mining and related industries in the Belgian part of the catchment.

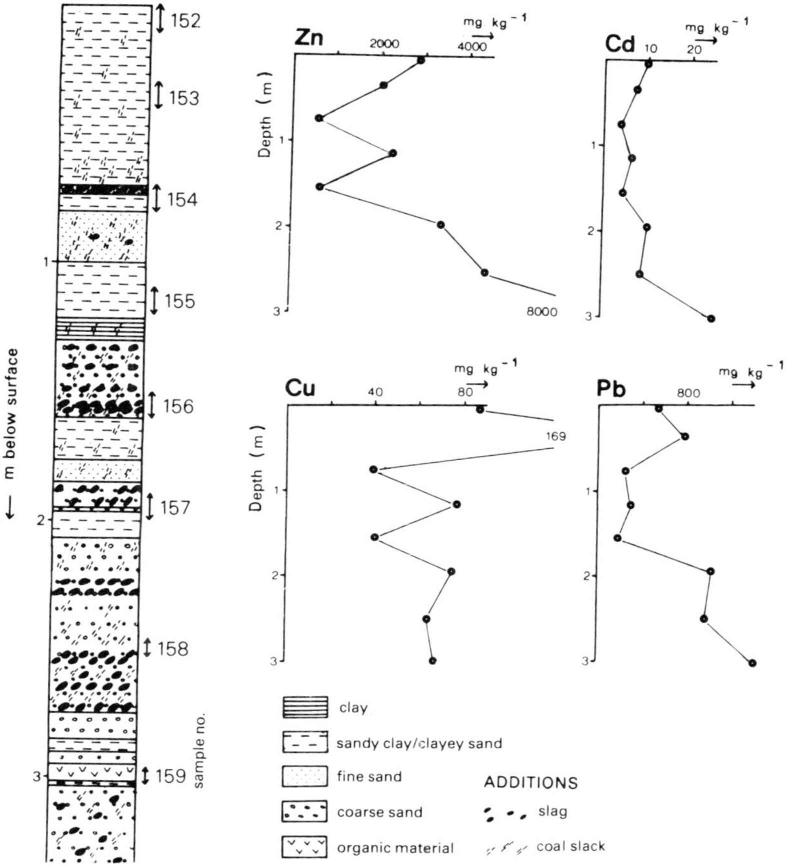

FIGURE 2. Profile composition and pollution of an abandoned meander of the Meuse near the confluence with the Geul

TABLE 13. Mean heavy metal contents (in mg kg^{-1}) of former and recent channel deposits

Location	Age (years)	n	Zn	Pb	Cd	Cu
Former channel	350	1	154	84	1.0	27
Former channel	120	1	8000	1406	24.7	67
Gravel pit	10	2	1660	530	27.0	160
Present channel	1	5	1600	390	21.8	90

CONCLUSIONS

All the floodplain soils of the Meuse that have a frequency of flooding of at least once per 250 years are polluted to some degree. The concentrations of zinc, cadmium, lead and, to a lesser extent, of copper, are particularly high. More than 8000 ha of the soil in the Limburg part of the Meuse valley is moderately to severely polluted. Approximately 3500 ha of this area, i.e. the soils that are inundated at least once every two years, are polluted severely to extremely severely. There seems to be a strong relationship between the degree of pollution of the uppermost layer of the soil and the soil type. The youngest soils, which are rich in lime and organic matter, appear to be the most polluted. This is fortuitous, because under such conditions the heavy metal uptake by crops is minimal.

The heavy metal pollution of the Meuse sediments began more than 350 years ago. The peak of pollution was reached about 100 years ago, coinciding with the zenith of ore mining in the Belgian part of the catchment. After erosion during high water the polluted river sediments that have been deposited on the floodplain form a secondary source of pollution that can have a strong and longlasting effect on the quality of the river silt.

REFERENCES

Bradley, S. B. and Cox, J. J. 1986. Heavy metals in the Hamps and Manifold Valleys, North Staffordshire, U.K.: distribution in floodplain soils. *The Science of the Total Environment*, **50**, 103–128.

Forstner, U. and Wittmann, G. L. 1983. *Metal Pollution in the Aquatic Environment*, Springer-Verlag, Berlin, 486 pp.

Lewin, J., Davies, B. E. and Wolfenden, P. J. 1977. Interaction between channel change and historic mining sediments. In Gregory, K. J. (Ed.), *River Channel Changes*, Wiley, Chichester, 353–367.

Salomons, W. and De Groot, A. J. 1978. Pollution history of trace elements in sediments, as effected by the Rhine River. In Krumbein, W. E. (Ed.), *Environmental Biochemistry, Volume 1*, Ann Arbor Sc. Publ., 149–162.

Salomons, W., Van Driel, W., Kerkdijk, H. and Boxma, R. 1982. *Help! Holland is plated by the Rhine (environmental problems associated with contaminated sediments)*, Effects of Waste Disposal on Groundwater and Surface Water, Proceedings of the Exeter symposium, July, IAHS Publ. No. 139.

Scheffer, F. and Schachtschabel, P. *Lehrbuch der Bodenkunde*, Ferdinand Enke Verlag, Stuttgart.

Van Mulken, L. 1949. Enige gegevens ove de vroegere loop van de Maas bij Elsloo. *De Maasgouw*, **63**.

Rang, M. C., Kleijn, C. E. and Schouten, C. J. 1985. Bodemverontreiniging met zware metalen in een deel van het winterbed van de Maas. *KNAG, Geografisch tijdschrift*, **4**.

Rang, M. C., Kleijn, C. E. and Schouten, C. J. 1986. Historic changes in the enrichment of fluvial deposits with heavy metals. In *Monitoring to detect changes in water quality series. Publication of the International Association of Hydrological Sciences*, **157**.

Rang, M. C. and Schouten, C. J. 1987. Bodemverontreiniging in het Maasdal: geen zaak van vandaag of gisteren. *Natuurhistorisch Maandblad*, **76**(4).

Historical Change of Large Alluvial Rivers: Western Europe
Edited by G. E. Petts
©1989 John Wiley & Sons Ltd

9

Palaeoecological Analyses of Large Rivers: Some Principles and Methods

CLAUDE AMOROS

Laboratory of Freshwater Ecology (U.A. CNRS 367), Université Claude-Bernard, Lyon 1, F–69622 Villeurbanne Cedex, France

and

GEERT VAN URK

Institute of Inland Water Management and Waste Water Treatment, Rijkswaterstaat, NL–8200 AA Lelystad (P.O. Box 17), The Netherlands

INTRODUCTION

Palaeoecology may be defined as the study of the remains of plants and animals in sedimentary deposits to assess their historic distribution and abundance, and to use this information to describe environmental changes. Any freshwater environment is subject to change. This may either be a natural succession, such as the development of abandoned river channels into terrestrial ecosystems, extending over no more than a few centuries, or the long-term evolution of a fluvial system as a whole, extending to geological timescales. Human activities have disrupted the natural processes of change in river systems and the extent of disruption may be assessed by palaeoecological methods. Furthermore, if attempts are made to achieve ecological management of river systems and to restore, or at least minimize, human impacts on rivers, it is essential to conceptualize possible original conditions (Hynes, 1986). Palaeoecological methods may be used to provide such baseline data.

The ecology of large rivers in temperate zones is only poorly understood. For example, it is often impossible to test the theories of the 'River Continuum Concept' (Vannote *et al.*, 1980), because human intervention has had such enormous effects. Data are required urgently to describe the historical dynamics of large rivers and this is true especially in a European context.

The palaeoecological approach has been proven useful in the study of the development of lakes (Frey, 1974; Berglund, 1986) and might be equally useful in the study of river systems. Reconnaissances into the possibilities of palaeoecological research were started on the Upper Rhône in 1982 and on the Lower Rhine in 1983. The primary objectives were somewhat different in these cases. For the Rhône, the main issues were the development of predictive scenarios for the effects of civil engineering projects (Bravard *et al.*, 1986), the study of rejuvenation processes within ecological successions (Amoros *et al.*, 1987) and the conservation of relict ecosystems. For the Lower Rhine, interest in palaeoecological research arose from the necessity to formulate ecological objectives to mitigate the effects of river engineering works (Smit and Van Urk, 1987). In the case of the Rhône, Cladoceran remains have proven to be useful indicators of ecological succession in the abandoned channels; in the Rhine, emphasis was laid on the study of macroinvertebrates, especially Chironomid remains.

Methodological aspects of sampling and of the use of the groups mentioned in the reconstruction of ecological conditions in the past are discussed below. For an insight into ecosystem functioning in the past, studies of a broad spectrum of organisms, including algae and macrophytes, are essential, but studies of this kind are still in a preliminary phase on studies of the Rhône and the Rhine. Some general principles of investigations on the groups will also be discussed.

SAMPLING STRATEGY

In palaeoecological investigations, information is derived from analyses of sediment cores. The ideal source of information is series of dated sediment cores. In river systems, in contrast to lakes, special problems are met in obtaining such series, not least because sedimentation patterns in river systems change more rapidly than those in lakes, due to changing channel morphology. At average discharges, sedimentation of the suspended matter carried by the

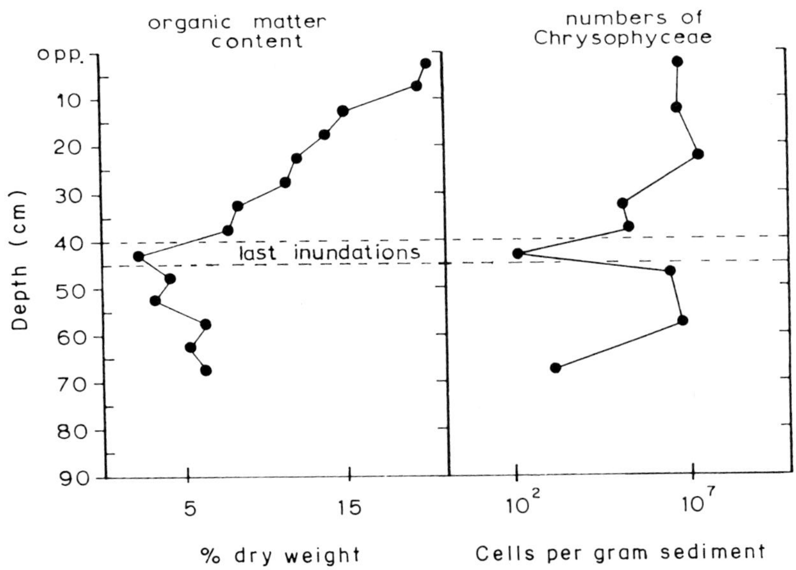

FIGURE 1. Organic matter content (percentage) and numbers of Chrysophyceans (Algae) in a sediment core from the 'Schoonrewoerrdse Wiel' in the Lower Rhine River (Chrysophyceans are autochthonous in the Wiel)

river occurs in the river's delta or estuary, including the freshwater tidal area, and in backwaters which are in open connection with the main channel. At peak discharges, river silt is deposited on the floodplain, in both terrestrial and aquatic habitats such as oxbow-lakes. Allochthonous river deposits in floodplain waters can be distinguished from autochthonous deposits by the low content of organic matter, as demonstrated for surficial sediments in the Rhône floodplain by Rostan *et al.* (1987). In rarely inundated floodplain lakes, anomalies in the organic matter content in a sediment core may indicate flooding events accompanied by the deposition of allochthonous material. Normally, the amount of organic matter in sediments from stagnant waters decreases with depth according to a first order process (Avnimelech *et al.*, 1984). In a Pb-210 dated sediment core from a stagnant water in the former floodplain of the Lower Rhine, the 'Schoonrewoerdse Wiel', the organic matter content in the upper 0.40 m of the core, follows approximately such a pattern (Figure 1). Below a depth of 0.40 m, anomalies of the organic matter content are observed. According to the sedimentation rates established by Pb-210 dating, the depth of 0.40–0.45 m coincides with the period of the last inundations of the area which occurred in 1809 and 1820 (Van de Ven, 1987). The sediment layers which are low in organic matter content, represent river sediment deposited during these inundations, but may be mixed with autochthonous sediments. Although the age of these sediments is known, the use of these sediments in palaeoecological investigations is generally not recommended because these are deposited at exceptional conditions and the composition of the sediment transported at peak discharges may be different from that transported at normal discharges. This can be easily demonstrated for the heavy metal content of the suspended matter (Figure 2), but might also apply for the amount of remains of organisms in the suspended matter. In palaeoecological research into conditions in the main channel, it seems therefore preferable to use sediments deposited at normal discharges in the deltaic of estuarine regions or in backwaters. Sedimentation rates can be very high at such locations and sedimentation patterns may shift with time (for an example in the Lower Rhine, Klink, this volume).

It is clear that in such situations a single sediment core cannot give a continuous record of the material carried by the river and that Pb-210 dating as applied to lake sediments is not possible. The requirements of the 'constant rate of supply' calculation model (Oldfield

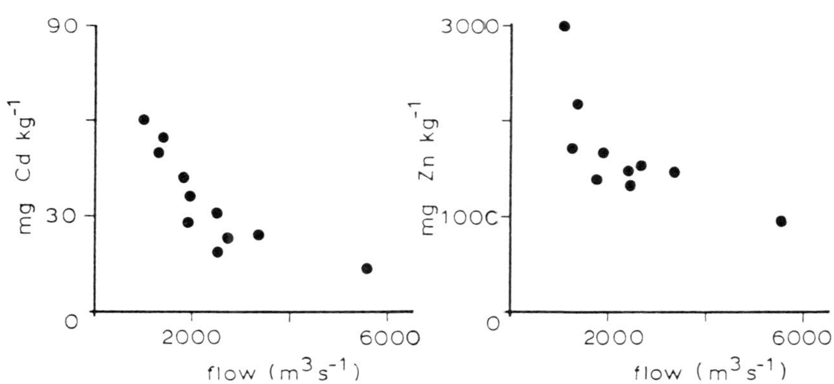

FIGURE 2. Relationship between the heavy metal content of the suspended matter of the Rhine River and discharge during 1978

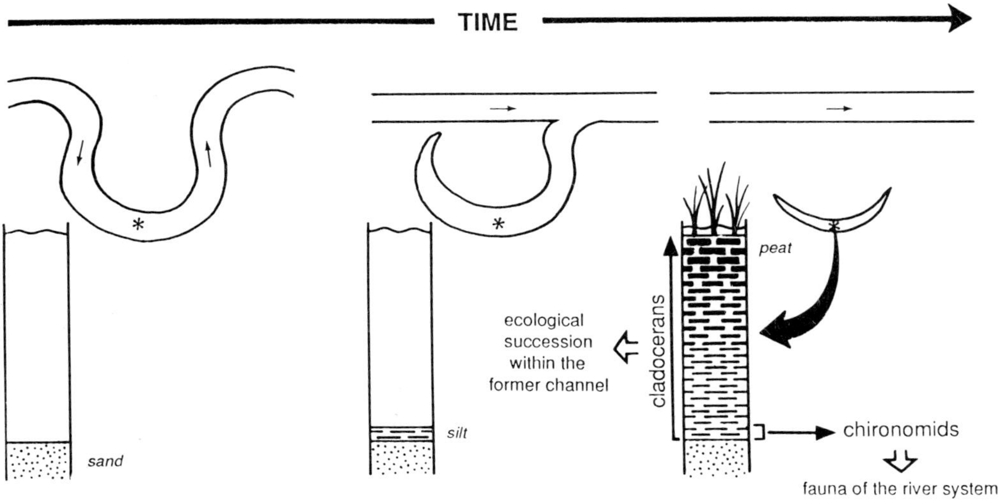

FIGURE 3. Schematic illustration of the sampling strategies (hypothetical situation)

and Appleby, 1984) are not fulfilled, whereas the assumption of a constant initial concentration cannot be validated without knowledge of sediment transport in previous periods.

The simplest method to obtain series of sediment samples of different age is to use historical data on the development of backwaters along a river, as can be derived from ancient maps. When a channel diversion occurs, flow conditions show an abrupt change which is reflected in the grain size distribution of the sediment. As long as the abandoned channel is not completely isolated, deposition of river silt will take place. Sediment cores from abandoned channels generally show a sudden change from sand and gravel substrates to fine sediments. It can be assumed that the lowest layer of fine sediment was deposited shortly after the channel diversion and remains of organisms characteristic of running waters will reflect the conditions in the main channel at the time of diversion. When isolation of the abandoned channel proceeds and ecological succession takes place, the relative amounts of autochtonous material in the sediments will increase and remains of organisms in the upper layers of the core will reflect this succession. This scheme of analysis of sediment cores from abandoned channels is illustrated in Figure 3.

ECOLOGICAL INDICATORS

Plant remains

Palaeoecological studies of lake sediments often use plant remains to elucidate both general changes of climate and land use at the drainage basin scale (Averdieck, 1983; Ammann and Tobolski, 1983; Prentice, 1985), and changes of lake water quality and productivity (Battarbee, 1986; Cronberg, 1986) which may or may not be related to such general changes.

Pollen

Palynology yields regional information about climatic changes, overall vegetation of the floodplain, and some human interventions such as deforestation or agriculture (Starling and Crowder, 1981). Because pollen are transported both by the air and the water, they are mixed and cannot afford precise data on the vegetation of each type of waterbody. Pollen of hydrophyte species could be used in palaeoecological studies of river systems, but more often these species reproduce vegetatively and their pollen are less abundant than helophyte and terrestrial pollen.

Diatom frustules

Among the remains of organisms in river sediments, diatom frustules are very common, as in lake sediments. Diatoms usually are the dominant group among the benthic algae and in the plankton of large rivers (Hynes, 1970). Analyses of Diatom communities have been proven to be useful in palaeoecological studies of lakes, in the assessment of eutrophication processes as well as acidification. Also in rivers, Diatom communities give information on water quality conditions. A review of pollution studies using freshwater diatoms is given by Patrick (1977). Ecological data and pollution tolerances of individual species were listed by Lowe (1974). Pollution studies in large rivers were carried out by Descy (1979) in the Meuse, and by Lange-Bertalot and Lorbach (1979) in the Rhine. There seems thus to be enough background information for trying to assess changes in a river system, notably changes in river water chemistry, from changes in subfossil Diatom communities. However, no study in this field has been published to our knowledge.

Algal pigments

In addition to Diatom frustules, Chrysophycean scales and colony-forming green algae such as *Pediastrum* can sometimes be found in sediment cores, but little is known about the relevance of these groups in palaeoecological studies. The analysis of sedimentary algal pigments is another method to assess shifts in the occurrence of different algal groups. The principles of this approach were discussed by Swain (1985). Concentrations and ratios of chlorophylls and total carotenoids are determined by allochthonous inputs, production and preservation conditions.

Preservation conditions are generally good in river sediments, due to a high rate of burial of pigments into reducing sediments. But on the basis of analyses of total chlorophylls and total carotenoids, it may be difficult to distinguish between allochthonous inputs containing pigments from terrestrial plants, and pigments from algae from the river itself and floodplain water bodies. Separation and quantification of individual carotenoids can be carried out by chromatographic techniques, enabling a distinction between carotenoids from macrophytes and those from most algal groups. An extensive study of sedimentary carotenoids in sediments from Swiss lakes by means of thin-layer chromatography was carried out by Zullig (1982). However, high-pressure liquid chromatography seems a more promising analytical tool for quantification of algal carotenoids.

Plant macrofossils

A third category of plant remains in sediments are the macrofossils: seeds, fruits, leaves etc. Birks (1980) has reviewed the palaeoecological importance of studying these macrofossils. He considers fluviatile sediments as a favourable environment for preservation of even the more fragile types of plant macrofossils such as leaves.

Preliminary investigations of Rhine sediment samples yielded only some hardly identifiable fragments of leaves and seeds. Only in some sediment cores from the freshwater tidal area of the Rhine, the large number of wood fragments in sandy sediments were conspicuous, and may be an indication that the snag habitat must have been very important (Rijkswaterstaat, unpublished results). The quantity of plant macrofossils found in surficial recent sediments is not related to the large quantities of plant fragments found in drift-nets in the Lower Rhine and it is clear that more data are needed on the transport and deposition of this material before the importance of plant macrofossils in palaeoecological studies of rivers can be assessed.

Cladoceran remains

Cladocera are microscopic crustaceans (size generally smaller than 1 mm), which proliferate in stagnant freshwater. Their exoskeletons and their shells are preserved in fine sediments (Figures 4 and 5). More than 90 species occur in European inland waters, within diversified habitats. Some species, such as *Daphnia longispina* or *Bosmina coregoni*, swim in open water (zooplankton); others colonize different kinds of aquatic plant communities (*Sida crystallina, Simocephalus vetulus, Graptoleberis testudinaria. . .*); and yet others live on the bottom, among mud, organic detritus or on sandy substrate (*Ilyocryptus sordidus, Leydigia leydigi. . .*).

The water types investigated using Cladoceran remains are limited to stagnant waters, such as oxbow-lakes, floodplain pools, marshes and backwaters. Although they occur in running waters, they are not used to reconstruct the populations of the lotic biotopes because the water flow washes away the Cladoceran remains and mixes the different populations, making the interpretations hazardous. It is more reliable to confine the palaeoecological studies to subfossil populations remaining within unreworked sediment assessed by granulometry; such decanted sediments deposited within quiet and standing water in the Rhône River alluvial floodplain exhibit a characteristic texture (Bravard, 1983).

The investigated timescales extend over the range of decades to centuries according to the persistence of the ecosystems of the former river channels. For example, on the Upper Rhône River, France, the complete development (from running water to terrestrial ecosystem required several decades to a century for the narrow former braided arms, and up to 2–3 centuries for the large and deep former meanders (Bravard *et al.*, 1986). According to their reproductive pattern, the Cladoceran populations can respond rapidly to changes of their environment: the individuals reproduce quickly by parthenogenesis; when the environmental conditions fit the requirements of a species, each breeding yields 5–10 juveniles (and sometimes more) that can reproduce themselves after 1–2 weeks. So, the populations increase exponentially and may give rise to several hundred individuals per litre and even more than one thousand in highly productive ecosystems. The density of the Cladoceran

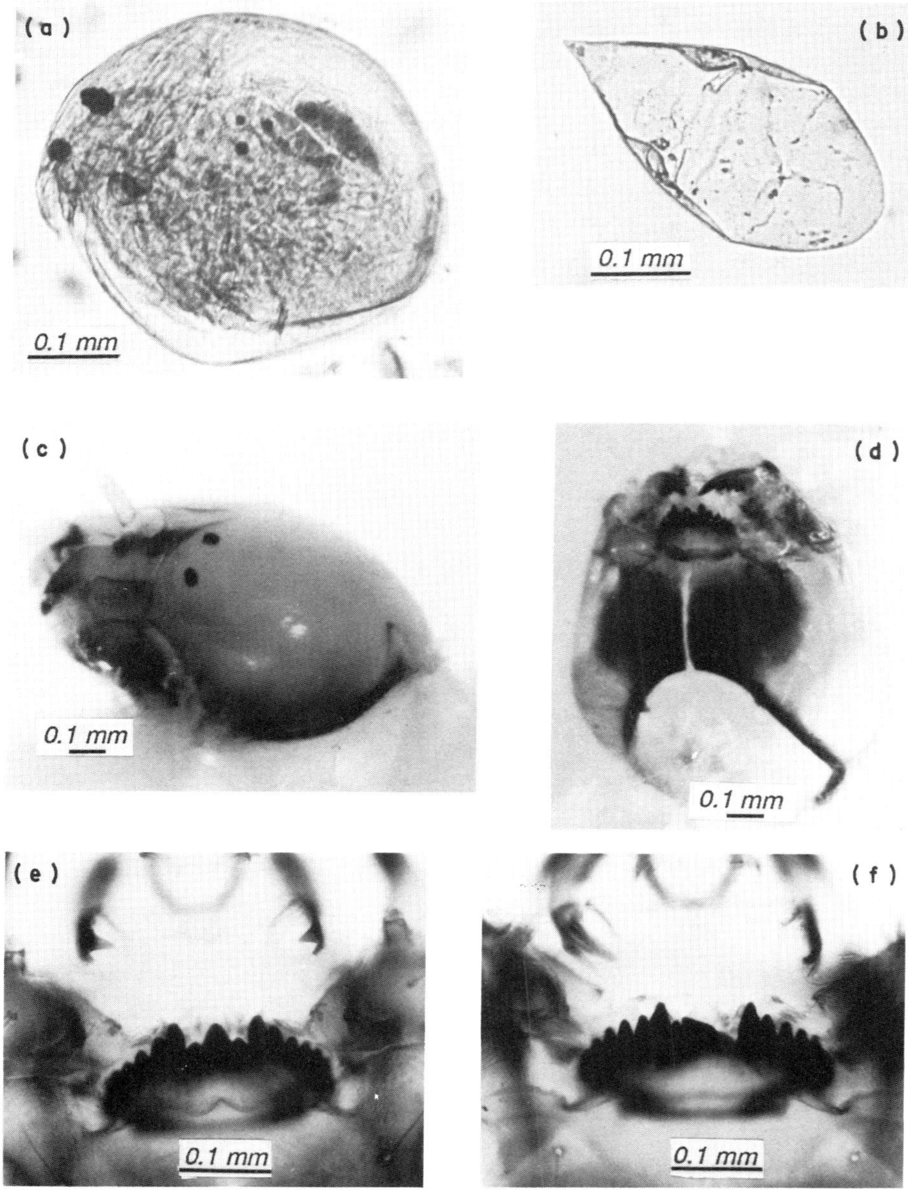

FIGURE 4. Living Cladoceran (a) and the remains of a headshield (b); head capsule of Chironomid larvae (c = lateral view; d = ventral view), its mentum (e) and an example of deformity (f)

remains sorted from the unreworked sediments is related to these abundances: we recovered up to 448 headshields, belonging to only two Cladoceran families, from 0.1 cm^3 fresh sediment in a cored from a former side arm of the Rhône River (Lône de la Coche, Serrières de Briord, Ain Dept., unpublished data).

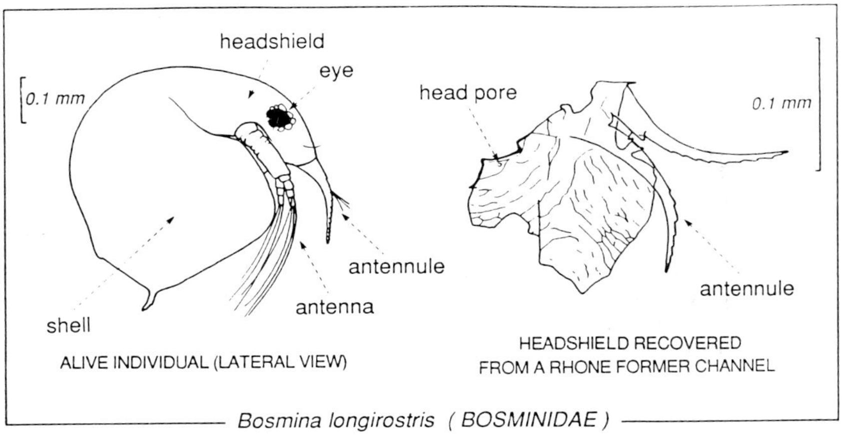

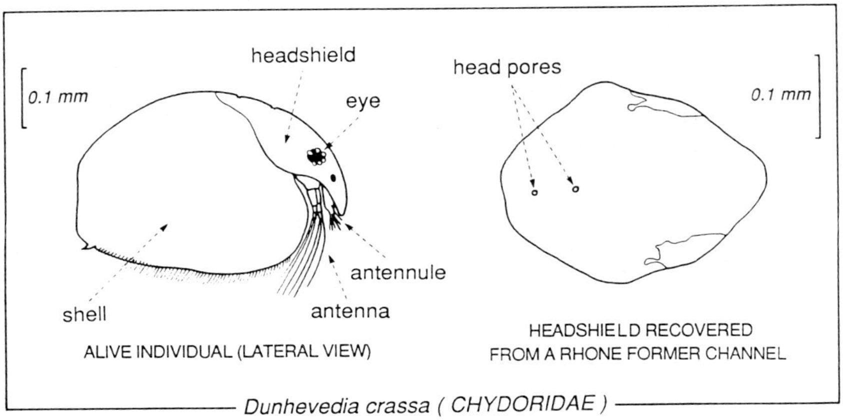

FIGURE 5. Living Cladocerans and remains of their headshields with detail of the head pores

The reconstruction of past ecological conditions is based on the interpretation of the succession of Cladoceran populations, namely the comparison of past species assemblages with present ones living in waterbodies the ecological conditions of which are known. For this, it is necessary to assess the information afforded by this taxa and to know the reliability and constraints of the method.

Reliability and constraints

The method is derived from previous palaeolimnological works but problems are different. Palaeolimnologists studied Cladoceran successions within one lake to detect the changes of the biotic functioning of the lake ecosystem depending or not on surrounding changes; river palaeoecology is based on comparative studies carried out on several ecosystems that change much more quickly than lakes. Moreover, the waterbodies of the former channel undergo morphogenesis and have ecological functions that differ from lacustrine ones. Thus there is a need for an elaboration of proper reference patterns.

Preservation and identification to species The first studies using Cladoceran remains were carried out by Frey (1958) based upon the discovery of disarticulated pieces preserved within lake sediments by Mesjatsev (1924). In 1960, Frey noticed that all the Cladoceran remains were not quantitatively preserved: in most families, except Bosminidae and Chydoridae, remains may be infrequent and badly preserved due to variable biodegradation. The differential preservation potentials among Cladocerans results from different chitin constitution: each polymer may occur within several crystalline configurations some of which are more hydrated and are better preserved (Deevey, 1964). Frey (1960) demonstrated that all the Chydoridae species living in Lake Mendota occurred as subfossils within the sediments. This preservation ability was assessed by Frey (1962, 1976, 1979), De Costa (1964) and Deevey (1964). For this reason, only the Bosminidae and Chydoridae remains will be considered within the river palaeoecological investigations. These remains include all the disarticulated chitinous parts: antennules, mandibles, parts of trunk limbs, postabdomen, postabdomen claws, shells, ephippiums, and headshields. Most of these parts cannot be confidently identified for all species but the headshields of Bosminidae and Chydoridae can be used to recognize all species except *Alona rectangula* and *Alona guttata*, which are often not separated (Frey, 1959, 1962b; Goulden and Frey, 1963; Korinek, 1971). The identification is based on the shape and the ornamentation of the headshield and the number, the size and the location of the headpores (Figure 5).

Morphological and ecological stability Cladocera is an old and primitive Crustacean taxon known from the Permian. The present species have not shown any morphological change for 11 000 years (Frey, 1976); herein, morphological changes do not concern cyclomorphosis, that is cyclical and reversible morphological variability related to periodical changes of the environment (seasonal cycles). The ecological requirements of Cladoceran appear stable, at least during the Holocene. Frey (1967) has shown that the present Cladoceran assemblages of Northern temperate lakes were established after the last glaciation. Moreover, in 77 Danish lakes, Whiteside (1970) listed the same Chydorids species that were earlier recovered by Frey (1962) in Danish lacustrine sediments from the Eemian interglacial, with relative abundances roughly similar to the present ones.

Widespread distribution and ecological requirements Cladocera can be widespread over large areas. Beyond their parthenogenesis, these animals can reproduce sexually (gamogenesis). When their environment becomes unfavourable (higher temperature, lower oxygen content, decrease of food amount, crowding), parthenogenic females produce males. The eggs laid after copulation do not develop immediately but remain in a resting stage until their environment becomes favourable. These eggs are protected by the ephippium that results from a transformation of the mother shell. Within this shelter, the resting eggs are capable of withstanding drying and freezing so they may be transported by birds or insects that move out from one waterbody to another one (Schlichting and Sides, 1969); Proctor (1964) and Mellors (1975) have shown that resting eggs within their epphipium are sometimes viable even after passing through the digestive tract of fishes, ducks and herons. The fertilized eggs permit not only the species perpetuation but their passive distribution too. So Cladoceran species may colonize new waterbodies such as recently cut off channels of rivers.

This advantage leads to a methodological constraint: if all species can be introduced in all waterbodies, then listing the occurring species is not sufficient. Palaeoecological investigations need quantitative data in order to put forward species assemblages that would fit different types of waterbodies and/or successional stages. Indeed, if resting eggs are widely and randomly distributed, only the eggs of those species with ecological requirements fitting the environmental conditions of one place could develop into large populations; if neonates of species unadapted to this habitat hatch, they would face severe conditions, constraining their development and reproduction, so that they would die or produce only very small populations. Quantitative studies show that some species have a large ecological tolerance and can colonize many types of waterbodies but another ones are more restricted then these latter ones may characterise particular habitats (Sramek-Husek, 1962; Flossner, 1964; Whiteside, 1970; Amoros, 1980; Amoros and Chessel, 1985).

A few of the Chydorid species appear to be good indicators of ecological conditions (Frey, 1979) but they are generally scarce and cannot afford by themselves enough information on the whole ecosystem. So, the data from the literature have to be used with caution for the interpretation of the past populations (Birks *et al.*, 1978; Whiteside *et al.*, 1980). For reconstructing past ecological conditions it seems better to compare the species assemblages that occurred along the succession with reference patterns elaborated from species assemblages occurring in present-day ecosystems within the same river system.

Construction of reference patterns Before constructing reference patterns it is necessary to test the reliability of the information afforded by the analysis of the Cladoceran remains recovered from the sediments of former river channels. Like other Arthropods, each individual generates several exuviae, and hence exoskeletal fragments, due to moulting before death. Nevertheless, it is generally assumed from a practical point of view that each species undergoes about the same number of moults, and consequently that each species contributes approximately the same number of remains to the sediment (Goulden, 1966; Frey, 1986).

For lakes, the exoskeletal remains have been shown to be passively transported 'offshore' and deposited in deep water sediments. Mueller (1964), De Costa (1968), Goulden (1969), Binford (1982) and Davis and Ford (1982) have demonstrated that all samples of surficial 'offshore' sediments yield the same percentage composition of Cladoceran remains including littoral species (associated with macrophytes or sediments) as well as planktonic ones. This means that the relative abundance of remains in the sediments of the deepest point of the lake integrates the distribution of the various species according to habitat preference; consequently, a single 'offshore' core of lake sediments is sufficient for analysing real changes over time of the overall community in the lake.

All these results were obtained within lakes, so it was necessary for the present purpose to test these features within former river channels. For this, Amoros and Jacquet (1987) studied, in the present situation, four waterbodies from two former Rhône channels. It was demonstrated that analyses of only the Bosminid and Chydorid populations yield practically the same information as an analysis of the total population of Copepoda and Cladocera about the development stage of aquatic ecosystems after they have been abandoned by the river. Moreover, the percentage composition of the Bosminid and

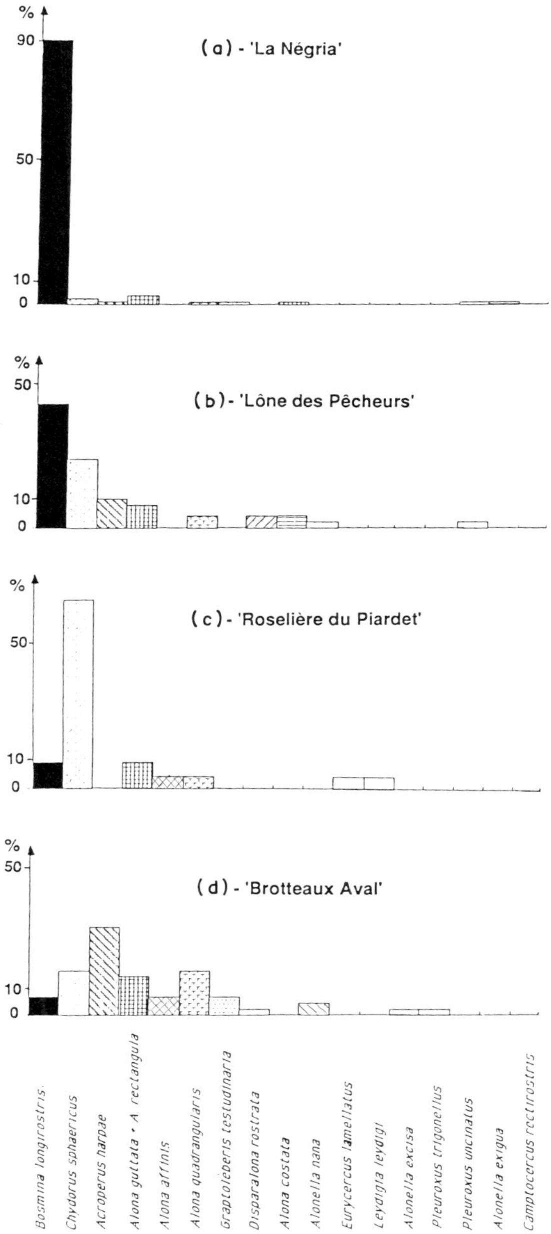

FIGURE 6. Example of Cladoceran species assemblages in four habitat types from the Upper Rhône River: (a) = deep open waterbody (max. depth 5 m) fringed by littoral belts with *Nuphar lutea* then *Carex elata*, (b) = open water associated with areas colonized by aquatic plant communities (*Myriophyllo–Nupharetum* type) and littoral areas overgrown by *Phragmites australis*, (c) = former channel undergoing terrestrialization (max. depth 0.2 m) almost entirely overgrown by *Phragmites australis*, (d) = shallow former meander of a tributary (the Ain River) colonized by several aquatic plant communities (*Ranunculus circinatus*, *Chara vulgaris*, *Potamogeton natans*, *Sparganium emersum*, *Groenlandia densa*, etc.)

Chydorid remains taken from the surficial sediments at the deepest point of each former channel strongly ressembles the average composition of living populations sampled at several stations during one full year. Therefore, Bosminid and Chydorid remains appear to afford adequate information concerning each phase of ecological succession that occurs in former river channels.

Within the Rhône River system, reference patterns of Bosminid and Chydorid assemblages are being constructed using data from 65 former channels of the Rhône River and 20 former channels of a tributary (unpublished data). For example, species assemblages that describe four habitat types may be compared:

— Assemblage from a deep open waterbody (maximum depth 5 m) fringed by littoral belts with *Nuphar lutea* then *Carex elata* (Figure 6a);
— Assemblage from a heterogenous waterbody with open water areas plus areas colonized by aquatic plant communities (*Myriophyllo–Nupharetum* type) and littoral areas overgrown by *Phragmites australis* (Figure 6b);
— Assemblage from a former channel undergoing terrestrialization (max. depth 0.2 m), almost entirely overgrown by *Phragmites australis* (Figure 6c);
— Assemblage from a shallow former meander (max. depth 1 m) of a tributary (the Ain River), the major part of which is colonized by different aquatic plant communities (*Ranunculus circinatus*, *Chara vulgaris*, *Potamogeton natans*, *Sparganium emersum*, *Groenlandia densa*. . .) (Figure 6d).

The interpretation of the results is not always easy because of the occurrence of intermediate stages between such typical habitats. Moreover, the meaning of specific assemblages is not universal and must be defined in relation to the studied river system (Bournaud and Amoros, 1984). The Bosminid and Chydorid assemblages yield information about the physiography of the waterbody such as the ratio between the open water area and the macrophyte area, the kinds of aquatic plant communities, and the degree of terrestrialization process. Other information may be extracted from these remain analyses; it concerns the eutrophication degree of the ecosystem (Hoffman, 1978). For example, Figures 6a, b and c correspond to eutrophic waterbodies while Figure 6d illustrates a mesotrophic one.

Material and methods

Since the techniques were previously described in detail (Jacquet *et al.*, 1984; Frey, 1986), here we will focus on sampling strategy. Because the Cladoceran remains mostly accumulate in the deepest part of a waterbody, the coring sites have to be located in the former talweg. This location is selected according to historical maps, geomorphological and granulometric data (if a cross-profile is available) and the present distribution of wetland and/or terrestrial plants. The former talwegs may be delineated by remote sensing and interpreting vegetation maps; since plant communities are closely related to the level of water table, they reveal the microtopography.

The sediment cores are transported to laboratory and stored in a cold room (0–5 °C). Samples of 0.1 cm^3 of wet sediment are recovered along the core. Each sample is constituted by the addition of two 0.05 cm^3 subsamples, which are both taken at the same depth in the same core. These small samples generally contain a small number of headshields but the time saved in sorting them, allows analysis of a greater number of sediment layers within one core, and/or the comparison of several former channel successions. This strategy uses very small but numerous samples in order to emphasize successional trends along the core rather than precise information about population structures at each stratigraphic level. For this purpose, numeric data (i.e. the number of headshields belonging to each species of Bosminids and Chydorids) are analysed by the means of computer techniques such as 'Factorial Analysis of Correspondence' (Lebart and Fenelon, 1971; Diday *et al.*, 1982). Plotting the coordinates from each factorial axis lead to a graphic partition of the succession into phases characterized by species assemblages. This procedure will be illustrated by the presentation of a case study.

Case study

A sediment core (1.80 m depth) was taken from the talweg of a former anastomosed side-arm of the Upper Rhône River, France ('Le Piardet', river-km 94). These fine sediments

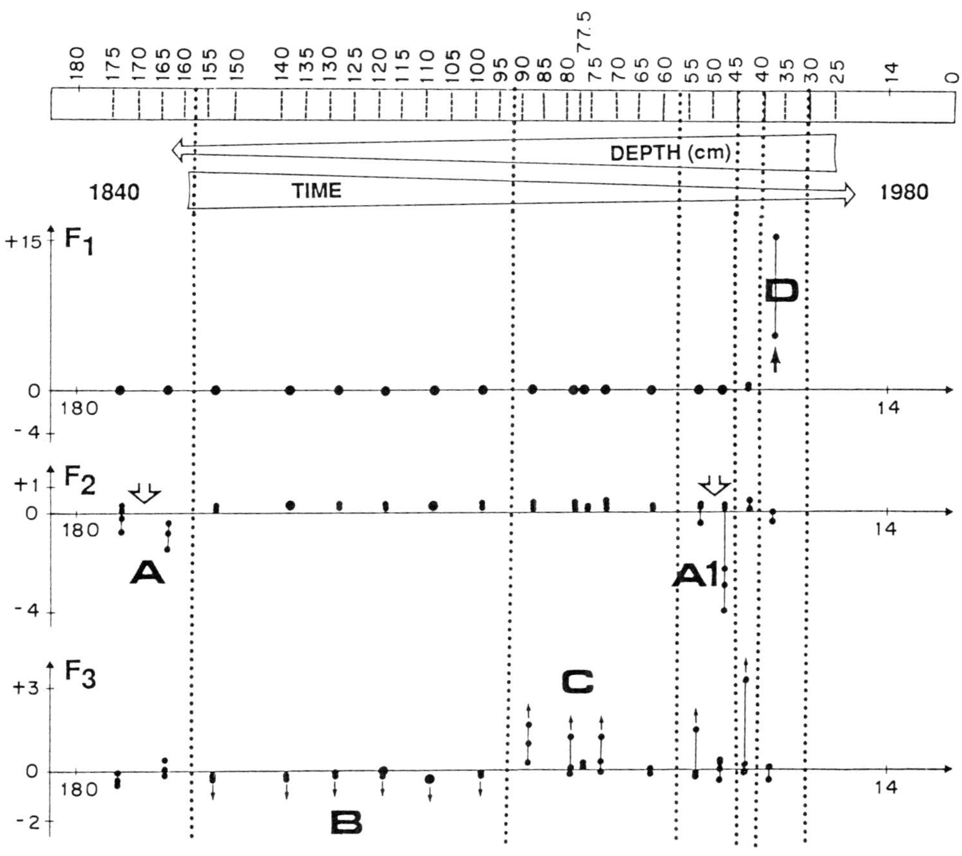

FIGURE 7. Cladoceran succession in a sediment core from a former channel of the Upper Rhône River: sample coordinates obtained from the Factorial Analysis of Correspondences, plotted in relation to their respective depth (F1, F2 and F3 = coordinates on the first three axes of the analysis; A, A1, B, C and D = the successional phases separated by the analysis) (modified from Jacquet *et al.*, 1984)

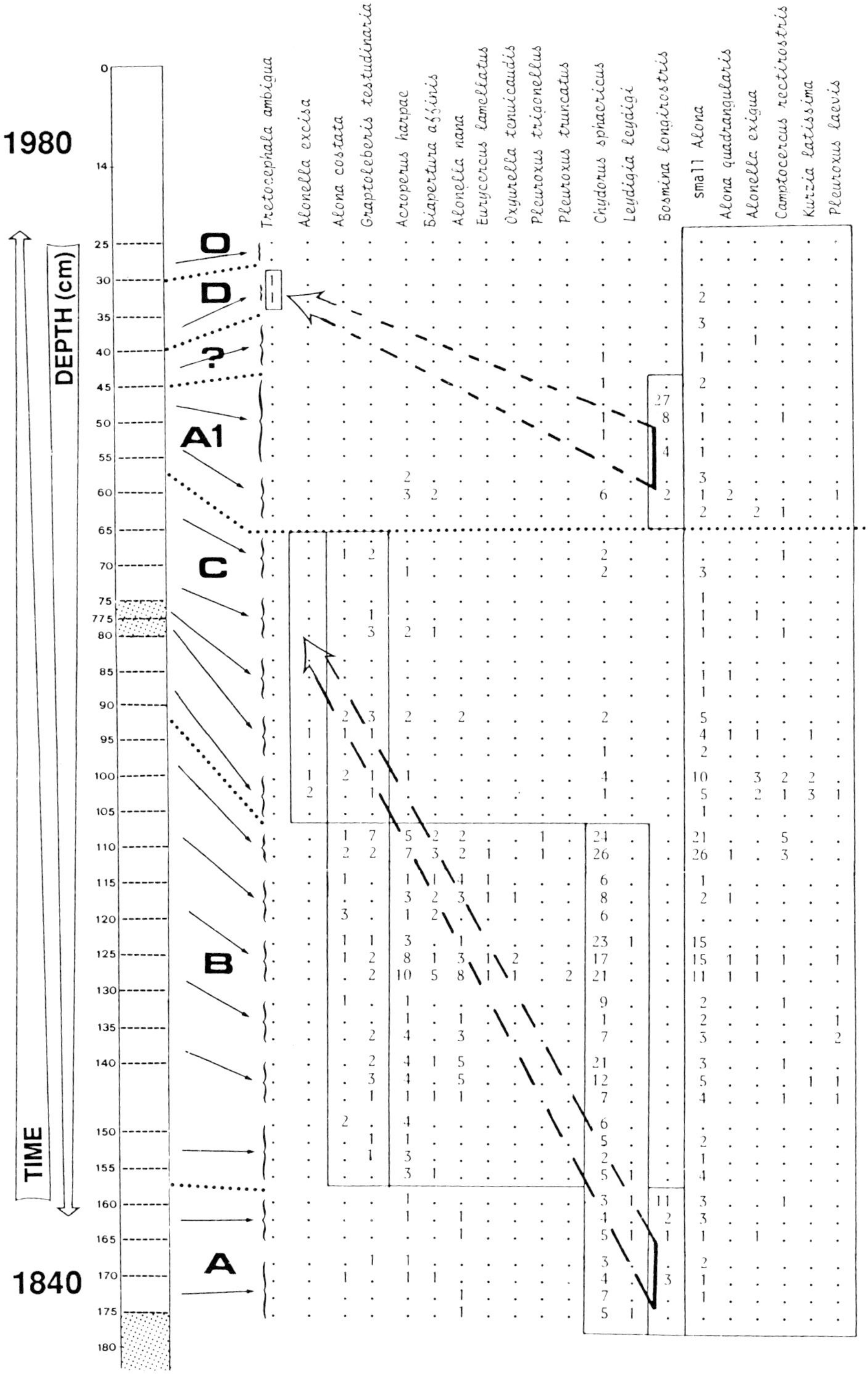
1980
1840
TIME
DEPTH (cm)
O
D
?
A1
C
B
A
Tretocephala ambigua
Alonella excisa
Alona costata
Graptoleberis testudinaria
Acroperus harpae
Biapertura affinis
Alonella nana
Eurycercus lamellatus
Oxyurella tenuicaudis
Pleuroxus trigonellus
Pleuroxus truncatus
Chydorus sphaericus
Leydigia leydigi
Bosmina longirostris
small Alona
Alona quadrangularis
Alonella exigua
Camptocercus rectirostris
Kurzia latissima
Pleuroxus laevis

were deposited on a coarse sand layer that corresponds to sedimentation within a running water channel. So, the transition from coarse sand to fine sediment (silt) is related to the side-arm cut off and dated to about 1840 according to the historical maps (Bravard, personal communication). This case study was used by Jacquet *et al.* (1984) to assess the reliability of small volume (0.1 cm^3) samples; for this purpose, 2–4 samples were taken at the same depth in the same core. The data were analysed by Factorial Analysis of Correspondence and the coordinates of each sample on the first three factorial axes are plotted according to their depth (Figure 7). The first factorial axis sorts the 0.35–0.40 samples from the rest of the core (phase D); the second factorial axis identifies phases A and A1 both constituted by samples with negative coordinates on this axis; and the third axis separates phase B (samples with negative coordinates) and phase C (samples with positive coordinates). This partition into successional phases, resulting from the data analysis, is drawn on Figure 8, which lists the species composition of each sample and supports the ecological interpretation.

The species assemblage of each phase is compared to those from ecosystems presently occurring in the same sector of the Upper Rhône River. After the anastomosed side-arm has been cut off, phase A indicates the occurrence of standing open-water with planktonic community (*Bosmina longirostris*) and the settlement of few aquatic plants. Then, the succession of phases B and C corresponds to the development of aquatic plant communities, which colonized the whole waterbody. Phase A1, similar to A (open-water with planktonic community), may be interpreted as a 'rejuvenation' of the ecosystem related to higher water level within this sector of the Rhône River at the beginning of the 20th century (Bravard *et al.*, 1986). This rejuvenation phase was followed by a rapid terrestrialization of the former channel because phase D indicates a very small and temporary waterbody, then no Cladoceran remains were recovered from the upper sediment layers. This surficial sediment corresponds to the present status of the former channel, which is flooded only a few days a year (plant community with *Phragmites australis* and *Convolvulus sepium* surrounded by *Alnus glutinosa*).

All the successional phases of this anastomosed side-arm refer to a eutrophic waterbody, while oligotrophic to mesotrophic stages (open water community with *Bosmina longispina*) occur within the ecological succession of a braided channel naturally abandoned by the Rhône River (Jacquet, 1986). The successional pattern of the former anastomosed channel differs in another way from that of the braided ones, which exhibit oscillating changes between open-water communities and phytophilous ones. In contrast, the Cladoceran succession studied by Jurasz and Amoros (in preparation) along a 6 m core from an abandoned Rhône meander, demonstrate a regular and progressive evolution of the ecosystem from the open-water stage to the terrestrial one, without either oscillation or rejuvenation. Such palaeological studies based on Cladoceran remains can afford adequate information on either natural or impacted development of ecosystems. The first results have provided evidence of different successional patterns in relation to geomorphological processes and the 1880s embankments of the Rhône River and have contributed to investigations of the historical changes of this fluvial system (Roux *et al.*, this volume).

FIGURE 8. (*opposite*) Cladoceran succession in a sediment core from a former channel of the Upper Rhône River: successional phases and species ordinated by the Factorial Analysis of Correspondences according to their abundances (from Jacquet *et al.*, 1984)

Chironomid remains

In river sediments, can be found remains of various insect groups, such as mayflies (Ephemeropterans), caddisflies (Trichopterans) and midges and blackflies (Dipterans: *Chironomidae* and *Simuliidae*). Chironomid remains often are numerically dominant and the composition of the chironomid fauna is a very good indication of past ecological conditions. This is because the chironomids are a group very rich in species which mostly live in a characteristic habitat. For Europe more than 2000 species are described, but the real number of species is probably lower as there are many double descriptions and is estimated to be about 1200–1500. A species may inhabit different types of water such as lakes or rivers but within all water types it is confined to the same habitat. Thus, some species inhabit sandy substrates, others occur only on stones etc. Species composition of the chironomid fauna in a sediment core yields information about the habitats of the waterbody from which the sediment is sampled. The relative numbers of remains of a particular species gives an indication of the relative importance of its habitat.

Preservation and identification to species

In chironomid identification, the head capsule is the most important part of the larval body (Figures 4 and 9); in some cases, the appendages of the last abdominal segment are also important. Larval skins are seldom preserved as a whole and generally only the head capsules are found in the sediments. In most cases the head capsules suffice for identification; the arrangement of teeth of the mentum is especially important, as is the mentum, which is one of the most chitinized parts of the head capsule.

One problem is to assess whether remains of individuals from different habitats have the same chance of entering the current and being deposited in the fine sediments in the river system. Generally it can be said that the material transported by the river will contain subfossil remains of organisms from erosional zones, so any sediment samples will contain mixtures of remains of different ages and from different habitats.

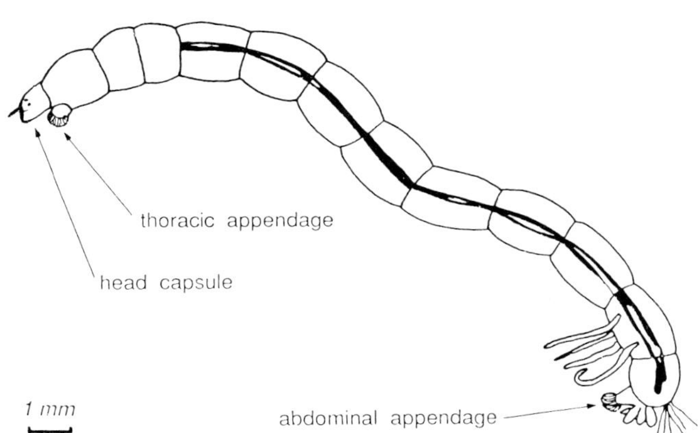

FIGURE 9. Living chironomid larvae

Morphological stability

The form of the mouth parts is clearly related to food items taken by the larvae and morphological stability is adequate for most palaeoecological investigations. Two points need to be discussed here: these are the existence of sibling species and the occurrence of mentum deformities in chironomids.

Especially in the genus *Chironomus* there are species which show hardly any morphological differences in the head capsules. These species can only be distinguished by karyological analysis of the giant chromosomes of the salivary glands. Species segregation is effected by differences of the swarming behaviour of the adults. These species are probably segregated only recently in evolutionary terms, and because they still occupy nearly the same habitat, the existence of these sibling species forms no special problem in reconstructing conditions in the past. An interesting phenomenon is the occurrence of deformities particularly in larvae of the genus *Chironomus*. Investigations in lakes (Warwick, 1986) have shown that the incidence of deformities has increased in the 20th century. The incidence of deformities may be related to the degree of chemical pollution of the sediments (Wiederholm, 1984) but few experimental data to support this hypothesis exist; Cushman (1984) reported a dose-dependent incidence of mentum deformities in *Chironomus decorus* in a semifield experiment with coal-tar derivatives.

The laboratory experiments, Kosalwat and Knight (1987) showed that copper could induce deformities of the epipharyngeal plates of *C. decorus* in addition to effecting growth and development. Their study suggests that the type of deformities might be pollutant specific. If this were true, investigations of deformities in subfossil *Chironomus* remains could reveal effects of pollution even if the presence of specific pollutants could not be proven because of their instability or mobility.

Reference patterns

Reconstruction of ecological conditions in the past from remains of organisms requires a reference pattern. The basic assumption in palaeoecological studies of rivers is that all rivers have some basic features in common. This can be easily demonstrated for stony headwater streams in which the fauna shows a great similarity, but is difficult to prove for the lower courses of large rivers which have generally been largely influenced by human activities. From the few rivers which are in a relatively unspoiled state it can be concluded that some characteristic habitats are present in large alluvial rivers. These are:

—Areas of shifting sand;
—The so-called 'snag habitat' consisting of branches and trees fallen into the river;
—The hyporheic zone (interstitial water flowing under the gravel armour layer); and
—Eroding clay banks.

The ecology of shifting sand habitats has been reviewed by Barton and Smith (1984) and that of the hyporheic zone by Williams (1984) in the same volume. In both habitats larval chironomids form an important fraction of the biomass. Many organisms spend only part of their life cycle in the hyporheic zone. In shifting sand habitats, a number of highly-specialized psammorheobiont chironomids occur, belonging to genera such as *Beckidia*, *Cyphomella*, *Chernovskiia*, *Robackia* and *Saetheria* (Barton and Smith, 1984). Although the taxonomy of this group has to be clarified, its presence in sediment samples can be easily recognized. The same can be said for lignicolous chironomids, such as *Symposiocladius lignicola*, from the snag habitat.

The fraction of these specialized species in the total chironomid fauna gives an indication of the areas occupied by their habitats. This applies also to other shifts in the species composition so far as the habitat requirements of the species involved are known. The ideal situation would be the availability of a data set of comparable samples from different rivers more or less influenced by man. As palaeoecological research in rivers is only in the beginning, no such data set is available and only rough comparisons between different rivers can be made. Nevertheless, comparisons can be made between samples from different periods in one river (in this case the Rhine). Classification of samples can be carried out by any of the techniques outlined by Pielou (1977). The ecological significance of differences in faunal composition established in this way can only be derived from studies of habitat requirements in the present day situation.

Most of the specialized chironomid species are very rare now and little is known about their niches in the system. This lack of knowledge forms a major limitation to the interpretation of the data. With regard to changes in deformity incidence in chironomids, lack of knowledge of the causative agents is the major limitation. Investigations at sites with no known sources of pollution indicate that the background level of deformity incidences may be below 1 per cent. The same applies for most samples of subfossil material in palaeoecological studies. Any significant increase of deformity incidence above this background level indicates an environmental change. Genetic drift in a population leading to higher background levels of deformity incidences is highly unlikely, as deformities impair development of the larvae.

Material and methods

A range of sampling equipment may be used to obtain sediment samples for analysis of subfossil chironomid remains. As the number of head capsules per volume unit of sediment may be low and the lowest layers of fine sediment in cores from backwaters and former channels must be used, heavy equipment suitable for obtaining long, large diameter sediment cores is recommended for sampling. The efficiency of different methods of recovering subfossil chironomid head capsules from sediments was investigated by Walker and Paterson (1985). Sieves having mesh sizes greater than 100 μm lose many of the smaller head capsules and a distorted picture of species composition may result. In the case of river sediments, sorting of the untreated wet sediments is impractical because of the relatively low numbers of head capsules. The best compromise is sieving through a 100 μ mess and sorting subsamples from the sieve under a stereomicroscope until a sufficient number of head capsules is recovered. For practical purposes, a total number of at least 300 head capsules results in a good overview of species composition. For investigations on a particular problem, e.g. the number of deformities in 4th instar *Chironomus* larvae which have large head capsules, the procedure may be modified.

The subfossil head capsules are identified to species by comparison with the head capsules of larvae which were sampled alive. Correct identification of chironomid larvae often requires the rearing of some larvae to adults, as the taxonomy is mainly based on male adult morphological characters. For this reason, identification of remains of species for which no comparative material is available (e.g. those species which are now extinct from the river) should be considered as provisional.

Outline of results of analysis of chironomid remains in the Rhine

The preliminary results of palaeoecological studies on macroinvertebrates in the Rhine are discussed in full detail by Klink (this volume). Some main trends are discussed here to demonstrate some principles of the methods used. A characteristic feature of the present chironomid fauna in the Lower Rhine is the dominance of species living on the stones of the groynes and the paucity of bottom-dwelling species. The chironomid fauna of the sandy bottoms in the Lower Rhine consists mainly of only one species *Polypedilum breviantennatum* (see Van Urk and Smit, this volume). Analyses of chironomid remains in sediments have shown that many of the psammorheobiont chironomids from shifting sand habitats must have been present in the Lower Rhine. These species are found especially in samples from the 18th century and before. The same applies to the lignicolous species *Symposiocladius lignicola*. It remains to be assessed whether the disappearance of these species coincided with channel constriction or with the deforestation of the floodplain.

In the Lower Rhine, deformity incidences in *Chironomus* larvae were already above background levels as early as the 18th century. This could be demonstrated in the sediment cores from the 'Schoonrewoerdse Wiel' (Figure 10). *Chironomus* head capsules from sediment layers deposited before 1800 showed about 4 per cent deformities; no deformities were found in *Chironomus* larvae from layers deposited after the isolation of the 'wiel' from the river. However, deformity incidences in subfossil materials from recently deposited sediment layers never reach the values found for living larvae. The maximum value in subfossil material is about 25 per cent, whereas living larvae deformity incidences of more than 40 per cent may be found. The difference may be caused by the presence

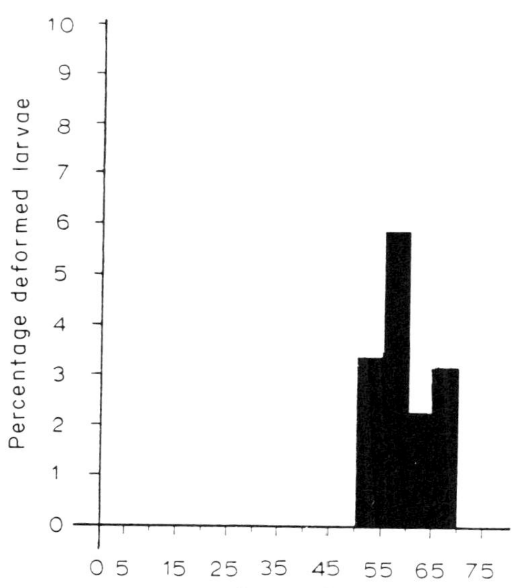

FIGURE 10. Percentages of deformed *Chironomus* larvae in relation to depth in a core from the 'Schoonrewoerdse Wiel' (see Figure 1 for other data on this core). (Deformity incidences established by Klink)

of allochthonous remains in the subfossil material. To solve problems like this, the mechanisms of transport, deposition and preservation of chironomid remains in the river should be investigated in more detail. This may be carried out by analysis of drift net samples taken in different seasons and at different flow conditions.

The species richness of the chironomids forms both an advantage and a limitation of their use in palaeoecological studies. The main advantage is that nearly all habitats contain specialized species; the main limitation is that the taxonomy of some groups needs further clarification and that the habitat requirements of some species, especially those that are now very rare, are poorly known.

CONCLUSION

The aim of palaeoecological studies is the reconstruction of past ecological conditions and historical changes of ecosystems, using remains of plants and animals preserved in fine sediments. Their value has been demonstrated in studies of lakes. Recent research carried out on the Rhône and Rhine systems show that palaeoecological techniques can be applied to river sediments as well.

Palynology addresses chiefly the terrestrial part of the alluvial floodplain; sediments from cut off channels constitute a convenient matrix for preserving the pollen in chronological sequence. Analysis of plant remains such as Diatom frustules, algal pigments and macrophyte remains could probably afford fuller information about aquatic ecosystem functioning and developments.

Cladoceran and Chironomid remains are abundant in unreworked sediments from backwaters and cut off channels of large rivers. These taxa are very diversified and can afford adequate information about the functioning and the development of the aquatic ecosystems before and after human impacts. These palaeoecological analyses, carried out in connection with historical investigations, granulometry and chemistry of sediments (organic matter, heavy metals), play a key role in understanding the mechanisms responsible for ecosystem changes and then in the constructin of predictive scenarios.

Dating of sediments is a special problem in palaeoecological river studies; Pb-210 dating is possible when the allochthonous component in the sediment is negligible but cartographic sources are most widely applicable. Generally historical data and granulometric analyses of the sediments provide adequate information on the period and the conditions of deposition. Sedimentation rates within river floodplains are much more rapid than those in lakes. The problem of sediment reworking, e.g. by bioturbation, which may impair time resolution in studies of lake sediments, is therefore probably only a minor one. The supply of allochthonous material into river sediments is believe to constitute a more important problem.

The reconstruction of ecosystems for which no present-day counterparts exist, such as the natural ecosystems within the man-impacted downstream course of large rivers in Europe is a particular problem. In such circumstances, the interpretation on ecosystem functioning must be totally derived from the sediments and can be only fragmentary. Where it is possible, comparisons of the palaeoecological results (diachronic analyses) with the community composition of present-day ecosystems occurring at different stages of succession (synchronic analyses), lead more rapidly to an understanding of the process involved in ecosystem development. Both diachronic and synchronic investigations are needed for assessment of past impacts and better environmental management of large river systems. We are just beginning to investigate the numerous and fast-changing ecosystems within river floodplains, and to develop or adapt methods for the analysis and the interpretation of the data. The availability of historical data and the occurrence of past human impacts on the large European rivers such as the Rhine or the Rhône would provide a promising experimental field for comparative palaeoecology.

REFERENCES

Ammann, B. and Tobolski, K. 1983. Vegetational development during the Late Würm at Lobsigensee (Swiss Plateau). Studies in the Late Quaternary of Lobsigensee 1. *Revue Paléobiol.*, **2**, 221–227.

Amoros, C. 1980. Structure et fonctionnement des écosystèmes du Haut-Rhône français: XV. Comparaison des peuplements de Cladocères et Copépodes de deux anciens méandres d'âges différents. *Acta Oecol., Ecol. Gen.*, **1**(2), 193–208.

Amoros, C. and Chessel, D. 1985. Les peuplements de Cladocères (Crustacés), descripteurs du fonctionnement hydrologique des bras-morts fluviaux. *Annls. Limnol.*, **21**(3), 227–240.

Amoros, C. and Jacquet, C. 1987. The dead-arm evolution of river systems: a comparison between the information provided by living Copepod and Cladocera populations and Bosminidae and Chydoridae remains. *Hydrobiologia*, **145**, 333–341.

Amoros, C., Rostan, J. C., Pautou, G. and Bravard, J. P. 1987. The reversible process concept applied to the environmental management of large river systems. *Environmental Management*, **11**(5).

Averdieck, F. R. 1983. Palynological investigations of the sediments of ten lakes in eastern Holstein. *Hydrobiologia*, **103**, 225–230.

Avnimelech, Y., McHenry, J. R. and Ross, J. D. 1984. Decomposition of organic matter in lake sediments. *Environmental Sciences and Technology*, **18**, 5–11.

Barton, D. R. and Smith, S. M. 1984. Insects of extremely small and extremely large aquatic habitats. In Resh, V. H. and Rosenberg, D. M. (Eds), *The Ecology of Aquatic Insects*, Praeger publishers, New York, 456–483.

Battarbee, R. W. 1986. Diatom analysis. In Berglund, B. E. (Ed.), *Handbook of Holocene Palaeoecology and Palaeohydrology*. Wiley, Chichester, 869 pp.

Berglund, B. E. 1986. *Handbook of Holocene Palaeoecology and Palaeohydrology*, Wiley, Chichester, 869 pp.

Binford, M. W. 1982. Ecological history of Lake Valencia, Venezuela: interpretation of animal microfossils and some chemical, physical and geological features. *Ecol. Monogr.*, **52**(3), 307–333.

Birks, H. H. 1980. Plant macrofossils in Quaternary lake sediments. *Ergebnisse der Limnologie*, **15**, 1–60.

Birks, H. H., Whiteside, M. C., Starr, D. M. and Bright, R. C. 1976. Recent palaeolimnology of three lakes in northwestern Minnesota. *Quat. Res. (N.Y.)*, **6**, 249–272.

Bournaud, M. and Amoros, C. 1984. Des indicateurs biologiques aux descripteurs de fonctionnement: quelques exemples dans un système fluvial. *Bull. Ecologie*, **15**(1), 57–66.

Bravard, J. P. 1983. Les sédiments fins des plaines d'inondation dans la vallée du Haut-Rhône (Approche qualitative et spatiale). *Rev. Géogr. alpine*, **71**(4), 363–379.

Bravard, J. P., Amoros, C. and Pautou, G. 1986. Impacts of civil engineering works on the succession of communities in a fluvial system: a methodological and predictive approach applied to a section of the Upper Rhône River. *Oikos*, **47**(1), 92–111.

Cronberg, G. 1986. Blue-green algae, green algae and chrysophyceae in sediments. In Berglund, B. E. (Ed.), *Handbook of Holocene Palaeoecology and Palaeohydrology*, Wiley, Chichester, 869 pp.

Cushman, R. M. 1984. Chironomid deformities as indicators of pollution from a synthetic coal-derived oil. *Freshwater Biology*, **14**, 179–182.

Davis, R. B. and Ford, M. S. 1982. Sediment focusing in Minor Lake, New Hampshire. *Limnol. Oceanogr.*, **27**, 137–150.

De Costa, J. 1964. Latitudinal distribution of Chydorid Cladocera in the Mississippi Valley based on their remains in surficial lake sediments. *Invest. Indiana Lakes and Streams*, **6**, 65–101.

De Costa, 1968. Species diversity of Chydorid fossil communities in the Mississippi Valley. *Hydrobiologia*, **32**, 497–512.

Deevey, E. S. 1964. Preliminary account of fossilization of zooplankton in Rogers Lake. *Verh. Internat. Verein. Limnol.*, **15**, 981–992.

Descy, J. P. 1979. A new approach to water quality estimation using diatoms. *Nova Hedwigia*, **64**, 305–323.

Diday, E., Lemaire, J., Pouget, J. and Testu, F. 1982. *Eléments d'Analyses de Données*, Ed. Dunod, Paris, 462 pp.

Flossner, D. 1964. Zur Cladoceren Fauna des Stechlin-Gebietes. 2. Okologische untersuchungen über die litoralen Arten. *Limnologia, (Berlin)*, **2**, 35–103.

Frey, D. G. 1958. The late glacial Cladoceran fauna of a small lake. *Arch. Hydrobiol.*, **54**, 209–275.

Frey, D. G. 1959. The taxonomic and phylogenetic significance of the head pores of the Chydoridae (Cladocera). *Int. Rev. Ges. Hydrobiol.*, **44**, 27–50.

Frey, D. G. 1960. The ecological significance of Cladoceran remains in lake sediments. *Ecology*, **41**, 684–699.

Frey, D. G. 1962. Cladocera from the Eemian interglacial of Denmark. *J. Palaeontol.*, **36**, 1133–1154.

Frey, D. G. 1962. Supplement to: the taxonomic and phylogenetic significance of the head pores of the Chydoridae (Cladocera). *Int. Rev. Ges. Hydrobiol.*, **47**(4), 603–604.

Frey, D. G. 1967. Cladocera in space and time. *Proc. Symp. Crustacea*, **1**, 1–9.

Frey, D. G. 1974. Palaeolimnology. *Mitt. Int. Verein. Limnol.*, **20**, 95–123.

Frey, D. G. 1976. Interpretation of Quaternary Palaeoecology from Cladocera and midges and prognosis regarding usability of other organisms. *Can. J. Zool.*, **54**, 2208–2226.

Frey. D. G. 1979. Cladocera analysis. In *Palaeohydrological Changes in the Temperate Zone in the Last 15000 Years*, UNESCO–IGCP, 227–257.

Frey, D. G. 1986. Cladocera analysis. In Berglund, B. E. (Ed.), *Handbook of Holocene Palaeoecology and Palaeohydrology*, Wiley, Chichester, 869 pp.

Goulden, C. E. 1986. La Aguada de Santa Ana Vieja: an interpretative study of the Cladoceran microfossils. *Arch. Hydrobiol.*, **62**, 373–404.

Goulden, C. E. 1969. Interpretative studies of Cladoceran in lake sediments. *Mitt. Int. Verein. Limnol.*, **17**, 43–55.

Goulden, C. E. and Frey, D. G. 1963. The occurrence and significance of lateral head pores in the genus *Bosmina* (Cladocera). *Int. Rev. Ges. Hydrobiol.*, **48**, 513–522.

Hofmann, W. 1978. Analysis of animal microfossils from the Grosser Segeberger See (F.R.G.). *Arch. Hydrobiol.*, **82**(1–4), 316–346.

Hynes, H. B. N. 1970. *The Ecology of Running Waters*, Liverpool University Press, 555 pp.

Hynes, H. B. N. 1986. Foreword. In Davies, B. R. and Walker, K. F. (Eds), *The Ecology of River Systems*, W. Junk Publishers, Dordrecht.

Jacquet, C. 1986. *Evolution des anciens chenaux du Rhône et de l'Ain: écologie rétrospective basée sur les restes Cladocères*, Thesis Univ. Cl. Bernard, Lyon 1, France, 194 pp. + 90 fig.

Jacquet, C., Amoros, C. and Auda, Y. 1984. Evolution des écosystèmes aquatiques abandonnés par les fleuves: recherches méthodologiques sur l'utilisation des restes de Cladocères en écologie rétrospective. *Archiv. für Hydrobiol.*, **102**(1), 73–89.

Korinek, V. 1971. Comparative study of head pores in the genus *Bosmina* Baird (Crustacea, Cladocera). *Vestn. Cesk. Spol. Zool.*, **35**(4), 275–296.

Kosalwat, P. and Knight, A. W. 1987. Chronic toxicity of copper to a partial life cycle of the midge *Chironomus decorus. Archives of Environmental Contamination and Toxicology*, **16**, 283–290.

Lange-Bertalot, H. and Lorbach, K. D. 1979. Die Diatomeenbesiedlung des Rheins in Abhangigkeit von der Abwasserbelastung. *Archiv. für Hydrobiologie*, **87**, 347–363.

Lebart, L. and Fenelon, J. P. 1971. *Statistique et Informatique Appliquées*, Ed. Dunod, Paris, 426 pp.

Lowe, R. L. 1974. Environmental requirements and pollution tolerance of freshwater diatoms. *EPA*, **670/4–74–005**, Cincinnati Ohio, vi + 334 p.

Mellors, W. K. 1975. Selective predation of ephippial Daphnia and the resistance of ephippial eggs to digestion. *Ecology*, **56**(4), 974–980.

Mesjatsev, I. I. 1924. Die fossile Fauna der Seen in Kossino. *Arb. Biol. Sta. Kossino*, **1**, 16–26.

Mueller, W. P. 1964. The distribution of Cladoceran remains in surficial sediments from three northern Indiana lakes. *Invest. Indiana lakes and streams*, **6**, 1–63.

Oldfield, F. and Appleby, P. G. 1984. Empirical testing of 210 Pb-dating models for lake sediments. In Haworth, E. Y. and Lund, J. W. G. (Eds), *Lake Sediments and Environmental History*, Leicester University Press, 93–124.

Patrick, R. 1977. Ecology of freshwater diatoms and diatom communities. In Werner, D. (Ed.), *In Biology of Diatoms*, Blackwell Scientific Publications, Oxford, 284–332.

Pielou, E. C. 1977. *Mathematical Ecology*, Wiley, New York, 385 pp.

Prentice, I. C. 1985. Pollen representation, source area, and basin size: toward a unified theory of pollen analysis. *Quat. Res.*, **23**, 76–86.

Proctor, V. W. 1964. Viability of Crustacean eggs recovered from ducks. *Ecology*, **45**, 656–658.

Rostan, J. C., Amoros, C. and Juget, J. 1987. The organic content of the surficial sediment: a method for the study of ecosystems development in abandoned river channels. *Hydrobiologia*, **148**, 45–62.

Schlichting, H. E. and Sides, S. L. 1969. The passive transport of aquatic microorganisms by selected Hemiptera. *Journal of Ecology*, **57**, 759–764.

Smit, H. and Van Urk, G. 1987. Het herstel van de ecologische waarden van de Rijn: over de zalm en ecologische doelstellingen. *H20*, **20**, 427–430.

Sramek-Husek, R. 1962. Die mitteleuropaischen Cladoceren- und Copepoden Gemeinschaften und deren Verbreitung in den Gewassern der CSSR. *Sbornik VSCHT Techn. Vody*, **6**(1), 99–133.

Starling, R. N. and Crowder, A. 1981. Pollen in the Salmon River system, Ontario, Canada. *Rev. Palaeobot. Palynol.*, **31**, 311–334.

Swain, E. B. 1985. Measurement and interpretation of sedimentary pigments. *Freshwater Biology*, **15**, 53–75.

Van de Ven, G. P. 1987. Jan Blanken en de waterstaat. In *De Physique Existentie dezes Lands*, Jan Blanken (1755–1838), Ed. Rijksmuseum, Amsterdam, 59–93.

Vannote, R. L., Minshall, G. W., Cummins, K. W., Sedell, J. R. and Cushing, C. E. 1980. The river continuum concept. *Canadian Journal of Fisheries and Aquatic Sciences*, **37**, 130–137.

Walker, I. R. and Paterson, C. G. 1985. Efficient separation of subfossil Chironomidae from lake sediments. *Hydrobiologia*, **122**, 189–192.

Warwick, W. F. 1980. Chironomidae (Diptera) responses to 2800 years of cultural influence: a palaeolimnological study with special reference to sedimentation, eutrophication and contamination processes. *Canadian Entomologist*, **112**, 1193–1238.

Whiteside, M. C. 1970. Danish Chydorid Cladocera: modern ecology and core studies. *Ecol. Monogr.*, **40**, 79–118.

Whiteside, M. C., Bradbury, J. P. and Tarapchak, S. J. 1980. Limnology of the Klutlan moraines Yukon territory Canada. *Quat. Res.*, **14**(1), 130–148.

Wiederholm, T. 1984. Incidence of deformed chironomid larvae (Diptera: Chironomidae) in Swedish lakes. *Hydrobiologia*, **109**, 243–249.

Zullig, H. 1982. Untersuchungen über die Stratigraphie von Carotinoiden in geschichteten Sedimente von 10 Schweizer Seen zur Erkundung fruherer Phytoplankton-Entfaltungen. *Schweizerische Zeitschrift für Hydrologie*, **44**, 1–98.

Historical Change of Large Alluvial Rivers: Western Europe
Edited by G. E. Petts
©1989 John Wiley & Sons Ltd

10

The Lower Rhine Geomorphological Changes

GEERT VAN URK and HENK SMIT

Rijkswaterstaat Institute for Inland Water Management and Waste Water Treatment, P.O. Box 17, NL-8200 AA Lelystad, The Netherlands

INTRODUCTION

The evolution of the Lower Rhine during the Holocene has been described by Brunnacker (1978) and Van der Woude (1981). A characteristic feature was the formation of alluvial fills indicating cyclic changes of channel morphology from strongly meandering to relatively straight channels. The sequence of changes is similar to that of other European rivers in a similar climatic zone (Starkel, 1983). Human activity disrupted the natural sequence of channel change. The first recorded historical evidence of human impact on the Lower Rhine dates back to the Roman era. The best-known example of Roman hydraulic engineering is the 'Drusus canal' (after the general Drusus), but its exact location is unknown. The canal may have been an artificial connection between the Rhine and (Oude) IJssel rivers. More tangible proof of Roman engineering is the inscription on a gravestone referring to 'Carvio ad molem' (Herwen at the dam). This 'mole' is believed to have influenced water distribution at the Rhine bifurcation in such a manner that the IJssel became an important branch of the Lower Rhine (Van Dalen, 1972; Van der Schrier, 1978). It may be assumed, however, that the Lower Rhine retained an almost natural character until the early Middle Ages. To judge by the extent of the river's deposits, the floodplain must have been tens of kilometres wide, bound only by the ridges of hills formed in the ice age (Figure 1).

With the development of agriculture in the Middle Ages the construction of dykes began. Generally, dykes were built on natural levees, as close to the river's channels as possible. Old meanders now generally are situated on the landside of the dykes (Figure 1a). In most areas embanking was completed around 1450; since that time, the area inundated during the floods has been confined to a narrow zone only a few kilometres wide along either side of the river. As a result, sedimentation of river silt, which formerly occurred all over the floodplain, concentrated on the river's forelands which were thus raised in height relative to the land behind the dykes. Despite all efforts to keep the dykes in a good condition, frequent inundations of the land behind the dykes took place mainly owing to the formation of ice dams in the river. The river was eventually regulated in the 19th century. This involved channel rectification and fixation, followed by channel constriction.

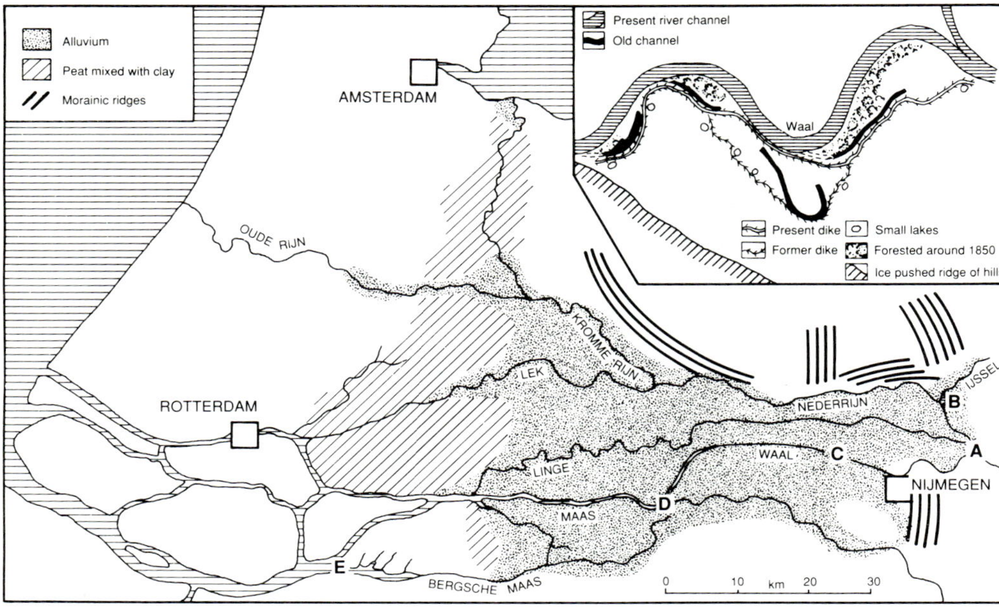

FIGURE 1. Holocene fluvial deposits of the Rhine's alluvial plain in the central part of the Netherlands. Before the embankment of the rivers, the peat area was only flooded at high river discharges. After embankment, the peat soil set to below normal river levels. The insert shows a map of the Ooypolder. The old channels in the forelands result from a channel correction in 1649. The meander on the inside of the present dyke was cut off in the 14th century, but was not embanked before the end of the 19th century. Several 'wielen' (small lakes) resulting from dyke bursts mainly in 1809 and 1820 are present near the dykes

Interest in the ecological consequences of river engineering and in the need to minimize damage did not arise until recently, when the restoration of the Rhine became an important issue (Smit and Van Urk, 1987). This chapter considers the morphological consequences of river engineering and discusses its possible ecological consequences. It is based, first, on a study of river maps made by the Survey Department of Rijkswaterstaat since 1829 (Boode, 1979), secondly, on analyses of sediment cores from the river, and thirdly on the ecological investigations carried out in the Waal.

CHANNEL RECTIFICATION IN THE LOWER RHINE

The Rhine bifurcations

The first major hydraulic engineering works since the Roman era were undertaken at the Rhine bifurcations in the early 1700s. Until then, channel corrections had been carried out and groynes, dams and revetments built, but they served only local purposes, such as the protection of dyke sections. A characteristic example is the bend correction in the Upper Waal in the 17th century; the former course of the Waal can be derived from Figure 1a (see also Van Heiningen, 1971). This bend correction probably influenced the water distribution at the bifurcation of the Waal and (Oude) Rijn. At the end of the 17th century, the Waal carried by far the greatest part of the Rhine discharge, probably over 95 per cent. Water supply to the Nederrijn and IJssel was so small that navigation became difficult: this is reflected in the shifting of navigation with timber rafts of up to 300 m in length from the Nederrijn to the Waal (Leemans, 1981).

The engineering works at the bifurcations of the Rhine were to ensure a better distribution of the total Rhine discharge among its three branches. The works are described in detail

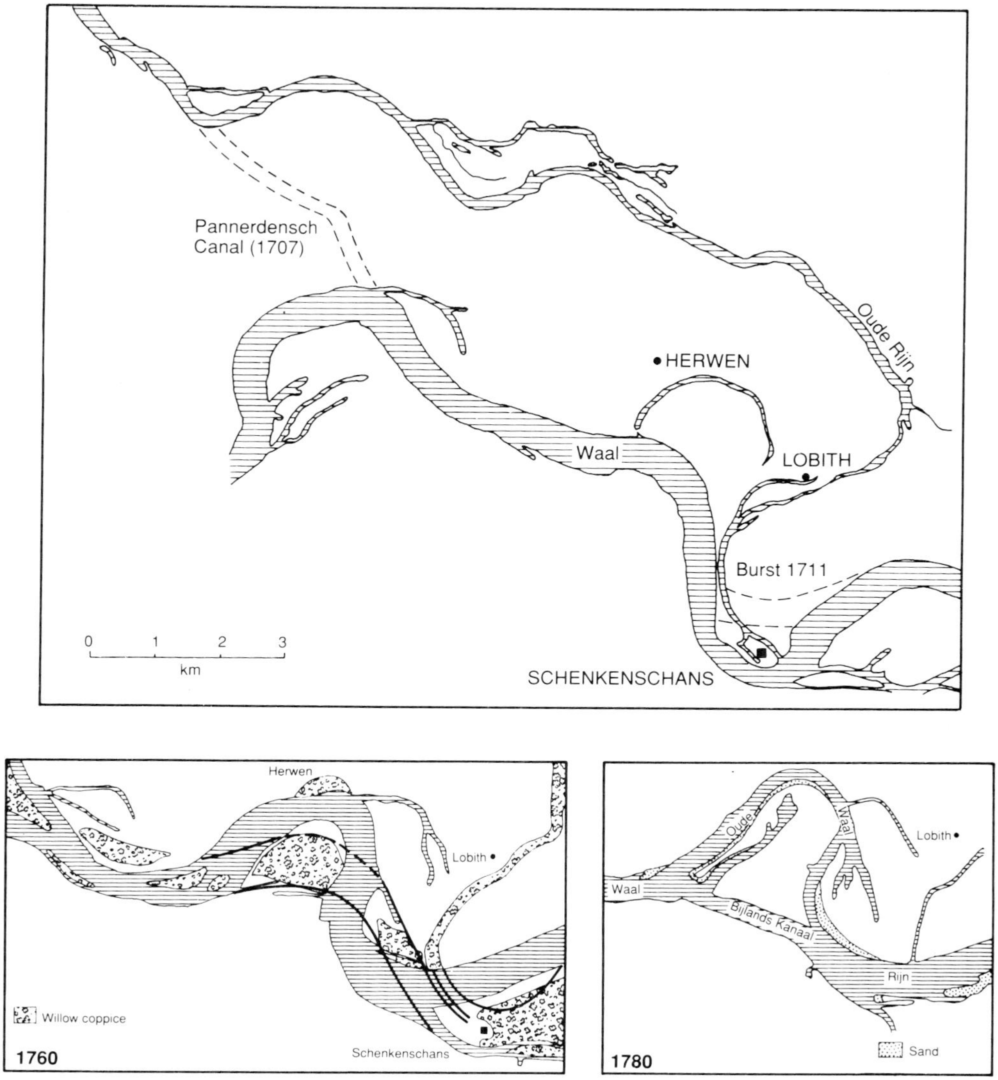

FIGURE 2. The situation near the first Rhine bifurcation at the end of the 17th century above, and, below, the Waal meander near Herwen around 1760 (left) also showing the situation after the construction of the Bijlandsch Kanaal around 1780

by Van de Ven (1976) and a general survey will suffice here. A sketch map of the situation by the end of the 17th century is given in Figure 2. At the first bifurcation, which was a strategically important point at the frontier between the Netherlands and the Duchy of Kleve (later to become a part of Germany), there is a military fort, Schenckenschans. As the head of the Oude Rijn was practically barred by sands, a new connection was dug between the Waal and Oude Rijn near Pannerden, several kilometres downstream of Schenckenschans. This so-called 'Pannerdensch Canal' functioned for the first time in 1707. However, the situation was still far from stable: in 1711 the Rhine broke through the dykes near Schenckenschans and formed a new meander (Figure 2b) which by the middle of the 18th century had scoured out the dykes near the village of Herwen (the 'Carvium' of the Romans). At last, but too late for the village, this meander was cut off by the construction of the 'Bijlandsch Kanaal', completed in 1780 (Figure 2c). By then, the church of Herwen had already been pulled down and the rest of the village had vanished into the river (Van Petersen, 1978). Around 1775, the bifurcation of the Nederrijn and the IJssel was reconstructed by digging a new channel in the upper part of the IJssel. From then, the IJssel received about one third of the discharge of the Pannerdensch Kanaal which in turn received about one third of the total Rhine discharge. The remaining two thirds of the flow was carried by the Waal. Stabilization of the situation was achieved by moles, constructed under supervision by Brunings who was one of the first engineers to carry out accurate flow measurements in the river (Van Schaik, 1984). Until the 1960s, the river bed of the Oude Rijn continued to function as a spillway for excess water, causing numerous floods in the region (Van Petersen, 1978).

River engineering in the 19th century

The works at the bifurcations improved water distribution among the Rhine branches, but floods continued to be hazardous. Administrative boards or water authorities, with responsibility for integrated river engineering, did not exist in the Netherlands until the 19th century. Individual landowners built bank revetments and groynes to protect their land from erosion and to increase sedimentation. However, the irregular array of groynes and the presence of many sand bars in the channel, not only impeded flow but particularly led to the formation of ice dams. A formal ban on the irregular construction of groynes proposed by the States of Gelderland in 1602 and finally adopted in 1715 (Van Heiningen, 1971), had little effect because of the lack of supervision. In 1809 and 1820, large areas were inundated (Van de Ven, 1987) and in 1821 a Government committee was appointed to make proposals for the solution of the problem. In 1825, the committee completed its report which was published in 1827 (Van de Ven, 1987). Meanwhile, prominent engineers such as Blanken, Goudriaan and Krayenhoff had published their different views on the best solution. Their recommendations varied from the creation of new spillways by the partial removal of dykes, to the building of sluices, and to the construction of a totally new canal from the entry of the Rhine in the Netherlands to the IJssel (Van Petersen, 1978). These drastic plans were never carried out; the solution finally adopted was outlined in a report submitted by inspectors Ferrand and Van der Kun of the newly formed Rijkswaterstaat. It proposed the removal of the sand bars and the constriction of the river's channels to accelerate flow. Only in the tidal part of the Rhine, was a totally new channel, the Nieuwe Merwede, constructed (see Van Urk, 1984).

TABLE 1. Survey of bend rectifications in the main channel of the Lower Rhine

Year	Location	River (km)	Channel shortening (km)
1649	Ooy	871–876	*c.* 4.0
1655	Hurwenen	929–932	*c.* 5.0
1677	Rees	834	*c.* 3.0
1780	Bijland	862	*c.* 2.0
1784	Wesel	812	*c.* 1.5
1788	Xanten	821	*c.* 5.5
1819	Grieth	845	*c.* 2.0

Other modifications to the river system included the complete separation of the Rivers Waal and Meuse at the turn of the century. Until this time, the narrow strip of land between the two rivers at Heerewaarden (Figure 1, D) was not embanked, allowing the free interchange of floodwater. To improve run off conditions, the Meuse was finally diverted to the Hollandsch Diep by the digging of the Bergsche Maas (Figure 1, E).

Recently, the capacity of the IJssel has been increased by the rectification of two bends. The construction of weirs in the Nederrijn and Lek caused the greater part of the discharge of the Pannerdensch Kanaal to be diverted to Lake IJssel, a large freshwater reservoir (see De Gaay and Blokland, 1970). A summary of channel rectifications in the main channel of the Lower Rhine is given in Table 1.

A CASE STUDY OF CHANNEL CONSTRICTION

As stated earlier, channel constriction was the solution finally adopted to increase the discharge capacity of the Lower Rhine. The general principles of channel constriction are discussed by Jansen *et al.* (1979). The method of channelization can best be demonstrated in a series of maps of a stretch of river which was relatively unaffected before work started. This excludes the bifurcations that were rechannelled in the 18th century, although it is a region for which the most extensive series of maps exists (Koeman, 1983).

Because the wide River Waal has a stronger tendency to form multiple channels than the much narrower, strongly meandering River IJssel, the effects of channel constriction on the Waal have been greater. Therefore, a stretch of the Waal, west of the city of Nijmegen, was chosen (Figure 1, C). The oldest map of this stretch, reproduced in Van Heiningen (1971), dates back to 1572, but lacks sufficient detail. A 18th century map made by triangulation, gives a clear picture of the unregulated river with multiple channels separated by sand bars which were partly overgrown with trees, probably willows (Figure 3a; Figure 4a). In the forelands, several old channels existed; some of them were still in open connection with the main channel, others were isolated at normal discharges. Because the channels run parallel to the river, they cannot be old meanders: they must have been formed by sand bars moving in the river. The forelands were only partly cultivated: large areas were covered with marsh vegetation and woods, probably mainly willow coppices.

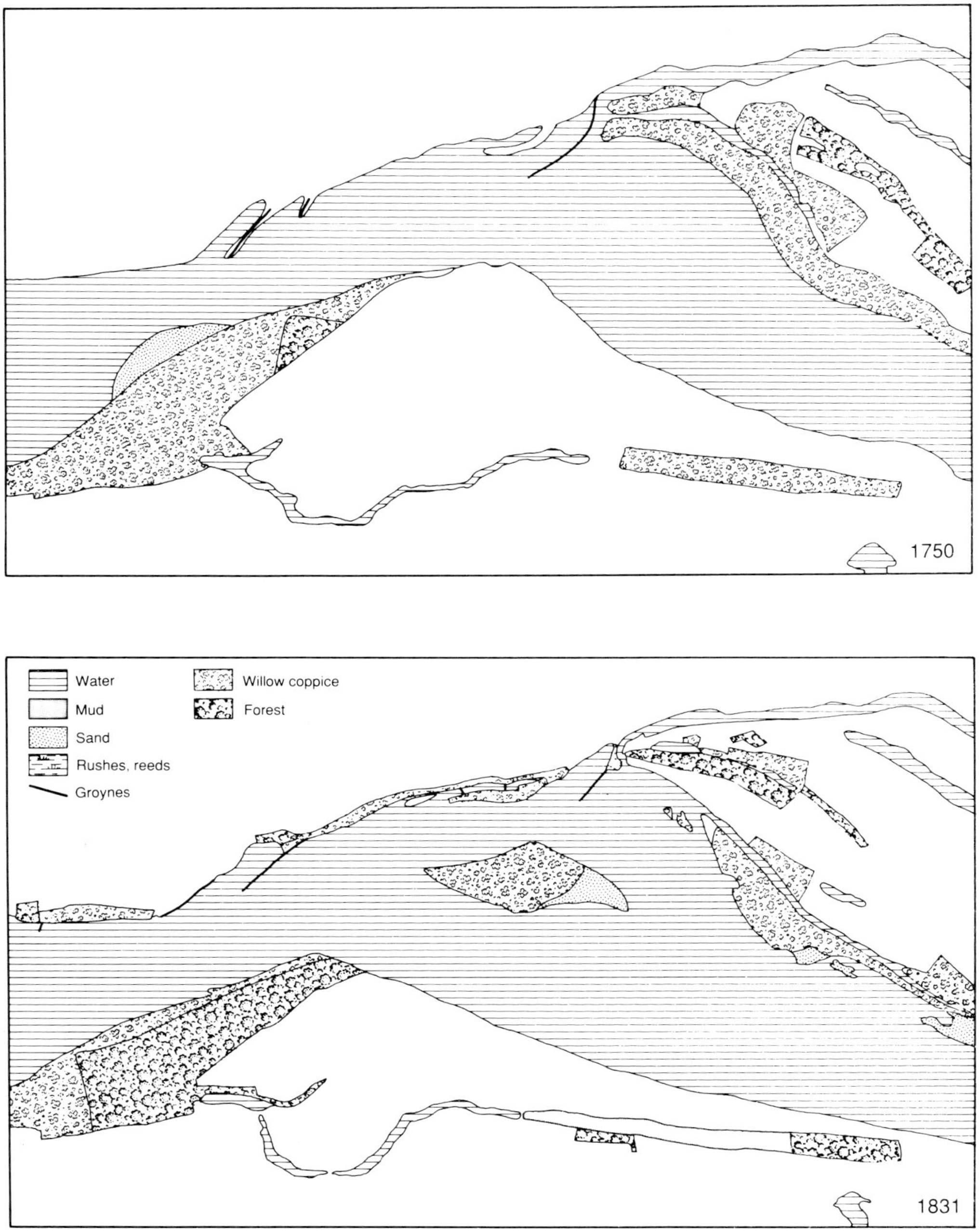

FIGURE 3. The Waal near river km 899–901 showing channel changes over a period of about 200 years. About 1830 (bottom left) the main channel had been diverted southward, a sand bar had developed between the two channels, and some dyke sections were protected by groynes. By 1870 (top right) the northern channel had been dammed. The dam is protected by short groynes; and

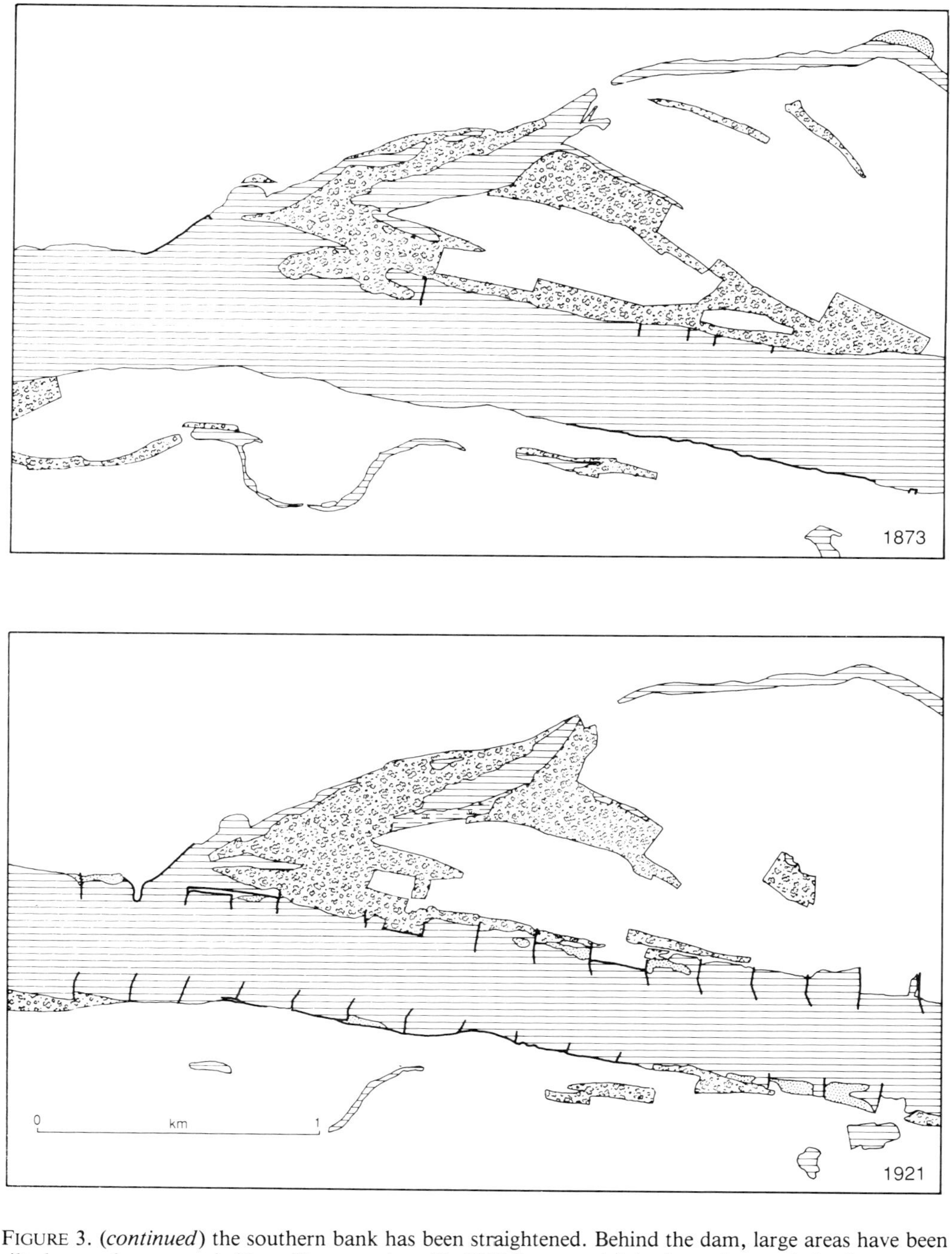

FIGURE 3. (*continued*) the southern bank has been straightened. Behind the dam, large areas have been silted up and are occupied by willow coppices. By 1920 (bottom right), channel constriction by groynes had been completed. The open water area had decreased and older willow coppices had been removed, but new willow coppices had developed in areas recently silted up

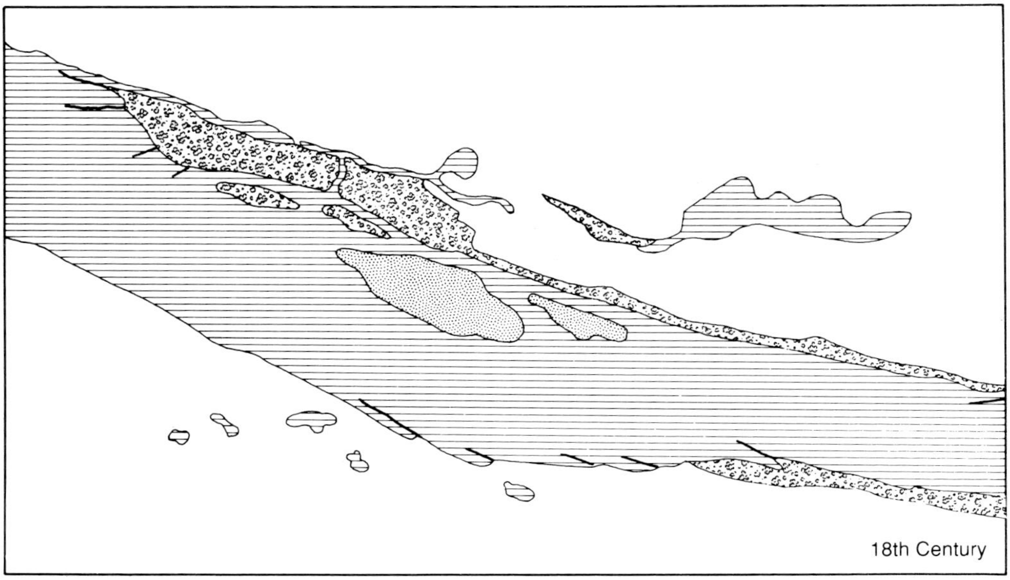

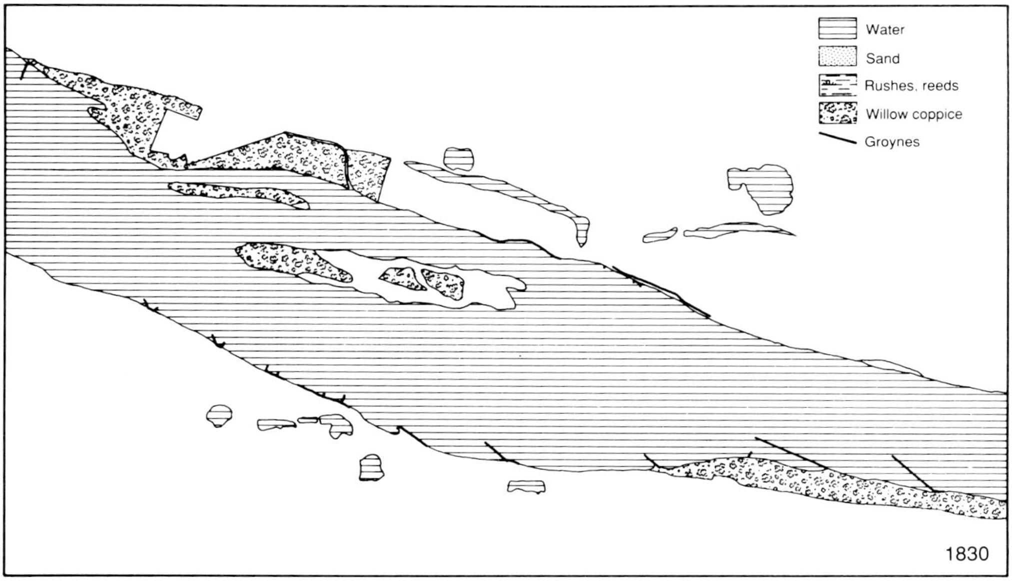

FIGURE 4. The Waal near river km 892–894 showing channel changes since the 18th century. By 1830 (bottom left) the sand bars had moved and the current had been diverted to the north bank (= right bank) by long groynes. By 1870 (top right) the northern channel had become the main channel. The large sand bar (called the 'Loenense Wel') had moved towards the south bank (= left bank). The smaller sand bars have disappeared for the greater part; a bank revetment was constructed on

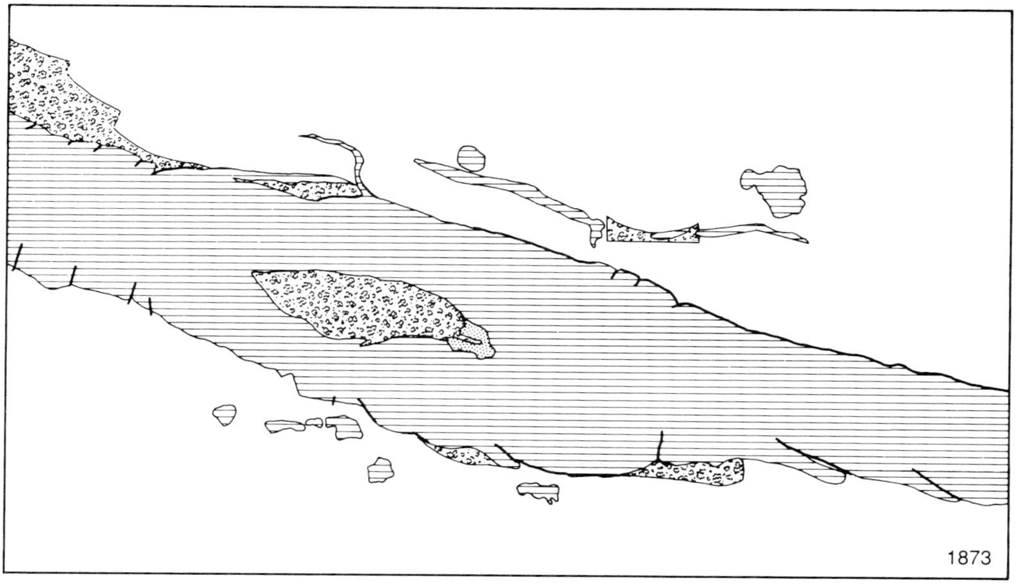

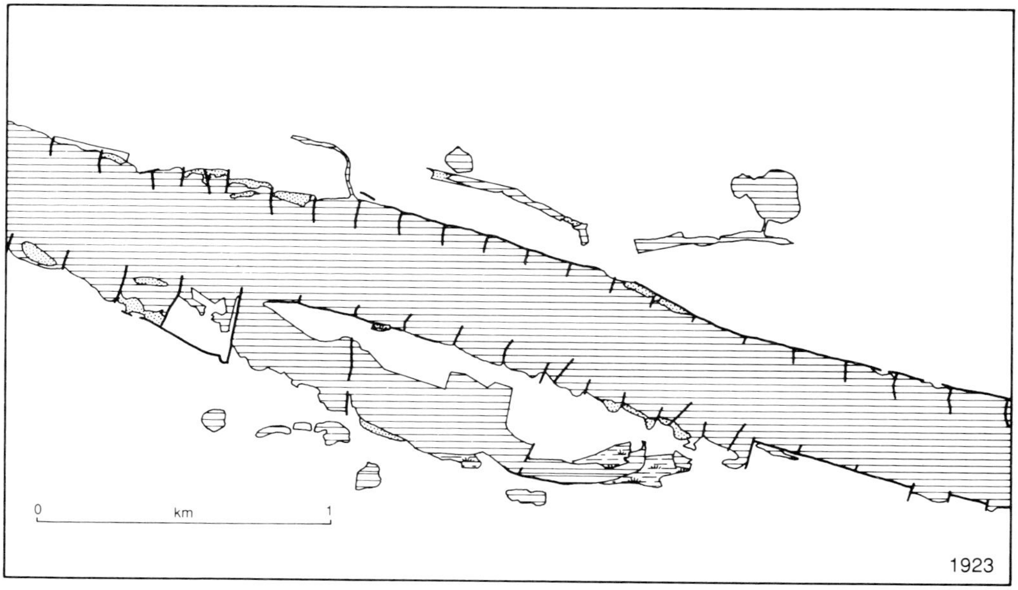

FIGURE 4. (*continued*) the north bank. By 1920 (bottom right) the 'Loenense Wel' had been excavated and the sand was used to dam the southern channel upstream of the 'Loenense Wel'. Channel constriction has been completed and some new, longer, groynes were already constructed in the second phase of channel constriction to a final width of about 260 m

TABLE 2. Water level lowering (m) in the German reach of the Lower Rhine (data from Eschweiler, 1952)

River (km)	Locality	Period 1820 1850	1850 1875	1875 1900	1900 1925	1925 1950
688	Cologne	−0.09	0.0	0.27	0.37	−0.01
744	Dusseldorf	0.0	−0.04	0.30	0.66	0.30
781	Ruhrort	0.0	0.26	0.49	0.70	0.78
814	Wesel	0.55	0.23	0.12	0.79	0.79
837	Rees	0.35	0.19	0.13	0.31	0.83
852	Emmerich	0.28	0.24	0.12	0.15	0.33

Morphological changes

The various operations carried out to transform the unregulated River Waal into a single constricted channel are shown in Figures 3 and 4. What the old maps do not show is the major degradation process going on in the river. Galay (1983) reviews the various causes of progressive bed degradation. The degradation process in the Lower Rhine is a complicated one; the data given by Eschweiler (1952) on the course of the degradation process between Cologne in Germany and the Dutch–German border are summarized in Table 2. Near the entry point of the Lower Rhine in the Netherlands, bed degradation occurred between 1930 and 1960; the bed lowering in this period was about 1.0 m. Since 1960, the river bed has, once again, achieved an equilibrium depth (CHR, 1978). The causes of bed degradation in the Lower Rhine are: the shortening of river bends (see Table 1), the dredging activities to maintain a navigable depth, the mining activities underneath the river bed, regulation of tributaries and channel constriction itself.

The depth contours on the 19th century river maps are insufficiently accurate to show but the most conspicuous bed forms. In the Waal, these were downstream moving sand bars which were removed at channel constriction. Today there is still considerable bed material transport and 'dune' bed forms of about 1 m high and some 10 m long can be found. Near the banks, the bed configuration is largely determined by the 19th century groynes. The flow pattern in groyne-fields is dominated by eddies causing localized bed scour. When large vessels, especially push tows, pass by, the flow pattern around the groynes may change and the flow near the head of the groyne may even be reversed. Furthermore, the waves produced by large vessels cause considerable erosion of the banks locally and in many places, the banks had to be protected by stones. Today, the bed material in a groyne-field usually consists of well-sorted medium sand with about 70 per cent, by weight, between 250 μm and 500 μm. At sites which are more exposed to the current, the bed material is less well sorted and an armour layer of gravel may be found. At these sites, over 25 per cent of the sediments are coarser than 2.0 mm and about 57 per cent is between 125 μm and 500 μm. The amount of clay and silt in the river bed is insignificant in this stretch of the River Waal; the sediments contain no coarse organic material nor any significant amounts of fine detritus.

CHANGES IN THE SUSPENDED LOAD AND SEDIMENTATION PATTERNS

Although a significant bed load of sand is transported down the Lower Rhine, the suspended load is more important. Changes in the bed load are difficult to assess, because of the

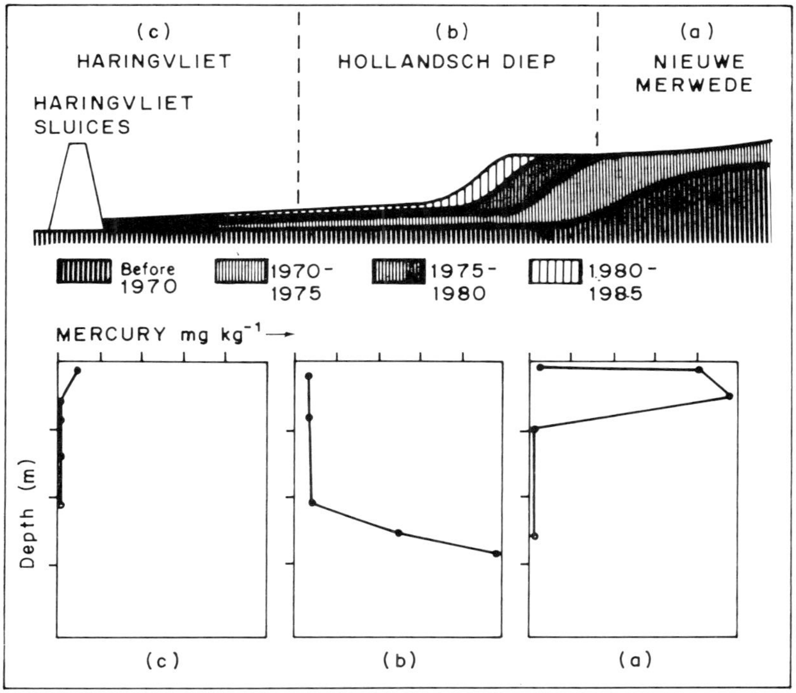

FIGURE 5. Sedimentation pattern in the former estuary of the Rhine. The sedimentation zone moved seaward in a delta-like manner. Concentration profiles of mercury in sediment cores from different sections of the sedimentation area are also shown (data from Rijkswaterstaat, Directie Benedenrivieren, 1987)

inaccuracy of measurement. Nevertheless, between 1870–1881 and 1970–1981 average annual suspended loads have been reduced from 4.2×10^9 kg yr^{-1} to 3.4×10^9 kg yr^{-1} (Santema, 1952; Dijkzeul, 1982). In the 19th century measurements, water samples were filtered over asbestos, so if a systematic analytical difference is suspected, one would expect today's methods to yield higher values. The difference here has been attributed to the canalization of the most important tributaries (Santema, 1952).

Sedimentation of the suspended material only takes place in the river's lowest reaches. In the past, some fine material was deposited in the groyne-fields too, but owing to increased navigation this is no longer so. Before 1970 much of the suspended material was trapped by the vegetation on the sand flats in the freshwater tidal area and made possible the eventual reclamation of polders (Van Urk, 1984). Figure 5 shows the sedimentation patterns after the closure of the Haringvliet in 1970, which has considerably reduced the tidal currents. Sedimentation concentrates in the deepest parts of the main channel and the deposition rate has been as high as 0.5 m per year as the delta advances (Rijkswaterstaat, Directie Benedenrivieren, 1987).

The Lower Rhine is seriously polluted by heavy metals and organic micropollutants, which for the greater part adsorb onto the suspended matter. In recent years, reductions of the loads of various heavy metals have been achieved (Dijkzeul, 1982), but the Rhine sediments still contain large quantities of pollutants carried by the river in previous years. Their concentration profiles in sediment cores can be quite different. Sediments deposited between 1970 and 1975 still contain very high concentrations. Toxic metals are ecologically 'available' in the near surface sediments towards the head of the delta but elsewhere the sediment deposited around 1970 is covered by top layers of relatively clean inorganic silt (Figure 5b; Rijkswaterstaat, Directie Benedenrivieren, 1987).

The amount of suspended material deposited on the floodplain is only about 5–10 per cent of the total suspended load. The concentrations of pollutants in the floodplain soils are related to flood frequencies: sites flooded only at very high discharges contain lower concentrations of heavy metals, for example, than sites already flooded at medium-high discharges (Institute of Soil Fertility, preliminary report); this results from the different composition of the suspended matter at various river discharges (Salomons and Förstner, 1984).

ECOLOGICAL CONSEQUENCES OF MORPHOLOGICAL CHANGE

Floodplain vegetation

The vegetation on the floodplain is determined, in the first place, by land use. Until the 18th century cultivation of the floodplain, particularly areas that were difficult to access, was minimal so that locally, notably along the IJssel, populations of beavers were still present (Van Wijngaarden, 1966). In Figure 1a it can be seen that until the middle of the 19th century land between the main channel and abandoned channels was forested. Other non-cultivated areas were those which were frequently flooded. In Figure 3 and 4 it can be seen that in the 19th century the lower ground along the Waal was covered by marsh vegetation and forest, probably willows, the only type of tree to withstand frequent flooding (Doing, 1962; Kop, 1961).

The fixation of the channels facilitated the cultivation of the floodplain. Initially areas of floodplain were enclosed by low dykes to prevent summer flooding (so-called 'summer dykes') to create meadows or hayfields.

Forests were removed also because they were considered obstacles at peak discharges; trees and hedges increase the hydraulic roughness of the floodplain (Klaassen and Van der Zwaard, 1974). Works to improve flow had already been carried out near the Lek around 1750 and near the IJssel around 1800 (Bardet, 1987). Particularly along the Nederrijn/Lek it was necessary to keep a tow-path free from obstacles, as this was the main trade route from Amsterdam to Cologne (Leemans, 1981). By the middle of the 19th century, the floodplains of the Waal were the least cultivated compared with those along the other branches (Kaartboek van Gelderland 1843), but since 1850, the forested area has been reduced by approximately 75 per cent (Jongman and Leemans, 1982). The vegetation on the floodplain must have been influenced also by the change in water balance. In addition to the summer dykes, cultivated areas were usually drained by ditches.

Figure 6 shows that present water levels in the river itself are lower than 19th century levels, partly owing to bed degradation and partly to varying discharge characteristics. As marsh vegetation is particularly sensitive to water level changes, this may be an additional cause of the disappearance of some plants such as reed and rushes. Only in the lowest

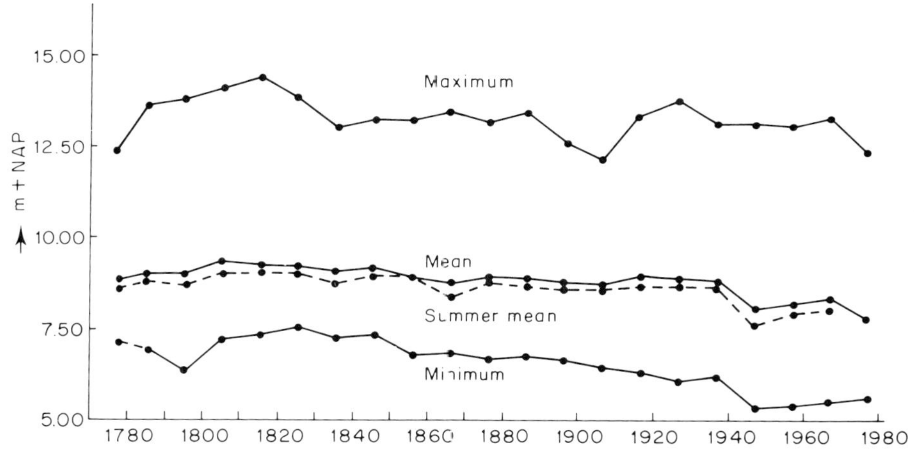

FIGURE 6. Ten-year maximum, mean and minimum water levels at the gauge at Nijmegen for the period 1780–1970

river reaches, a small zone of emergent vegetation is still found along the banks. Due to bank erosion, the banks of the Waal are, for the greater part, barren.

Benthic communities in the main channel

It is clear that river engineering works have had a marked influence on the characteristics of the mineral substrates in the river. However, the supply of organic matter is a key factor in the development of benthic communities in running waters. The 'River Continuum Concept' (Vannote *et al.*, 1980) suggests that benthic communities in large rivers should thrive mainly on phytoplankton and fine detritus from upstream. Sedell and Froggatt (1984) and Triska (1984), however, stress the importance of river/floodplain interactions; they argue that wood debris and other organic matter from the floodplain are essential for habitat diversity.

In the Lower Rhine, the amount of macrophyte debris, either from terrestrial or aquatic sources, is insignificant at the present time. In the IJssel, the amount transported is well below 1 per cent of the total organic material in the suspended matter. This situation must have been different once given the presence of such macrophyte debris in all types of older deposits in the Nieuwe Merwede/Hollandsch Diep area where large numbers of sediment cores have been investigated (see Figure 5). Coarse organic material is a scarce food source for benthic invertebrate communities in the river today. As the gravel and sand substrates contain little fine organic matter, the only available food source for non filter-feeding animals inhabiting such substrates would be benthic algal production. The fairly uniform depth of about 3 m at low water levels prevents light from reaching the bottom, except near the banks, the light extinction coefficient being about $2.0\,m^{-1}$ at suspended matter concentrations of $40\,mg\,l^{-1}$. But near the banks the forces exerted by the wash of the waves are exceptionally strong and together with water level fluctuations make the river a severely stressed habitat. Thus, it is difficult to envisage the development of a diverse benthic community in the sand and gravel substrates of the Lower Rhine. The fauna in sand substrates of the Waal is extremely poor, consisting mainly of the chironomid *Polypedilum breviantennatum* (syn *Tripodura scalaenum*) (Kerkum and Van Urk, in preparation). The scarce fauna in the sand and gravel substrates in the Waal is in strong contrast to the rich insect fauna in small alluvial streams which have roughly the same types of mineral substrates but a much greater variety of coarse and fine organic material.

Remarkably, burrowing filter feeding animals like the mayfly larva *Palingenia longicauda* have also disappeared from the Lower Rhine. These species seem to prefer river stretches where eroding clay deposits in newly formed channels provide a stable substrate and this type of habitat is no longer available. Moreover, their long life cycle of two to three years make these species particularly vulnerable to pollution.

In the sedimentation zone, the presence of heavy metals and other pollutants in the sediments may be expected to have an impact on the bottom fauna. *Chironomus* larvae from these areas show high incidences of deformities of the mouth parts and delayed development (Van Urk and Kerkum, 1987).

Between the shifting sand and gravel habitats and the sedimentation zone, there is narrow transitional zone where fine organic matter is deposited temporarily at low discharges. In the transitional zone of the IJssel for instance, the bottom fauna is becoming more diverse due to improvements in water and sediment quality in recent years. The same applies to filter feeding animals on the stones of the groynes (Van Urk, 1984). In an unregulated channel with a great variety of flow velocities, habitats characterized by the temporary deposition of organic material occupy a much greater area than they do at present.

CONCLUSION

In the Lower Rhine channel fixation and constriction during the 19th century facilitated cultivation of the floodplain, thus allowing for the removal of the remnants of the natural vegetation. It may be assumed that this caused a drastic reduction of supply of organic matter from the floodplain to the main river. Instead of an alternation of transitional and depositional habitats over relatively short distances, today there are long stretches of sand and gravel habitats and a clearly defined sedimentation zone with only a narrow zone of transition. Physical constraints and a shortage of food supply limit benthic communities in the sand and gravel substrates, whereas pollutants have strongly impacted the benthos of the sedimentation zone. Water quality improvement is reflected in a more diverse fauna on the stones of the groynes and in the transition zone, but even if the pollution load of the Rhine was to decrease to negligible amounts, the existing channel morphology would prevent the restoration of a diverse river ecosystem. For this purpose, morphological adaptations will be required as well as recovery of floodplain functions. Plans in this direction are being developed, such as the 'Plan Ooievaar' for the restoration of floodplain forests. A balance needs to be found between safety demands for the dykes, navigation requirements and ecological management of the river basin.

REFERENCES

Bardet, J. 1987. Civiele techniek. In *De Physique Existentie dezes Lands, Jan Blanken (1755–1838)*. Ed. Rijksmuseum Amsterdam, 112–140.

Boode, M. F. 1979. *150 Jaar rivierkaarten van Nederland*, Meetkundige Dienst, Delft, Publicatie nr 30, 52 pp.

Brunnacker, K. 1978. Der Niederrhein im Holozan. *Fortschritte der Geologie Rheinland und Westfalen*, **28**, 399–440.

Commission de Hydrologie du Rhin 1978. *Le bassin du Rhin*, Text, tables, maps, ś-Gravenhage.

De Gaay, A. C. and Blokland, P. 1970. The canalization of the Lower Rhine, *Rijkswaterstaat Communications*, **10**, 24 pp.

Dijkzeul, A. H. J. 1982. De waterkwaliteit van de Rijn in Nederland in de periode 1970–1981. *RIZA nota nr*, **82–061**, Lelystad, 112 pp.

Doing, H. 1962. Systematische Ordnung und floristische Zusammensetzung niederlandischer Wald- und Gebuschgesellschaften. *Wentia*, **8**, 1–85.

Eschweiler, W. 1952. Die Wasserspiegel- und Sohlensenkung am Niederrhein. *Besondere Mitteilungen zum Deutschen Gewasserkundlichen Jahrbuch*, No. 7, Bundesanstalt für Gewasserkunde, Koblenz, 69 pp.

Galay, V. J. 1983. Causes of river bed degradation. *Water Resources Research*, **19**, 1057–1090.

Jansen, P. Ph., Van Bendegom, L., Van den Berg, J., De Vries, M. and Zanen, A. 1979. *Principles of River Engineering. The Non-tidal Alluvial River*, Pitman, London, 509 pp.

Jongman, R. H. G. and Leemans, J. A. A. M. 1982. *Vegetatie-onderzoek Gelderse Uiterwaarden*, Rapport Dienst Landinrichting en Landbouw van Gelderland, 99 pp + App.

Kaartboek van Gelderland 1843. Facsimile reprint 1971. Canaletto, Alphen aan de Rijn, 28 pp.

Klaassen, G. J. and Van der Zwaard, J. 1974. Roughness coefficients of vegetated floodplains. *Journal of Hydraulic Research*, **12**, 43–63.

Koeman, C. 1983. *Geschiedenis van de kartografie van Nederland*. Canaletto, Alphen aan de Rijn, 274 pp.

Kop, L. G. 1961. Walder und Waldentwicklung in alten Fluszbetten in den Niederlanden. *Wentia*, **5**, 86–111.

Leemans, W. F. 1981. De grote Gelderse tollen en de tollenaars. *Gelderse Historische Reeks*, **XIV**, Walburg Pers, Zutphen, 127 pp.

Rijkswaterstaat, Directie Benedenrivieren, 1987. *De waterbodem van het Noordelijk Deltabekken*, Dordrecht, 29 pp.

Salomons, W. and Förstner, U. 1984. *Metals in the Hydrocycle*, Springer-Verlag, Berlin, 349 pp.

Santema, P. 1952. *Beschouwingen over het slibtransport van de Bovenrijn en zijn vertakkingen*, Rijkswaterstaat, Directie Benedenrivieren, rapport, 17 pp + App.

Sedell, J. R. and Froggatt, J. L. 1984. Importance of streamside forests to large rivers: the isolation of the Willamette River, Oregon, USA, from its floodplain by snagging and streamside forest removal. *Verhandlungen internationaler Verein für theoretischen und angewandten Limnologie*, **22**, 1828–1834.

Smit, H. and Van Urk, G. 1987. Het herstel van de ecologische waarden van de Rijn: over de zalm en ecologische doelstellingen. *H20*, **20**, 427–430.

Starkel, L. 1983. The reflection of hydrologic changes in the fluvial environment of the temperate zone during the last 15,000 years. In Gregory, K. J. (Ed.), *Background to Palaeohydrology*, John Wiley and Sons, Chichester, 213–235.

Triska, F. J. 1984. Rol of wood debris in modifying channel geomorphology and riparian areas of a large lowland river under pristine conditions. A historical case study. *Verhandlungen internationaler Verein für theoretischen und angewandten Limnologie*, **22**, 1876–1892.

Vannote, R. L., Minshall, G. W., Cummins, K. W., Sedell, J. R. and Cushing, R. E. 1980. The river continuum concept. *Canadian Journal of Fisheries and Aquatic Sciences*, **37**, 130–137.

Van Dalen, A. G. 1972. *Rondom het Tolhuys aan Rijn en Waal*, Walburg Pers, Zutphen, 215 pp.

Van der Schrier, D. M. 1978. Langs IJssel, Vecht en Reest. In *Rivieren en beken in Overijssel*, Waanders, Zwolle, 13–33.

Van der Woude, J. D. 1981. *Holocene palaeoenvironmental evolution of a perimarine fluviatile area*, Thesis, Vrije Universiteit Amsterdam, 112 pp.

Van de Ven, G. P. 1976. Aan de Wieg van Rijkswaterstaat. Wordingsgeschiedenis van het Pannerdensch Kanaal. *Gelderse Historische Reeks*, **VIII**, Walburg Pers, Zutphen, 440 pp.

Van de Ven, G. P. 1987. Jan Blanken en de Waterstaat. In *de physique Existentie dezes lands, Jan Blanken (1755–1838)*. Ed. Rijksmuseum Amsterdam, 59–93.
Van Heiningen, H. 1971. *Tussen Maas en Waal*, Walburg Pers, Zutphen, 669 pp.
Van Petersen, J. W. 1978. *De waterplaag*, Walburg Pers, Zutphen, 160 pp.
Van Schaik, P. 1984. *Christiaan Brunings 1736–1805. Waterstaat in opkomst*, Walburg Pers, Zutphen, 88 pp.
Van Urk, G. 1984. Lower Rhine-Meuse. In Whitton, B. A. (Ed.), *Ecology of European Rivers*, Blackwell, Oxford, 437–468.
Van Urk, G. and Kerkum, F. C. M. 1987. Chironomid mortality after the Sandoz accident and deformities in Chironomus larvae due to sediment pollution in the Rhine. *Aqua*, **4**, 191–196.
Van Wijngaarden, A. 1966. De bever, Castor fiber L., in Nederland, *Lutra*, **8**, 33–52.

Historical Change of Large Alluvial Rivers: Western Europe
Edited by G. E. Petts
©1989 John Wiley & Sons Ltd

11

The Lower Rhine: Palaeoecological Analysis

ALEXANDER KLINK

Klink Hydrobiological Consultants, Hydrobiologisch Adviesburo, Boterstraat 28, 6701 CW Wageningen, The Netherlands

INTRODUCTION

This chapter reviews the first few years of palaeoecological river research, by means of insect remains, in the Netherlands. In 1983 the first floodplain samples from the River Rhine were analysed and since then, progressively, we have been building up biological evidence concerning the history of the river. As is symptomatic of any new development in science, numerous questions arise from the analysed material and very few answers can be given. Nevertheless, it has become clear that palaeoecological river research has considerable scope.

HISTORICAL DATA ON THE INSECT FAUNA OF THE RHINE

Data on the insects inhabiting the Rhine in the past are very scarce. From the most comprehensive record of riverine insects (Albarda, 1889) it is apparent that the characteristic riverine Ephemeroptera, Plecoptera, Odonata and Trichoptera were still present in the 19th century. The best example of the Ephemeroptera is *Palingenia longicauda*, which burrowed into the clay banks in large numbers: it was last recorded in 1907 (Mol, 1985). The Plecoptera in the Rhine included the genera Oemopteryx, Isoperla and Chloroperla. These stoneflies became extinct in the early decades of this century (Geijskes, 1940). *Gomphus flavipes*, a mud-dwelling dragonfly, was last collected in 1902 (Geijskes and Van Tol, 1983). The Trichoptera included, among others, *Cheumatopsyche lepida*, *Lepidostoma hirtum*, *Psychomyia pusilla* and many *Hydropsyche* species. The first two species have not been collected during the 20th century. The last record of *Psychomyia pusilla* dates back to 1933 (Fischer, 1934). Very little is known about the Diptera. Van Der Wulp (1877) notes that adult Simuliidae (only four species described) were very common on places near the rivers Rhine and Meuse. The chironomids mentioned in his work cannot be assigned to the typical river fauna.

From the early records it can be concluded that in the second half of the 19th century and the first few decades of the 20th century a dramatic decline was observed in the larger insects. The river Rhine today is deprived of Ephemeroptera (except *Caenis* species), Plecoptera, Anisoptera and Simuliidae. Apparently Trichoptera have been absent from the Rhine for a long period of time, too. From 1978 on, however, a remarkable return of *Hydropsyche contubernalis* and some other caddisflies has been observed (Van Urk, 1984).

CHEMICAL CHANGES IN THE RHINE

Evidence that early pollution of the Dutch rivers did not go unnoticed can be found in the following quotation from Velsen (1768):

> 'Our people have always been utterly indifferent to a vital part of this country, the rivers. We used them and abused them as we pleased, without any supervision whatsoever. If the rivers were to die tomorrow not a soul would care or be grieved. Instead of worshipping her as gods or saints, we treat them as sewers, in which anyone may freely dispose of his wastes. This is what has caused their abominable state.'

Most probably the pollution referred to above was predominantly organic. Large-scale mining activities did not start until the middle of the 19th century, and it was not until the 20th century that steel production became a major industry and pesticides based on mercury and arsenic were commonly used. In Figure 1 the history of heavy metal pollution is inferred from analyses of old river sediments and soil samples from a branch of the Lower Rhine. Organic pollution in former times has been deduced from the population

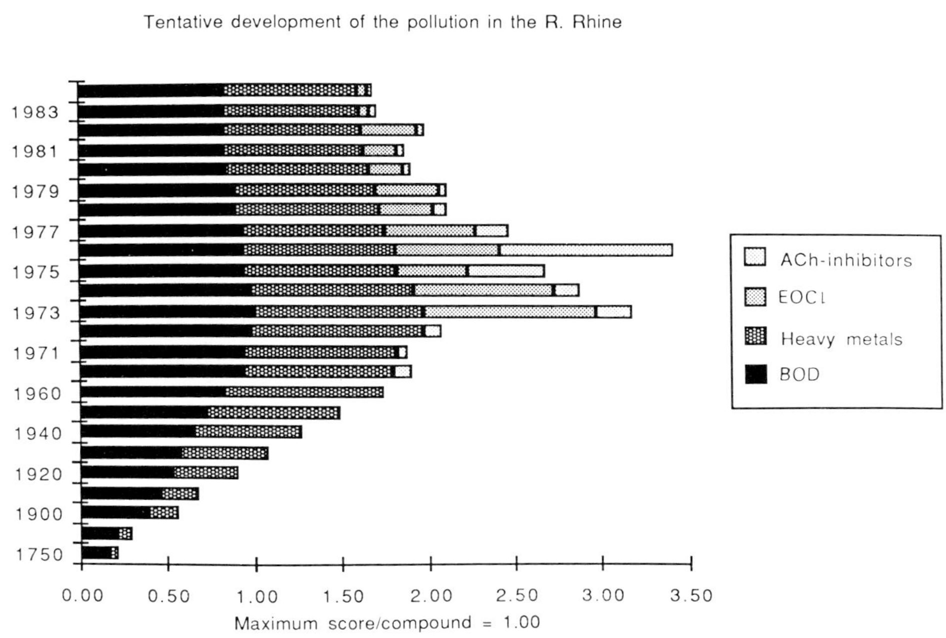

FIGURE 1. Tentative compilation of the pollution history of the Rhine derived from direct and indirect evidence

statistics of the Netherlands. The first decades of the 20th century were characterized by the onset of a steep rise in organic and heavy metal pollution in the Rhine.

Since the Second World War an entirely new branch of industry has added its own characteristic compounds to the environment. With the development of the petrochemical industry the family of chlorinated hydrocarbons (DDT, PCBs, drins and dioxins) have been released into the river. In the 1960s and 1970s acetylcholinesterase inhibitors were used as insecticides, resulting in concentrations in the Rhine of up to 50 $\mu g\,l^{-1}$ (Greve, 1980). The initial rise in chlorinated hydrocarbons (as EOCl) is not known as no analyses were made until 1973. In the mid nineteen seventies, however, concentrations as high as

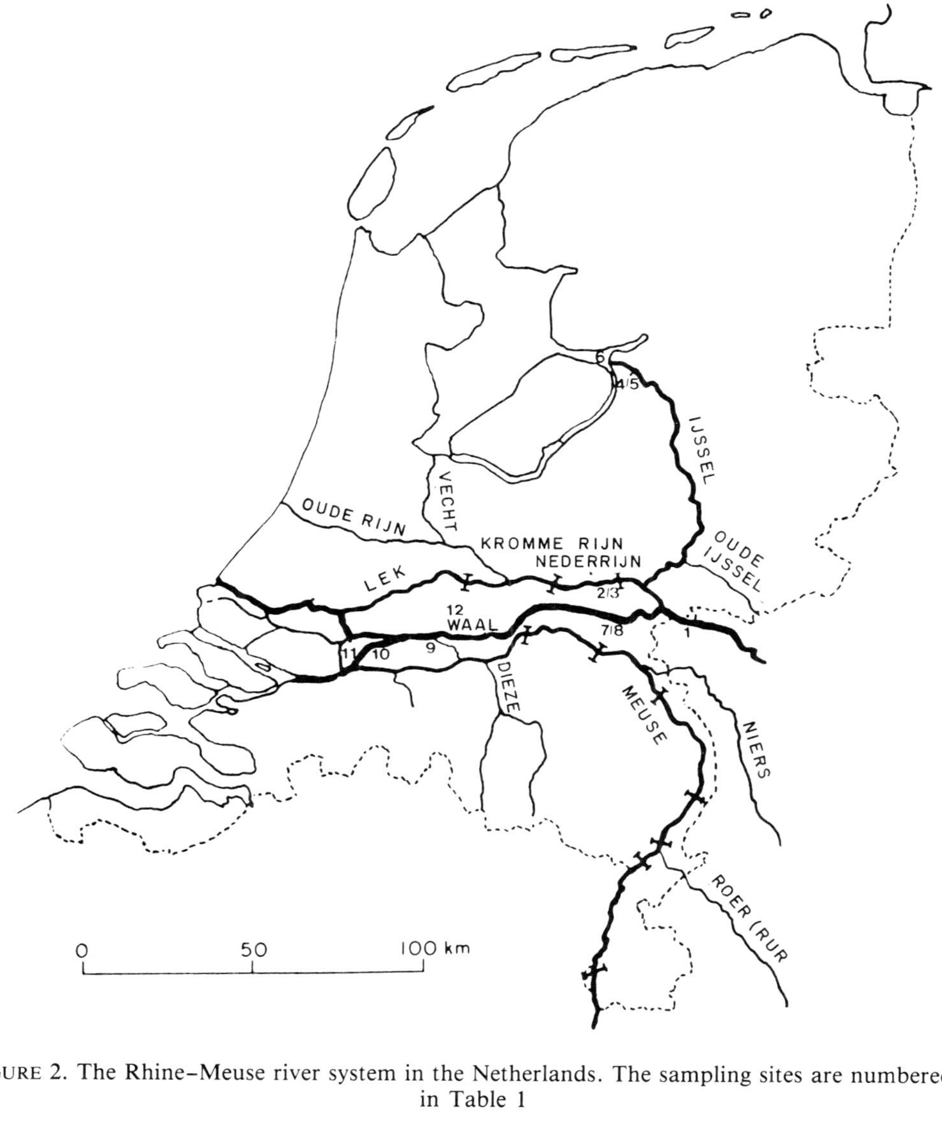

FIGURE 2. The Rhine–Meuse river system in the Netherlands. The sampling sites are numbered as in Table 1

TABLE 1. Sampling sites by type and date of sedimentation/sampling

Number	River	Type	Date
1	Rhine	Floodplain deposit	-1745
2	Nederrijn	Floodplain deposit	-1880
3	Nederrijn	Exuviae sampling	1981
4	IJssel	Sediment core	?
5	IJssel	Exuviae sampling	1986
6	IJssel	Sediment core	?
7	Waal	Exuviae sampling	1985
8	Waal	Suspended solids ($\geq 100\ \mu$m)	-1984
9	Boven Merwede	Floodplain deposit	1914
10	Nieuwe Merwede 9	Sediment core	?
11	Nieuwe Merwede 12	Sediment core	?
12	Schoonrewoerdse Wiel	Sediment core	1750–1984

$30\ \mu g\ l^{-1}$ were measured (Greve, 1980). Concurrently, the pollution of the Rhine reached its maximum. Subsequently, improvements in oxygen content and a decline in micropollutants have been observed. The return of caddisflies in the Rhine, notably *H. contubernalis* could be the result of this improvement (Becker, 1987).

Localities and methods

The sampling sites used in this survey of the Lower Rhine are shown in Figure 2 and listed in Table 1.

The sediment samples were collected in three different ways. First, sediment cores from the river-bed were taken by means of a mudcorer installed on a research vessel. The core from Schoonrewoerdse Wiel (nr. 12) was taken by means of scuba gear. Secondly, floodplain deposits were collected with hand auger equipment for soil research. The augers were drilled into the soil until a layer of coarse sand was met, indicating the former river-bed. Since these layers contain hardly any remains of invertebrates, the samples were taken from the silt layer directly on top of the sand. Thirdly, suspended solids were collected from the river itself by means of a driftnet (mesh $100\ \mu$m). To compare these data with the present situation, monthly exuviae collections were carried out at three sampling stations from May to October by means of a driftnet (mesh $500\ \mu$m).

The deposits were sieved over a mesh of $100\ \mu$m and the macroinvertebrate remains were handpicked under a magnification of 40–80 and preserved in ethanol 70 per cent. Only five insect orders were found in sufficient numbers for sample comparison. Distribution of aquatic Coleoptera remains was too local to take them into consideration, while no Odonata remains were found. Of the non-insect groups, mollusc shells appeared to be absent in all but a few samples, while Tricladida, Oligochaeta and Hirudinea remains were not found at all. The Tricladida and Hirudinea contain no chitinous parts that stay intact in the sediment. The chitinous parts of the Oligochaeta (chaetae and cuticular penis sheaths in some Tubificidae) must have been lost in sieving or overlooked during sorting with this magnification. Curiously no remains of Isopoda or Amphipoda were recognized.

The following groups are considered and identified according to the selected parts: Ephemeroptera (mandibles); Plecoptera (maxilla); Heteroptera (tergites); Trichoptera (frontoclypeus); and Diptera (headcapsule).

The remains of the insects were identified by means of the available literature on the various groups. As single references rarely provided complete identification by means of the selected parts a wide variety of special articles was consulted. The remains that could not be identified in this way were compared to the material in reference collections. If this was not successful, the taxa were drawn and given a tentative name (for example Chironomini gen. no. 1).

Abandoned main channels were selected for the dating of the floodplain deposits. As old river maps show, the filling of these channels is a process that may take no more than a few decades. The layer directly above the coarse sand (assumed to represent the former channel bed) was taken for analyses of the remains as described and dated according to the period that the river abandoned that particular course. The estimated age of the sediment remains is the youngest possible age because the sediment contains an unknown fraction of older resuspended deposits. The inferred age of the sediment, therefore, is an underestimation of the average age of the sediment. The sediment cores from the present river-bed have not been dated, because no old river maps could be applied to the coring sites, and no sequential isotope decay series has been found or is expected to be found in the fluctuating sedimentation and resuspension environment that take place in rivers.

The identified remains were treated as units. This means that a mandible of a mayfly was counted as one mayfly. Mathematically this is not correct, since one mandible stands for only one half of one of the 25 instars that mayflies usually have. The one part = one specimen approach was chosen instead of the mathematically correct alternative because if not the numbers of hemimetabolic insects would have been too low to differentiate in statistical analyses. The homometabolic insect remains were treated similarly except the Chironomidae in which especially half headcapsules of the Orthocladiinae were found, which were counted as half specimens. The exuviae collections were summarized over the period of sampling (May–October). Finally, the counts in the sediment samples and the exuviae collections were recalculated to a total of 500 specimens and $\ln(x+1)$ transformed for factor and cluster analyses. These analyses were done by the statistical package SYSTAT (Wilkinson, 1986).

RESULTS

A total of approximately 15 000 insect remains covering 167 taxa were found in 52 samples from sediment cores, floodplain deposits and coarse suspended solids, while exuviae collections contained a little over 13 000 specimens comprising 66 taxa. From the statistical analyses it appears that subsamples from the same sediment core are always more closely related to each other than to samples from other cores or floodplain deposits, even in cores of 3 m long. In addition to the sedimentation process in the abandoned channels referred to earlier, the analyses of the insect remains also show that sediment cores from the river-bed may represent only very brief periods of time.

The names of the taxa in Figure 3 represent separate clusters by their most differentiating taxon. The bold-faced characters indicate the status of the clusters according to historical and recent documentation (e.g. the *Nanocladius* cluster includes species that have invaded the rivers). The distance from the clusters to the origin is a relative scale of abundance.

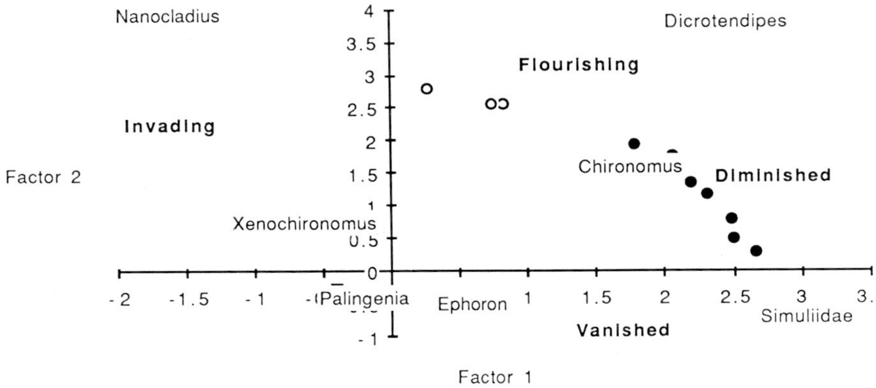

FIGURE 3. Changes of the insect fauna of the Rhine. The open circles are exuviae samples and the solid circles are samples from sediment cores and floodplain deposits

The first quadrant of Figure 3 contains taxa that are abundant in both sediment and exuviae samples. The *Dicrotendipes* cluster stands for *Dicrotendipes nervosus* and the *Cricotopus/Orthocladius* aggregate (not subdivided in the remains). They are under-represented in the sediment samples but are 'flourishing' so to speak in the present Rhine. In the exuviae collections *Cricotopus bicinctus* is by far the most abundant representative of the Cricotopus/Orthocladius aggregate. The Chironomus cluster includes *Chironomus* spp., *Cricotopus* sg. *Isocladius*, *Cryptochironomus* spp., *Glyptotendipes gr. pallens*, *Hydropsyche contubernalis*, *Parachironomus arcuatus*, Pentaneurini (not identifiable to a generic level) and *Polypedilum scalaenum*. This cluster stands for species that are still plentiful in the Rhine but not as abundantly as in the sediment samples. Despite the spectacular return of *Hydropsyche contubernalis* in the Rhine since 1978/1979 (Van Urk, 1984), their relative abundance in the sediments has not been equalled (yet).

The second quadrant consists of two discrete clusters of species which have largely vanished from the Rhine. The Simuliidae cluster includes (besides Simuliidae) *Microtendipes gr. chloris* and *Procladius* spp. Neither Simuliidae and *Microtendipes gr. chloris* were found in exuviae samples at all, while *Procladius* appeared to be extremely rare in the exuviae collections, in contrast to the sediment samples. The *Ephoron virgo* cluster differs from the Simuliidae cluster only in that the 30 taxa involved are less abundant. These taxa are frequently found in the sediment samples but are absent or extremely rare in the exuviae collections. The taxa concerned are mainly riverine Ephemeroptera and Trichoptera (e.g. Ephemera, Ephemerella, Baetis spp. Heptageniidae, Leptophlebiidae and *Brachycentrus subnubilus*, *Cheumatopsyche lepida*, *Hydropsyche pellucidula*, *Lepidostoma hirtum*, *Psychomyia pusilla*), the adults of which were all collected on the river banks in the last century (Albarda, 1889). Noteworthy is the occurrence of *Byssodon maculatum* (Simuliidae). It is uncertain if the species was ever caught alive in the Netherlands since the only encountered record may well refer to *Byssodon maculatum* and *Wilhelmia equina* (Van Der Wulp, 1877). According to Rubsow (1964), *Byssodon maculatum* inhabits large lowland rivers from Western France (River Seine) to the ultimate eastern part of the USSR (River Kolima). Even in the Mississippi this malicious biting blackfly has been collected.

The third quadrant contains only the Palingenia cluster. It contains 98 of the total of 167 taxa. The cluster lies close to the origin, which means that the species involved are not abundant and infrequently met in the dataset. Most of its species are extinct in the Rhine, including *Palingenia longicauda* itself, *Aphelocheirus aestivalis* and *Potamanthus luteus*. Species characteristic for large rivers are some sand-dwelling chironomid taxa (*Beckidia zabolotzkyi*, *Chernovskiia 'macrocera'*, *Paratendipes connectens 3 Lipina*, *Paratendipes intermedius*, *Robackia demeijerei* and *Potthastia gaedii*). They have not been collected alive in the Dutch rivers. Another chironomid (*Symposiocladius lignicola*) has recently been described (Cranston, 1982). The larvae burrow into submerged wood and are thus confined to snag habitats. The Trichoptera in this cluster are *Chimarra marginata*, Goeridae, Hydroptilidae, Micrasema spp., *Odontocerum albicorne*, *Oligoplectrum maculatum* and Rhyacophila sp. These taxa are not characteristic of lowland rivers and predominantly live upstream. It is uncertain if these species have actually ever inhabited the Lower Rhine.

The clusters in the fourth quadrant contain species that rarely occur in the sediment samples and can therefore be characterized as invaders. They all belong to the Chironomidae. The Nanocladius cluster includes Nanocladius spp., *Rheocricotopus chalybeatus*, Rheotanytarsus spp. and *Parachironomus longiforceps*. The Xenochironomus cluster includes the less abundant midges *Synorthocladius semivirens* and *Xenochironomus xenolabis*. Their absence in all but a few (recent?) sediment samples may be due to the physical changes taking place in the Rhine. Two of the most conspicuous changes are the increasing the water temperatures and the changing habitats.

Possible effects of long-term changes in water temperature

Climatic conditions have changed and thermal pollution has increased. This combination has so far caused a rise of 2.5 °C in the average temperature of the Rhine since 1908 (Department of Public Works, 1851–. . .; RIWA (Cooperating Rhine and Meuse Water Supply Companies), 1983). Also, observations of ice formation in the Rhine over the last two centuries (Figure 4) show that there has been a steady increase in the temperature of

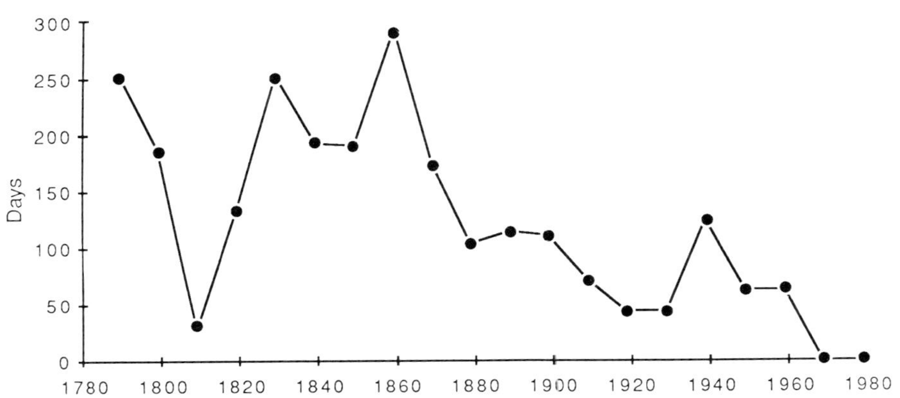

FIGURE 4. Number of days with ice in the Rhine at the German–Dutch border since 1780

the river since the middle of the last century. The documented invasion of the freshwater shrimp (*Atyaephyra desmaresti*: Decapoda) in the Rhine can be considered as biological evidence for rising winter temperatures (Van Den Brink and Van Der Velde, 1986). The species has a southern European distribution and was first collected in the Netherlands in 1915 (Redeke, 1936).

Other species for which the rise in temperature may have been beneficial are the 'invaders' *Rheocricotopus chalybeatus*, *Nanocladius rectinervis* and the *Rheotanytarsus* species. They are very abundant in the Rhine today, and can well add up to 60 per cent of the total insects in the exuviae collections. *Rheocricotopus chalybeatus* has a southern distribution. In Europe, the species does not occur in Scandinavia (Fittkau and Reiss, 1978), but is very common in the River Po in Italy (Rossaro, 1984) and the Llobregat in Spain (Prat *et al.*, 1984). The species has also been collected in Lebanon (Moubayed and Laville, 1983) and Syria (Reiss, 1986). *Nanocladius rectinervis* roughly follows the same distribution pattern as *Rheocricotopus chalybeatus*. In Europe, the genus Rheotanytarsus is restricted to the area south of Scandinavia (Fittkau and Reiss, 1978). The distribution area of *R. photophilus* is not clear. Another species (*R. rhenanus*) has recently been described by Klink (1983), who collected the species not only in Dutch rivers but also in the rivers Meuse in S. Belgium, Lahn (W. Germany), Loire (France) and Tisza (Hungary). Saxl (personal communication) collected material from the River Krems (Austria), and Moubayed (personal communication) caught the species in Lebanese rivers. No collection sites north of the Netherlands are known at the moment. Future research may produce evidence that the expansion of these species to the north is due to the rise in river temperatures. It should be more interesting, though, to trace species showing a reverse trend, which would allow us to monitor the negative effects of thermal pollution. The Rhine is unsuitable for such research, however, because of the other profound changes that have taken place.

Changing habitats in the Rhine and their effect on the insect fauna

The original habitats of the 'natural' River Rhine can be deduced at least to some extent from the insect remains and historical research. The river-bed must have been a gradient passing from coarse sands in the erosive zones to silt in the depositional areas. Stones did not occur in the natural river-bed. Part of the banks may have been covered with vegetation, with dead trees providing snag habitats. The biological importance of the snag habitat has been pointed out by Bencke *et al.* (1984). The historical struggle of man against wood debris in rivers has been described in a case study by Triska (1984).

The changes which have affected the Rhine habitats have been discussed in the previous chapter but in brief the following has happened. The natural forests along the Rhine must have been cleared long ago, but the low-lying parts of the floodplains were partly covered by stands of willow until the late 19th/early 20th century. Jetties consisting of basalt blocks were built from the second half of 19th century. Recently part of the banks are protected by stones to prevent erosion from the increasingly powerful cargo vessels. In consequence, the habitat distribution in the Rhine has changed completely. The vegetation and snag habitats have been replaced by heavy boulders and the gradients in the composition of the river-bed have disappeared because of the jetties.

The overall impact of changed habitats on the insect fauna can be derived from Figure 5. The sediments contain 20–55 per cent of bottom-dwelling insect remains, while

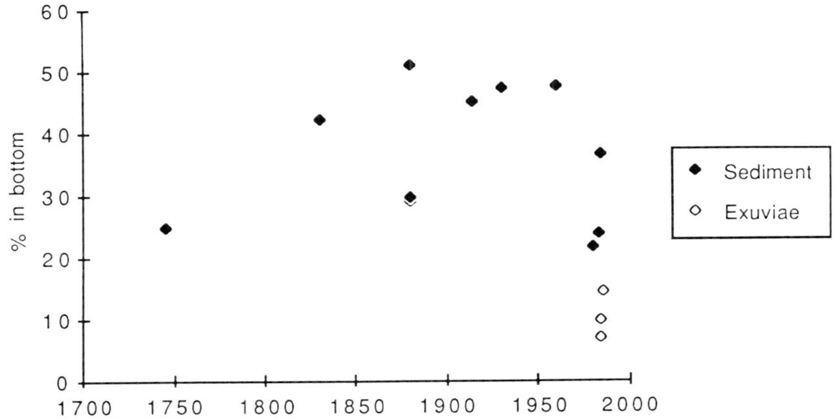

FIGURE 5. Relative abundance of benthic insect fauna in relation to the relative age: sediment (old), exuviae (recent)

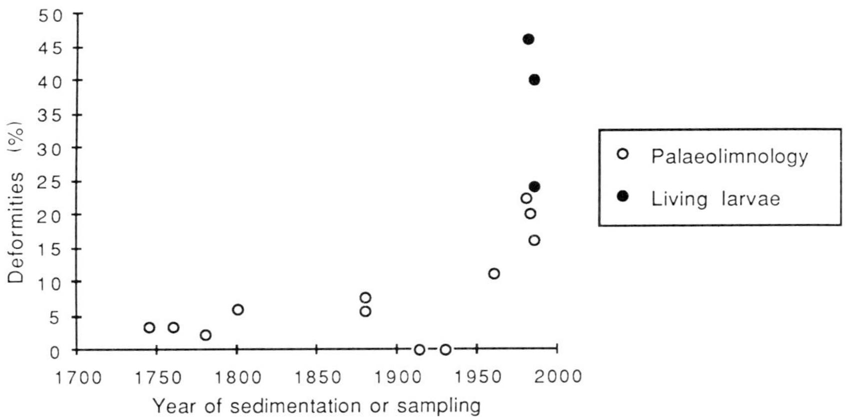

FIGURE 6. The incidence of deformities in the headcapsules of *Chronomus* larvae, in relation to inferred (probably underestimated) sediment age

in the recent situation not more than 15 per cent of the insect fauna lives in the bottom. The 'invader' species *Parachironomus longiforceps* and *Xenochironomus xenolabis* may have benefited from these altered habitats. The former live in colonies of Bryozoa (Ertlova, 1974). The latter inhabit sponges (Pagast, 1934). In the present Rhine both Bryozoa and Porifera are confined to the stones on the banks and jetties. To what extent the natural snag provides a suitable habitat for these colonies is still uncertain. Unlike the supposed thermophilic invaders, the distribution area of these midges and their hosts extends into Scandinavia.

Incidence of deformities in headcapsules of mud-dwelling *Chironomus* larvae

Although marked physical changes have taken place in the river-bed of the Rhine, mud habitats still exist and these are preferred by *Chironomus* larvae. Unfortunately heavy metals

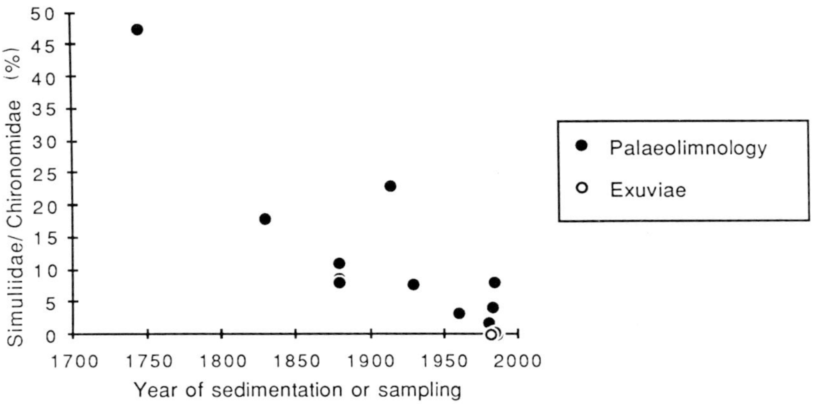

FIGURE 7. Changes in Diptera composition in the Rhine according to sediment samples and recent exuviae collections. Ages are probably underestimates

and other pollutants are accumulating in this sediment. According to Cushman (1984), Hamilton and Saether (1971), Warwick (1985) and Wiederholm (1984), the incidence of deformities in the headcapsules of *Chironomus* larvae is strongly propagated by xenobiotic pollutants. The most suspect compounds are heavy metals and chlorinated hydrocarbons. However, causal relationships between particular pollutants and deformities are not yet known. In the Rhine deformities of up to 40 per cent are not exceptional (Van Urk and Kerkum, 1986).

Comparison of the percentage of deformities in the sediment samples with recently collected living larvae, as shown in Figure 6, reveals a trend in the incidence of head deformities and the (underestimated) age of the sediment. Despite the lack of reliable sediment dating, it is obvious that deformities occurred as early as the 18th century, but that the sharp increase was only generated in this century. The rise in heavy metal concentrations (Figure 1) is strikingly similar to this rise in deformity rates. The relationship has yet to be proven by laboratory research, as concurrently with the rise in heavy metals the levels of other pollutants have been increasing too. Phenol pollution in the early 1900s, for instance, was traced because salmon meat tasted 'carbolic' (Van Drimmelen, 1982).

Changes in the Diptera fauna of the Rhine

One of the most striking discrepancies found between the sediment samples and exuviae collections is the absence of Simuliidae in the present Rhine (see Figure 7). The habitat of Simuliidae is solid substrate in the current, a habitat which is heavily propagated by the presence of jetties. The cause of their extinction is unknown but it could be related to the quality of the suspended solids, their food source. Another factor adversely affecting their existence is fluctuating current velocities. In the case of the Rhine, instantaneous fluctuations are generated by the wave action from boats, especially around the jetties. Future research on the sensitivity of Simuliidae to chemical and physical factors can provide valuable information on the threshold levels of these factors. Such information is crucial if the restoration of the Rhine ecosystem is to be attempted.

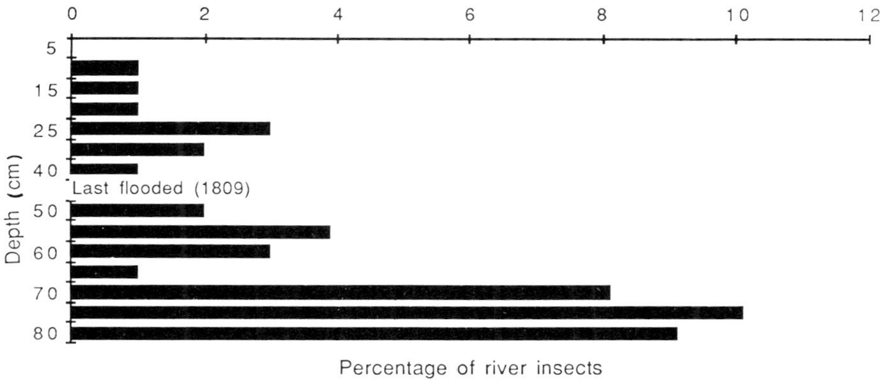

FIGURE 8. The percentage of typical river insects in the sediment core from L. Schoonrewoerdse Wiel. The 35 cm layer was discarded as no subfossil insects were found

FLOODPLAIN STRATIGRAPHY BASED UPON TWO CORES OF DIFFERENT ORIGIN

Description of a sediment core from a lake generated by a dykeburst

In the Netherlands most large rivers were embanked as early as the 13th century (Pons, 1957). When a river bursts through a short stretch of dyke, the result is a large eddypool which creates localized scour and leaves a small lake. In this way many dozens of lakes—called 'Wielen'—were generated by the Rhine. The Schoonrewoerdse Wiel (1573) is one of the larger ones, with an area of approximately 0.25 km^2 and a depth of 8 m. It was flooded several times in the past. It is not known how often this occurred, but in 1672 it was part of the Dutch 'waterlinie', a strip of land inundated as a defence against the army of Louis XIV of France. The Wiel has not flooded again since 1809, although it remained connected to a branch of the R. Rhine by means of a ditch. For the past 30 years it has been completely isolated.

In 1984 a sediment core (length 80 cm) was taken from the deepest part of the Wiel by means of scuba gear. One half of the core was dated by means of Pb-210, and its phytoplankton analysed. The other half was processed for insect analyses. The core covers the period from approximately 1750 to 1984. In Figure 8 the percentage of typical river insects is shown against the depth of the sample (depth 0–5 cm is the top). Despite a sharp decline in the riverine insects (Simuliidae, Heptageniidae and Hydropsychidae, as well as some rheophilic Chironomidae) the indirect influence of the river is still notable, even at a depth of 10 cm (1972). This core reveals that large areas of the western part of the Netherlands evolved by slow vertical accretion of Rhine sediments.

Description of a river deposit in the freshwater tidal belt in the western part of the Netherlands

In 1986 two sediment cores were taken from the Nieuwe Merwede (Figure 2; Table 1). The genesis of this tidal area is discussed by Van Urk (1984) and will only be summarized

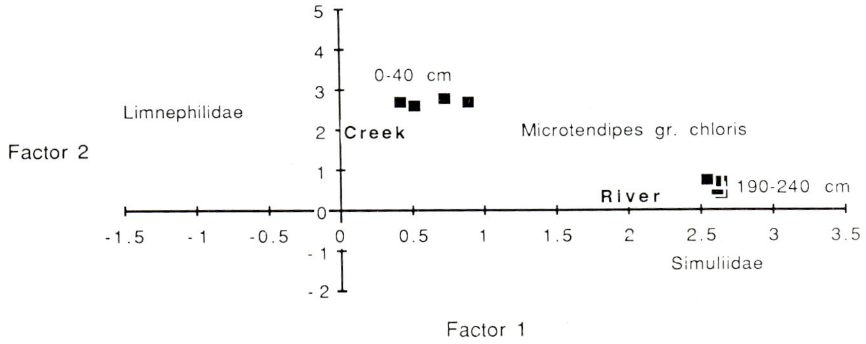

FIGURE 9. Factor analysis of the samples and insects in a sediment core from the Nieuwe Merwede

here. In 1421 there was a disastrous flood in the southwestern part of the Netherlands (the Saint Elizabeth flood). The River Merwede completely washed away a vast area of land. In the following centuries sedimentation progressed and resulted in the formation of a dense system of small and large creeks. A century ago this system was cut through by a newly dug river called the Nieuwe (New) Merwede. The sediment composition in the core is roughly as follows:

0–40 cm: predominantly coarse sand with pieces of wood
40–190 cm: peat and clay in varying quantities
190–240 cm: predominantly fine sand clay

In Figure 9 the insect analyses have been summarized by means of a factor analysis. The most striking feature of the analysis is the clear separation between the upper and lower part of the core. The peat in between hardly contains any insect remains. The Simuliidae cluster represents river-inhabiting taxa and apart from Simuliidae includes *Harnischia* sp., *Potthastia gaedii*, *Stenochironomus* sp., *Hydropsyche contubernalis*, *Polypedilum scalaenum* and a score of less abundant taxa. The Limnephilidae cluster includes *Micropsectra* sp., *Zavrelia pentatoma*, *Glyptotendipes pallens*, *Chaetocladius* spp., *Metriocnemus* with three species, *Bryophaenocladius* sp. and *Limnophyes* sp. The first two representatives in particular can reach high abundances in small slow-flowing streams in the eastern part of the Netherlands. The latter four genera (all Orthocladiinae) are (semi)terrestrial, which may be evidence that the creek was a very shallow one, with parts becoming dry temporarily. The cluster between these two extremes is either formed by taxa which occur in both large rivers and small streams (e.g. *Microtendipes gr. chloris*), or by taxa that cannot properly be identified in the samples (e.g. Pentaneurini).

According to geological studies (Zonneveld, 1960; Verbraeck, 1970; Verbraeck and Bisschops, 1980; Berendsen, 1982) peat formation and river deposits alternate in Holocene sediments in the western part of the Netherlands. The bulk of peat formed in the subboreal (2000–5000 BP) as part of the so-called Hollnd peat, which has a widespread lateral extension throughout the western part of the country. The age of the river deposit at the bottom of the core has not been established, but it is probably several thousands of years. (the peat has an age of 5140 ± 40 BP). The insects in this deposit may be considered as the inhabitants of the natural Rhine or Meuse. As pointed out above, in the natural Rhine the number of insects that inhabited the bottom compares to those that inhabited the solid substrate, which also is the case in these deposits with an even distribution over both types of habitats (45 per cent each). If its estimated age is correct, the natural swamp forests near the river must have consisted mainly of Alnus. On the lower parts, willow coppice must have grown. The higher parts were covered with Quercus, Ulmus, Fraxinus and Corylus (Zonneveld, 1960; Van Der Woude, 1981).

The tidal creek deposit at the top of the core is of unknown age but could date from after the Saint Elisabeth flood (1421) when the numerous smaller and larger tidal creeks were formed. This deposit reveals that these creeks used to be inhabited by caddisflies (Limnephilidae). This is an important fact, as Van Urk (1984) found a distinct absence of Trichoptera in these tidal creeks. From this evidence, it may be concluded that the absence of Trichoptera must be entirely due to river pollution, as the morphology of the creeks is still quite natural.

CONCLUSION

In this palaeoecological study the lack of a reliable dating method of river sediment cores was a major problem. For the time being this problem can be partly overcome by collecting floodplain deposits, with minimum ages determined from old maps. Palaeoecology in large rivers appears to be a powerful instrument, however, in generating the species composition of former insect faunas and in revealing the differences between a former and the present situation.

As a consequence of large-scale chemical and physical changes which have taken place in the Rhine, the insect fauna of the river has changed completely. Eighty to one hundred insect taxa have disappeared from the river over the last few centuries (see appendix). The changes in habitats documented in the literature are reflected by these changes in the insect fauna. Three important changes of individual taxa are:

— the propagation of species with a southern distribution possibly caused by the long-term rise in water temperature;
— a positive relationship between the deformities of headcapsules and the heavy metal content of the river sediment; and
— the decline and extinction of Simuliidae — once abundant — over the last few centuries.

This study can be considered as the initial phase of palaeolimnological river research in the Netherlands. The changes observed in the Rhine by means of sediment analyses still stand fairly isolated, since no direct links can be made with causal factors. Research needs to focus on establishing these links. A possible approach is to relate the biological characteristics in the sediment to the levels of persistent pollutants. Another line of approach would be to determine the sensitivity spectra of the extinct indicator species still living in other rivers. These spectra would help greatly if we want to restore the Rhine into a suitable ecosystem for these species.

ACKNOWLEDGEMENTS

The research was financed by the Institute for Inland Water Management and Waste Water Treatment and National Institute for Water Supply. The author is greatly indebted to Dr. G. Van Urk for the fruitful discussions and critical comments on the manuscript.

APPENDIX 1. LIST OF TAXA IN THE SEDIMENT AND EXUVIAE COLLECTIONS

Key: E = Eurytope
L = Lithon
LR = Lithorheon
P = Pelon
PS = Psammon
T = Terrestrial
V = Vegetation
? = Unknown

Taxon	Habitat	Palaeo	Recent
EPHEMEROPTERA			
Baetidae	LR	+	
Caenis sp.	L	+	+
Cloeon sp.	L	+	+
Ecdyonurus sp.	LR	+	
Ephemera sp.	PS	+	
Ephemerella sp.	LR	+	
Ephoron virgo	PS	+	
Heptagenia sp.	LR	+	
Leptophlebiidae	V	+	
Palingenia longicauda	P	+	
Potamanthus luteus	LR	+	
Raptobaetopus tenellus	PS	+	
Rhithrogena sp.	LR	+	
PLECOPTERA			
Nemouridae	LR	+	
Perlidae	LR	+	
HETEROPTERA			
Aphelocheirus aestivalis	LR	+	
TRICHOPTERA			
Athripsodes aterrimus	V	+	
Athripsodes sp.	V	+	
Brachycentrus subnubilus	LR	+	
Ceraclea sp.	LR	+	+
Cheumatopsyche lepida	LR	+	
Chimarra marginata	LR	+	
Cyrnus sp.	L	+	+
Ecnomus tenellus	L	+	+
Glossosomatidae	LR	+	
Goeridae	LR	+	
Hydropsyche angustipennis	LR	+	
Hydropsyche contubernalis	LR	+	+
Hydropsyche pellucidula	LR	+	
Hydropsyche saxonica	LR	+	
Hydropsyche sp.	LR	+	+
Hydroptilidae	V	+	
Lepidostoma hirtum	LR	+	
Leptoceridae	V	+	
Limnephilidae	PS	+	
Lype sp.	L	+	
Micrasema sp.	LR	+	

Taxon	Habitat	Palaeo	Recent
Molanna sp.	PS	+	
Mystacides longicornis	V	+	
Neureclipsis bimaculata	L	+	+
Notidobia ciliaris	PS	+	
Odontocerum albicorne	LR	+	
Oecetis sp.	V	+	
Oligoplectrum maculatum	LR	+	
Phryganea bipunctata	V	+	
Polycentropus flavomaculatus	LR	+	
Psychomyia pusilla	LR	+	
Rhyacophila sp.	LR	+	
Sericostoma sp.	PS	+	
SIMULIIDAE			
Byssodon maculatum	V	+	
Eusimulium sp.	LR	+	
Prosimulium sp.	LR	+	
Simuliidae gen. 1	LR	+	
CHIRONOMIDAE-TANYPODINAE			
Labrundinia sp.	?	+	
Pentaneurini	?	+	+
Procladius sp.	P	+	+
Tanypus sp.	P	+	+
DIAMESINAE			
Diamesa carpatica	LR	+	
Diamesa sp.	LR	+	
Monodiamesa bathyphila	PS	+	
Potthastia gaedii	P	+	
Potthastia longimana	P	+	
Prodiamesa olivacea	P	+	+
Syndiamesa sp.	LR	+	
ORTHOCLADIINAE			
Acricotopus lucens	V	+	
Brillia longifurca	LR	+	+
Brillia modesta	LR	+	+
Bryophaenocladius sp.	T	+	+
Cardiocladius fuscus	LR	+	+
Chaetocladius piger agg.	T	+	
Chaetocladius sp.	T	+	
Chaetocladius? sp.	T	+	
Corynoneurinae indet.	V	+	+
Cricotopus flavocinctus	LR	+	
Cricotopus sg. Isocladius	L	+	+
Cricotopus triannulatus	LR	+	+
Cricotopus trifasciatus	L	+	
Cricotopus/Orthocladius	LR	+	+
Diplocladius cultriger	V?	+	
Eukiefferiella claripennis agg.	LR	+	+
Eukiefferiella ilkleyensis	LR	+	
Eukiefferiella sp.	LR	+	+
Gymnometriocnemus sp.	T	+	+
Heterotrissocladius marcidus	PS	+	

continued

Taxon	Habitat	Palaeo	Recent
Hydrobaenus lugubris	T	+	+
Limnophyes sp.	T	+	+
Metriocnemus fuscipes	T	+	+
Metriocnemus hirticollis agg.	T	+	+
Metriocnemus terrester	T	+	+
Metriocnemus? sp.	T	+	
Nanocldius sp.	E	+	+
Orthocladius (Euorthocladius) rivulorum	LR	+	
Orthocladius (Euorthocladius) sp.	LR	+	+
Paracladius conversus	PS	+	+
Parakiefferiella bathophila	L	+	
Parametriocnemus stylatus	PS	+	
Paratrichocladius rufiventris	LR	+	+
Psectrocladius gr. sordidellus	V	+	+
Psectrocladius platypus	V	+	
Psectrocladius psilopterus	V	+	
Pseudorthocladius sp.	T	+	+
Pseudosmittia sp.	T	+	+
Rheocricotopus sp.	LR	+	+
Smittia gr. aquatilis	T	+	+
Symposiocladius lignicola	V	+	
Synorthocladius semivirens	V	+	+
CHIRONOMINI			
Beckidia zabolotzkyi	PS	+	
Chernovskiia 'macrocera'	PS	+	
Chironomini gen. 1	?	+	
Chironomus gr. plumosus	P	+	+
Chironomus gr. uliginosus	P	+	+
Cladopelma gr. laccophila	P	+	+
Cladopelma gr. lateralis	P	+	
Cryptochironomus sp.	PP	+	+
Cryptotendipes gr. holsatus	PS	+	
Demeijerea rufipes	L	+	
Demicryptochironomus vulneratus	PS	+	
Dicrotendipes gr. nervosus	L	+	+
Dicrotendipes notatus	L	+	
Einfeldia dissidens	P	+	
Endochironomus albipennis	P	+	+
Endochironomus gr. dispar	V	+	
Endochironomus tendens	P	+	
Glyptotendipes caulicola	V	+	+
Glyptotendipes gr. pallens	P	+	+
Glyptotendipes gr. signatus	V	+	
Glyptotendipes sp.	?	+	+
Harnischia sp.	P	+	+
Kiefferulus tendipediformis	P	+	
Kloosia pusilla	PS	+	+
Lipiniella arenicola	PS	+	+
Microchironomus tener	P	+	+
Microtendipes gr. chloris	PS	+	
Microtendipes rydalensis agg.	?	+	
Microtendipes tarsalis agg.	?	+	

Taxon	Habitat	Palaeo	Recent
Parachironomus arcuatus	L	+	+
Parachironomus longiforceps	L	+	+
Paracladopelma laminata agg.	PS	+	
Paracladopelma sp.	PS	+	
Paralauterborniella nigrohalteralis	?	+	
Paratendipes connectens 3 Lipina	PS	+	
Paratendipes gr. albimanus	PS	+	
Paratendipes intermedius	PS	+	
Phaenopsectra sp.	P	+	+
Polypedilum cf. *uncinatum*	?	+	
Polypedilum gr. bicrenatum	P	+	+
Polypedilum laetum agg.	L	+	+
Polypedilum nubeculosum	P	+	+
Polypedilum pedestre agg.	V	+	+
Polypedilum scalaenum	PS	+	+
Polypedilum sordens	V	+	+
Pseudochironomus prasinatus	PS	+	
Pseudochironomus sp.	PS	+	
Robackia demeijerei	P	+	
Stenochironomus sp.	V	+	
Stictochironomus sp.	PS	+	
Tribelos intextus	P	+	
Xenochironomus xenolabis	L	+	+
Zavreliella marmorata	V	+	
TANYTARSINI			
Cladotanytarsus gr. mancus	PS	+	+
Micropsectra sp.	LR	+	+
Paratanytarsus confusus	L	+	+
Paratanytarsus tenuis	L	+	+
Rheotanytarsus sp.	LR	+	+
Stempellina sp.	PS	+	
Tanytarsus gr. brundini	PS	+	
Zavrelia pentatoma	P	+	
Total taxa		167	66

REFERENCES

Albarda, H. 1889. Cataloque raisonné et synonymique des Neuroptères observés dans les Pays-Bas et dans les Pays limitrophes. *Tijdschrift voor Entomologie*, **32**, 211–375.

Becker, G. 1987. *Lebenszyklus, Reproduktion und ökophysiologische Anpassungen von Hydropsyche contubernalis, einer Köcherfliege mit massenvorkommen im Rhein*, Inaugural Dissertation Köln, 108 pp.

Bencke, A. C., Van Arsdall, Jr. T. C. and Gillespie, D. M. 1984. Invertebrate productivity in a subtropical blackwater river: the importance of habitat and life history. *Ecological Monographs*, **54**(1), 23–63.

Berendsen, H. J. A. 1982. De genese van het landschap in het zuiden van de provincie Utrecht. *Utrechtse geografische Studies*, **25**, 259 pp.

CBS (Centraal Bureau voor de Statistiek) 1983. *Statistisch zakboek 1983*, Staatsuitgeverij, 'S-Gravenhage, 385 pp.

Cranston, P. S. 1982. The metamorphosis of Symposiocladius lignicola (Kieffer) n.gen., n.comb., a wood-mining Chironomidae (Diptera). *Entomologica scandinavica*, **13**, 419–429.

Cushman, R. M. 1984. Chironomid deformities as indicators of pollution from synthetic, coal derived oil. *Freshwater Biology*, **14**, 179–182.

Dijkzeul, A. 1982. De waterkwaliteit van de Rijn in Nederland in de periode 1970–1981. *Report Government Institute for Sewage and Waste Water Treatment (RIZA)*, **82.061**, 112 pp.

Ertlova, E. 1974. Einige Erkenntnisse über Chironomiden (Diptera, Chironomidae) aus Bryozoen. *Biologia Bratislava*, **29**, 869–876.

Fischer, F. C. J. 1934. Verzeichnis der in den Niederlanden und dem Nachbargebiete vorkommenden Trichoptera. *Tijdschrift voor Entomologie*, **77**, 177–201.

Fittkau, E. J. and Reiss, F. 1978. Chironomidae. In Illies, J. (Ed.), *Limnofauna Europaea*, Fischer Verlag, Stuttgart, 404–440.

Geijskes, D. C. 1940. Verzeichnis der in den Niederlanden vorkommenden Plecoptera, mit einigen geschichtlichen, ökologischen und systematischen Bemerkungen. *Tijdschrift voor Entomologie*, **83**, 3–16.

Geijskes, D. C. and Van Tol, J. 1983. *De libellen van Nederland (Odonata)*, Koninklijke Nederlandse Natuurhistorische Vereniging, Hoogwoud, 368 pp.

Greve, P. A. 1980. Toxische microverontreinigingen in de Rijn. *H_2O*, *13*, 360–364.

Hamilton, A. L. and Saether, O. A. 1971. The occurrence of characteristic deformities in the chironomid larvae of several Canadian Lakes. *Canadian Entomologist*, 103, 363–368.

Hueck Van Der Plas, E. H. (Ed.) 1983. *Zware metalen in aquatische systemen. Geochemische inventarisatie biologisch onderzoek in Rijn en Maas en de daardoor gevoede bekkens 1. Geochemische inventarisatie*, Rapport Organisatie voor toegepast natuurwetenschappelijk Onderzoek TNO, 127 pp.

Klink, A. G. 1983. Rheotanytarsus rhenanus n.sp. A common midge of the lithorheophilic fauna in large lowland rivers (Diptera: chironomidae). *Entomologische Berichten*, **43**, 136–138.

Mol, A. W. M. 1985. Een overzicht van de Nederlandse haften (Ephemeroptera) 2. Overige families. *Entomologische Berichten*, **45**, 128–135.

Molt, E. L. 1961. Verontreiniging van het Rijnwater. In *Drinkwatervoorziening 13e vakantiecursus*, Moormans periodieke pers Den Haag, 46–71.

Moubayed, Z. and Laville, H. 1983. Les chironomides (Diptera) du Liban 1. Premier inventaire faunistique. *Annales Limnologica*, **19**(3), 219–228.

Pagast, F. 1934. Uber die Metamorphose von Chironomus xenolabis Kieff. eines Schwammparasieten (Dipt.). *Zoologische Anzeiger*, **105**, 155–158.

Pons, L. J. 1957. De geologie, de bodemvorming en de waterstaatkundige ontwikkeling van het land van Maas en Waal en een gedeelte van het Rijk van Nijmegen Verslagen Landbouwkundig Onderzoek (63.11). *Bodemkundige Studies*, **3**, 156 pp.

Prat, N. F., Puig, M. A., Gonzalez, G. *et al.* 1984. Llobregat. In Whitton, B. A. (Ed.), *Ecology of European Rivers*, Blackwell, Oxford, 527–552.

RIWA (Cooperating Rhine and Meuse Water Supply Companies) 1983. *Jaarverslag 1983. Deel A: de Rijn*, RIWA (Cooperating Rhine and Meuse Water Supply Companies), Report, 76 pp.

RIWA (Cooperating Rhine and Meuse Water Supply Companies) 1984. *Jaarverslag 1984. Deel A: de Rijn*, RIWA (Cooperating Rhine and Meuse Water Supply Companies), Report, 96 pp.

Government Institute for Sewage and Waste Water Treatment (RIZA) 1983. De waterkwaliteit van de Rijn in 1982. *Report Government Institute for Sewage and Waste Water Treatment (RIZA)*, **83-080**, 1–39.

Redeke, H. C. 1936. La crevette d'eau douce, Athyaephyra desmaresti (Mill.) dans les Pays-Bas. *Mémoires Musée Histoire nationale Belgique*, **3**(2), 227–231.

Reiss, F. 1986. Ein Beitrag zur Chironomidenfauna Syriens (Diptera: Chironomidae). *Entomofauna*, **7**(11), 153–168.

Department of Public Works 1851–present. *Tienjarige overzichten der waterhoogten en afvoeren*, Staatsuitgeverij's Gravenhage.

Rossaro, B. 1984. The chironomids of the Po River (Italy) between Trino Vercellese and Cremona. *Aquatic Insects*, **6**(2), 123–135.

Rubsow, I. A. 1964. 14. Simuliidae (Melusinidae). In Lindner, E. (Ed.), *Die Fliegen der palaearktische Region. Schweizerbart'sche Verlagsbuchhandlung Stuttgart Band*, **III-4**, 689 pp.

Salomons, W. and De Groot, A. J. 1977. Pollution history of trace metals in sediments, as affected by the Rhine River. *Publikaties Waterloopkundig Laboratorium Delft Hydraulica Laboratorium*, **184**, 21 pp.

Triska, F. J. 1984. Role of wood debris in modifying channel geomorphology and riparian areas of a large lowland river under pristine conditions: a historical case study. *Verhandlungen Internationale Vereinigung für theoretische und angewandte Limnologie*, **22**, 1876–1892.

Van Drimmelen, D. E. 1982. De visserij in het stroomgebied van de Rijn. Syllabus voor de cursus 'De Rijn, waterbron van Nederland'. Stichting postakademiale Vorming Gezondheidstechniek. *Cursus 'de Rijn Waterbron van Nederland*, **2.5**, 16 pp.

Van Urk, G. 1984. Lower Rhine-Meuse. In Whitton, B. A. (Ed.), *Ecology of European Rivers*, Blackwell, Oxford, 437–468.

Van Urk, G. and Kerkum, F. C. M. 1986. Misvormingen bij muggelarven uit Nederlandse oppervlaktewateren. *H_2O*, **26**, 624–627.

Van Den Brink, F. W. B. and Van Der Velde, G. 1986. Observations on the seasonal and yearly occurrence and the distribution of *Atyaephyra desmaresti* (Millet, 1831) (Crustacea, Decapoda, Natantia) in The Netherlands. *Hydrobiological Bulletin*, **19**(2), 193–198.

Van Der Woude, J. R. 1981. *Holocene palaeoenvironmental evolution of a perimarine fluviatile area*, Dissertation Vrije Univ. Amsterdam, 118 pp.

Van Der Wulp, F. M. 1877. *Diptera Neerlandica. De tweevleugelige insecten van Nederland*, Martinus Nijhoff, 's Gravenhage, 497 pp.

Velsen, C. 1768. *Rivierkundige verhandelingen. Afgeleid uit waterwigt- en waterbeweegkundige grondbeginselen. En toepasselijk gemaakt op den Rhyn, de Maas, de Waal, de Merwede en de Lek*, Harlingen, 3 pp.

Verbraeck, A. 1974. *Toelichting bij de geologische kaart van Nederland 1:50 000, blad Gorkum Oost (38 O)*, Rijksgeologische Dienst Haarlem, 140 pp.

Verbraeck, A. and Bisschops, J. H. 1980. *Toelichting bij de geologische kaart van Nederland 1:50 000, blad Willemstad Oost (43 O)*, Rijksgeologische Dienst Haarlem, 112 pp.

Warwick, W. F. 1985. Morphological abnormalities in Chironomidae (Diptera) larvae as measures of toxic stress in freshwater ecosystems: indexing antennal deformities in Chironomus Meigen. *Canadian Journal of Fisheries and aquatic Sciences*, **42**, 1881–1914.

Wiederholm, T. 1984. Incidence of deformed chironomid larvae (Diptera: Chironomidae) in Swedish lakes. *Hydrobiologia*, **109**, 243–249.

Wilkinson, L. 1986. *SYSTAT: The system for statistics*, Evanston, Illinois.

Zonneveld, I. S. 1960. De Brabantse Biesbosch. Een studie van bodem en vegetatie van een zoetwatergetijdendelta. A. Summary with text figures and tables. *Bodemkundige Studies*, **4**, 210 pp.

Historical Change of Large Alluvial Rivers: Western Europe
Edited by G. E. Petts
©1989 John Wiley & Sons Ltd

12

Changes in Fish Stocks and Fisheries: the Lower Elbe River

HEINO MÖLLER

Institut für Meereskunde, Dusternbrooker Weg 20, D-2300 Kiel, West Germany

INTRODUCTION

The large rivers of western Europe once sustained important fish stocks and supported valuable fisheries. Records of fish catches extend back for about one hundred years for many rivers and for longer in some. This paper illustrates the use of such historic documentary data sources in assessing changes of fish stocks and fisheries by considering the lower Elbe River. The Elbe rises in Czechoslovakia and crosses the German Democratic Republic; after 888 km it reaches the territory of the Federal Republic of Germany; after an additional 135 km it crosses the city centre of Hamburg. The 120 km long section from Hamburg to the river mouth near the Isle of Scharhörn is the subject of this review (Figure 1).

In 1916, Sterner (1916b) classified the Elbe as the European river most densely inhabited by fishes. The same author reviewed the remarkable variety of fishing methods which had developed there (Sterner, 1916c–e, g, 1918a–f). His scattered information later was compiled and updated by Schnakenbeck (1928). Duncker and Ladiges (1960) described the biology and distribution of the different fish species, mainly referring to studies made before 1935. Reports and memoirs of fishermen (Breckwoldt 1894; Holst 1931; Rübcke 1910, 1911, 1912, 1914) and reviews of professionals from administration and the fishing industry (Blankenburg, 1912; Lübbert and Wiese, 1949), and fisheries biologists (Schnakenbeck, 1937), today give us insights into the problems and prospects the Elbe fishery had to face at the turn of the century. They also document that the German high-sea fisheries mainly evolved from Elbe fishing villages like Finkenwerder and Blankenese.

During the second half of the last century, the Elbe fisheries faced a considerable boom, but already in the mid-1920s, fishing activities decreased dramatically. A parallel development is found in the number of scientific publications dealing with fish life in the Elbe.

Only few reliable catch statistics from the Elbe are available. Data presented by the 'Deutscher Seefischereiverein' for 1887–1919 (Table 5) are rough estimates only. At the time Sterner (1916b) had criticized them because their sources were not clear. In the few cases where counterchecks are possible, those data can be shown to have clearly underestimated the real catches made during that time. Landings at the two traditional

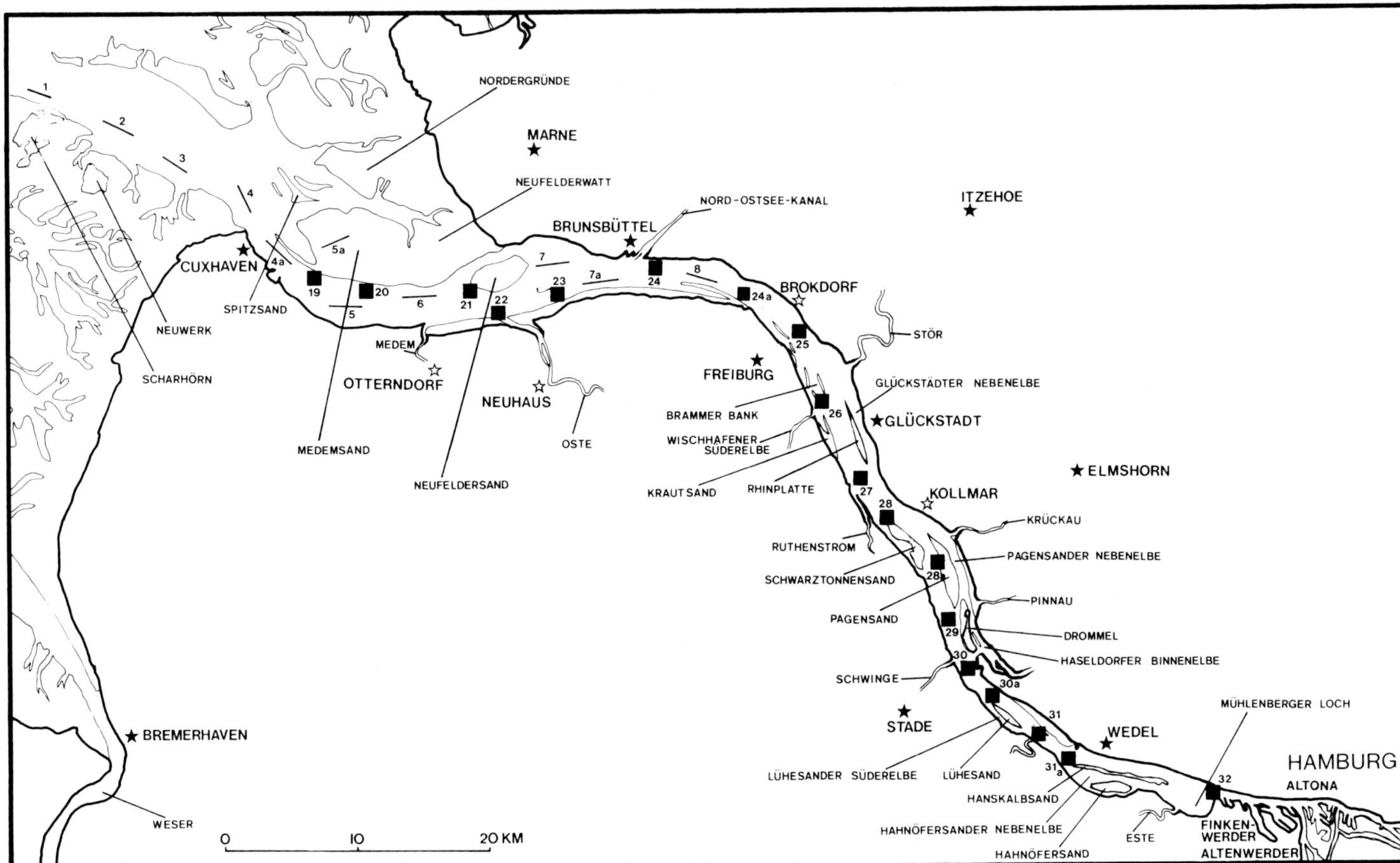

FIGURE 1. The lower Elbe with fishing stations used during the surveys in 1981–86. Lines: shrimp trawl, squares: anchor net

fish markets in Altona and Hamburg (Koops, 1959), which were united in 1937, and at the fish market in Cuxhaven (Kühl, 1976; Rönnau, 1987), established in 1908, where most anchor net catches have been landed since World War II, are incomplete as eel and flounder especially have traditionally been sold directly by fishermen. Market statistics therefore are considered to be only of limited value for the estimation of the economic significance of the Elbe fisheries.

The reasons for the decline in Elbe fisheries have been discussed increasingly since the green movement gained strength in Germany in the mid-1970s. Due to the lack of data and different ambitions, disputes among scientists developed. The idea was favoured that the decline was mainly due to industrialization and increased shipping (Kühl, 1976; Mann, 1968; Peters, 1978; Wilkens and Köhler, 1977). Fishermen blocked shipping lanes and environmentalists organized demonstrations in order to draw public attention to the problem. These actions contributed to the establishment of research and monitoring programs on river ecology; such programmes began at the end of the 1970s.

Riedel-Lorjé and Gaumert (1982) reviewed '100 years of Elbe research 1842–1943'. They compiled reports on the effects of human population increase, embankment, pollution, and shipping on river life. The hydrography of the Elbe, its present state and changes during the past, was reviewed by ARGE Elbe (1984). This governmental organization initiated continuous surveys of hydrochemical and hydrographical parameters in 1977 (Flügge, 1988). Its activities justify the statement that the lower Elbe today is one of the best studied rivers in Europe.

Conclusions made in this paper mainly are drawn from comparisons of published as well as of unpublished historical data with results from two recently made surveys on fish biology in the lower Elbe River (Möller, 1984, 1988). Monthly surveys were carried out from September 1981 to November 1982 and again from November 1984 to July 1986. Shrimp trawlers with usual commercial gear (two 11-m beam trawls) were chartered for fishing in the estuary, anchor net vessels (net opening 56–80 m^2) were chartered for fishing between Cuxhaven and Hamburg (Figure 1). Catches from these two vessels are not comparable, however. Quantitative data presented here therefore are restricted to results from catches of anchor nets, the most important fishing gear on the river today.

THE FISHES OF THE ELBE

Faunistic studies of freshwater fishes recently were carried out in Niedersachsen (Gaumert, 1981) and Schleswig-Holstein (Nellen and Dehus, 1985). Similar studies of the Elbe between these two reaches do not exist, although a faunistic study of fish in Hamburg was finished recently (Diercking, personal communication).

During 1851 hours of experimental fishing in 1981–1986, 66 fish species were caught, 35 of them belonging to the marine fauna. Eighteen species occurred with less than 10 specimens in the catches. The geographical border between marine and freshwater fauna shifts about stream-km 690 near Brunsbüttel. Clupeids and gobiids occasionally occur further upstream, while among the freshwater group the percids show the farthest distribution towards the open sea (Table 1). During single tides, the brackish water frontier may move approximately 20 km. Its position depends on the freshwater discharge, lying relatively far upstream during the dry summer season. Accordingly, the greatest extension of marine species can be observed in summer, while freshwater species occur farthest down-river in winter (Möller, 1987).

TABLE 1. Records of fish species and their distribution in the lower Elbe during experimental fishing in 1981–86. +: less than 10 specimens in all catches; *: introduced species

	Scharhörn						Cuxhaven								Kiel Canal						Glückstadt				Stade				Hamburg
Station no.	01	02	03	04	04a	19	05	05a	20	06	21	22	07	23	07a	24	08	24a	25	26	27	28	28a	29	30	30a	31	31a	32
Stream-kilometre	740								720						700						680				660				640
Marine species																													
01 *Clupea harengus*, herring	XX	XX	XX	XX	XX	XX	XX	XX	XX	XX	XX	XX	XX	XX	XX	XX	XX	XX	XX	XX	XX	XX							
02 *Sprattus sprattus*, sprat	XX	XX	XX	XX	XX	XX	XX	XX	XX	XX	XX	XX	XX	XX	XX	XX	XX	XX	XX	XX									
03 *Belone belone*, garfish +									XX																				
04 *Siphonostoma typhle*, deep-snouted pipefish			XX	XX		XX	XX			XX		XX	XX	XX			XX												
05 *Syngnathus rostellatus*, Nilsson's pipefish						XX	XX		XX		XX	XX	XX																
06 *Entelurus aequoreus*, snake pipefish +	XX			XX					XX																				
07 *Gadus morhua*, cod	XX	XX	XX	XX	XX	XX	XX	XX	XX	XX	XX	XX	XX	XX		XX	XX	XX											
08 *Trisopterus luscus*, bib +	XX	XX			XX																								
09 *Trisopterus minutus*, poor cod	XX	XX	XX	XX			XX		XX		XX	XX		XX		XX													
10 *Trisopterus esmarkii*, Norway pout +	XX		XX																										
11 *Merlangius merlangus*, whiting	XX	XX	XX	XX	XX	XX	XX	XX	XX	XX	XX	XX	XX	XX	XX														
12 *Pollachius virens*, saithe +		XX																											
13 *Raniceps raninus*, tadpole-fish +						XX																							
14 *Gaidropsaurus mediterraneus*, shore rockling	XX	XX	XX	XX																									
15 *Ciliata mustela*, five-bearded rockling						XX					XX																		
16 *Trachurus trachurus*, scad +	XX	XX	XX	XX					XX																				
17 *Hyperoplus lanceolatus*, greater sandeel	XX	XX	XX	XX	XX	XX	XX		XX	XX				XX															
18 *Callionymus lyra*, dragonet	XX	XX	XX		XX		XX																						
19 *Chelon labrosus*, thick-lipped grey mullet +																XX													
20 *Scomber scombrus*, mackerel +																		XX											
21 *Pholis gunnellus*, butterfish	XX	XX	XX	XX	XX	XX				XX				XX															
22 *Zoarces viviparus*, viviparous blenny	XX	XX	XX	XX	XX	XX	XX	XX	XX	XX	XX	XX	XX	XX	XX	XX	XX												
23 *Pomatoschistus* sp., goby	XX	XX	XX	XX	XX	XX	XX	XX	XX	XX	XX	XX	XX	XX	XX	XX	XX	XX	XX	XX	XX	XX		XX	XX	XX	XX	XX	XX
24 *Eutrigla gurnardus*, grey gurnard	XX	XX	XX	XX						XX																			
25 *Taurulus bubalis*, sea scorpion	XX	XX	XX	XX	XX	XX	XX	XX	XX	XX	XX	XX	XX	XX	XX	XX		XX											
26 *Agonus cataphractus*, hooknose	XX	XX	XX	XX	XX	XX	XX	XX	XX	XX	XX	XX	XX																
27 *Cyclopterus lumpus*, lumpsucker		XX	XX			XX	XX		XX		XX					XX													
28 *Liparis* sp., sea-snail	XX	XX	XX	XX	XX	XX	XX	XX	XX	XX	XX	XX	XX	XX		XX													

29 *Spinachia spinachia*, 15-spined stickleback +	XX
30 *Scophthalmus maximus*, turbot	XXXXXXXXXX XXXXXXXXXXXXXXXX XXXX
31 *Scophthalmus rhombus*, brill	XXXXXXXXXXXXXX XX
32 *Hippoglossoides platessoides*, long rough dab +	XX
33 *Limanda limanda*, dab	XXXXXXXXXXXXXXXXXXXXXXXXXXXX XX XX
34 *Pleuronectes platessa*, plaice	XXXXXXXXXXXXXXXXXXXXXXXXXXXXXXXXXXXX
35 *Solea solea*, sole	XXXXXXXXXXXXXXXXXXXXXXXXXXXXXXXXXX
Euryhaline species	
36 *Lampetra fluviatilis*, lampern	XXXXXXXX XX XX XXXX XXXXXXXXXXXXXXXXXXXXXXXXXXXXXXXX
37 *Petromyzon marinus*, lamprey +	XX XX XX
38 *Alosa fallax*, twaite shad	XXXXXX XXXXXXXXXXXXXXXXXX XXXX XXXXXXXXXXXXXXXXXXXXXX
39 *Salmo salar*, salmon +	XX XX XX XX XX XX XX
40 *Salmo trutta*, trout	XX XX XX XXXX XX XX XXXXXXXXXX XXXXXXXX
41 *Salmo gairdneri*, rainbow trout*	XX XX
42 *Coregonus lavaretus*, powan* +	XX XXXXXX
43 *Osmerus eperlanus*, smelt	XXXXXXXXXXXXXXXXXXXXXXXXXXXXXXXXXXXXXXXXXXXXXXXXXXXXXXXXXX
44 *Anguilla anguilla*, eel	XXXXXXXXXXXXXXXXXXXXXXXXXXXXXXXXXXXXXXXXXXXXXXXXXXXXXXXXXX
45 *Gasterosteus aculeatus*, stickleback	XXXXXXXXXXXXXXXXXXXXXXXXXXXXXXXXXXXXXXXXXXXXXXXXXXXXXXXXXX
46 *Platichthys flesus*, flounder	XXXXXXXXXXXXXXXXXXXXXXXXXXXXXXXXXXXXXXXXXXXXXXXXXXXXXXXXXX
Freshwater species	
47 *Acipenser ruthenus*, sterlet* +	XX
48 *Esox lucius*, pike +	XX XXXXXXXXXX
49 *Rutilus rutilus*, roach	XX XXXX XXXXXXXXXX
50 *Leuciscus leuciscus*, dace	XX XX
51 *Leuciscus cephalus*, chub	XXXX
52 *Leuciscus idus*, ide	XXXX XXXXXXXXXX
53 *Scardinius erythrophthalmus*, rudd	XX XXXXXXXXXX XXXX
54 *Aspius aspius*, asp	XX XXXXXXXXXXXX XXXXXXXX
55 *Alburnus alburnus*, bleak	XX
56 *Blicca bjoerkna*, silver bream	XXXX XXXX XXXXXX XXXXXXXX
57 *Abramis brama*, bream	XX XXXXXXXXXXXXXXXXXXXXXXXX
58 *Abramis ballerus*, zope	XX XXXXXX
59 *Carassius carassius*, Crucian carp	XX XX XXXXXXXXXX
60 *Cyprinus carpio*, carp* +	XX XXXX XX
61 *Misgurnus fossilis*, weatherfish +	XX XX XX
62 *Lota lota*, burbot +	XX
63 *Perca fluviatilis*, perch	XX XX XX XX XXXX XXXXXXXXXXXXXX
64 *Stizostedion lucioperca*, zander	XX XX XX XXXXXXXXXX XXXXXXXXXXXXXXXX
65 *Gymnocephalus cernuus*, ruffe	XX XX XXXX XXXXXXXXXXXXXXXXXXXXXXXXXXXX
66 *Pungitius pungitius*, 9-spined stickleback	XXXX XXXXXXXXXX XXXX

TABLE 2. Total catch in kg per one hour anchor net (80 m^2) fishing in the lower Elbe. Data are averages of 17 series between November 1984 and July 1986

Station	Jan	Feb	Mar	Apr	May	Jun	Jul	Aug	Sep	Oct	Nov	Dec	Average
No. of survey	1	0	2	1	2	2	2	1	1	1	2	2	
19	4		13	32	32	16	33	15	43	15	22	21	22
20	5		13	26	23	20	27	12	14	25	45	44	23
21	4		17	47	25	37	68	34	22	39	53	61	37
23	4	I	13	14	14	12	22	12	10	11	20	21	14
24	3	C	10	6	8	5	44	5	7	11	12	53	15
24a	11	E	25	5	48	19	42	36	17	33	46	48	30
25	2		5	2	7	5	37	15	8	16	17	18	12
26	10	C	10	2	65	12	45	50	31	31	23	15	27
28	17	O	7	1	6	6	58	33	41	35	41	9	23
28a	1	V	3	1	1	3	9	11	24	0	7	3	6
29	2	E	2	3	4	2	84	11	8	4	5	3	11
30	1	R	8	0	0	1	90	6	4	1	1	2	10
30a	2		1	1	5	0	23	15	6	13	7	5	7
31a	2		2	4	8	3	28	13	15	6	6	12	9
32	1		1	5	2	1	0	0	0	2	8	7	2
Average	5		9	10	16	9	41	18	17	16	21	21	17

The annual average fish biomass caught upstream from station 28 was only one third of that caught between stations 19–28. Lowest catches were made in Hamburg harbour (station 32). Although represented by few species, the euryhaline group contributed 65–99 per cent of the biomass at single stations. On average, of all anchor net catches made between Cuxhaven and Hamburg in 1984–1986, smelt (68 per cent), shad (24 per cent), and herring (4 per cent) contributed 96 per cent of the individuals. Ninety-two per cent of the weight consisted of smelt (56 per cent), shad (15 per cent), flounder (8 per cent), herring (7 per cent) and eel (6 per cent). Smelt was outnumbered by shad and eel only at stations 29–31a (Table 2).

Results from ichthyoplankton surveys indicate that smelt and shad spawn along the southern shore of the lower Elbe (Figure 2). Together they contribute to more than 95 per cent of the pelagic fish larvae in the freshwater region in spring and summer. No pelagic fish larvae are present in late autumn and winter (Dieckwisch, 1987). Smelt mainly spawns upstream from station 29; first larvae appear in April or May. Shad larvae are found about three weeks later with highest concentrations between stations 25 and 28. Spawning smelt and smelt eggs have also been caught in Hamburg harbour (Lillelund, 1961; Möller, 1984). A traditional winter fishery for smelt in the Oste estuary indicates that this tributary river also possibly might be used for spawning. High catches of flounder larvae are common in the River Este near Hamburg and to a lesser extent also in other tributary rivers, and this supports the view that this species occasionally spawns successfully in pure freshwater. Larvae of percids and cyprinids were found only in low quantities during recent surveys.

The principles of spawning migrations of smelt and shad, as well as of flounder migration, in relation to water temperature and oxygen conditions are well known (Möller, 1987). These migration patterns, together with the emergence of smelt and shad fry in early summer, explain much of the seasonal and regional variations in anchor net catches as

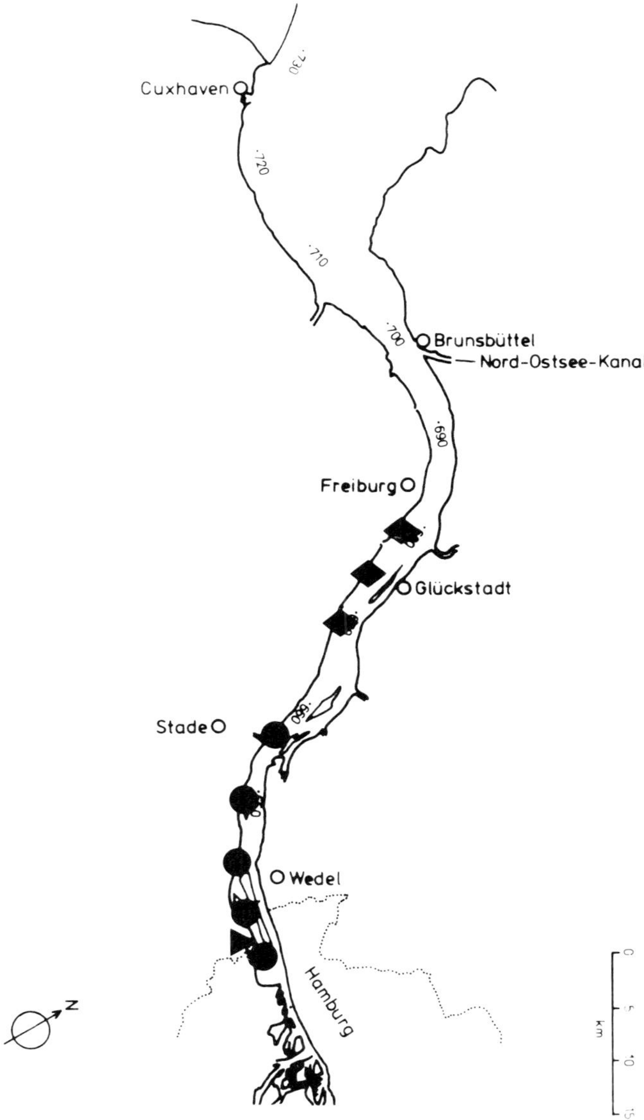

FIGURE 2. Major spawning grounds of shad (rhombs) and smelt (circles) and highest densities of flounder larvae (triangle) as estimated from ichthyoplankton surveys in 1985 and 1986 (Dieckwisch, 1987)

indicated in Table 2. The mass occurrence of young smelt and shad is mainly responsible for a quadrupling of the average catch from June to July with a maximum of 144 kg per hour at station 30 in July 1985. In August 1987 fishermen reported smelt catches of 3 t per tide and boat near station 28 (Röbcke, personal communication). The unusually high figure for station 26 in May is mainly made up by spawning shad (May 1986: 108 kg h^{-1}) which is caught only occasionally by anchor nets.

TABLE 3. Percentage of eight species dominating (by weight and number) in total anchor net catches. Data are averages of 17 series between November 1984 and July 1986

Station	*Clupea harengus*	*Gadus morhua*	*Alosa fallax*	*Osmerus eperlanus*	*Anguilla anguilla*	*Gasterosteus aculeatus*	*Platichthys flesus*	*Abramis brama*
				Weight				
19	26	6	3	44	2	0	16	0
20	10	11	4	50	2	1	14	0
21	14	5	2	66	3	1	11	0
23	11	2	6	66	3	1	7	0
24	19	0	9	45	6	8	9	0
24a	1	0	17	72	3	2	3	0
25	0	0	3	77	6	1	8	0
26	0	0	25	66	4	1	2	0
28	0	0	5	79	5	1	11	0
28a	0	0	18	64	5	1	8	2
29	0	0	61	29	6	0	3	0
30	0	0	74	17	5	1	2	0
30a	0	0	11	32	46	1	6	1
31a	0	0	16	29	37	1	3	6
32	0	0	0	48	18	1	25	3
				Number				
19	19	0	12	57	0	1	2	0
20	13	0	3	62	0	2	2	0
21	11	0	3	74	0	1	2	0
23	16	0	6	69	0	2	1	0
24	11	0	11	66	0	6	1	0
24a	1	0	3	91	0	3	0	0
25	0	0	3	95	0	1	1	0
26	1	0	1	97	0	1	0	0
28	0	0	7	92	0	0	1	0
28a	0	0	5	93	0	0	1	0
29	0	0	64	35	0	0	0	0
30	0	0	88	12	0	0	0	0
30a	0	0	55	39	4	1	0	0
31a	0	0	57	39	2	1	0	0
32	0	0	0	81	8	6	5	0

The majority of 0-group smelt and shad leave the river in autumn for feeding grounds in the Wadden Sea, while young eel and flounder—katadromous species—prefer freshwater feeding grounds. In the lower Elbe, they face considerable local differences in food supply. The biomass of zooplankton (mainly copepods and cladocerans) as well as of zoobenthos (mainly oligochaetes and amphipods) which may serve as fish food, is highest in the freshwater area but exceptionally low in Hamburg harbour. The biomass declines significantly in the central estuary and increases again further towards the open sea (Fiedler, 1988). These differences are due to effects of eutrophication near Hamburg and to adverse conditions due to tidal fluctuations of salinity in the estuary. This may explain why bottom-feeding fish like flounders are known to be well nourished and fat when caught in the freshwater area upstream from Freiburg, but in low body condition in the estuary (Möller, in preparation; Sterner, 1916c).

Changes in fish stocks

Sturgeon, allis shad, salmon, houting, zährte, and burbot were described as extinct by Wilkens and Köhler in 1977. One sturgeon of 36 cm length was recorded in July 1986 by a local fisherman near Krautsand (Röbcke, personal communication). Allis shad has never been reported from the lower Elbe, but a few specimens have been caught in the 19th century far upstream (Fric, 1859). Conflicting reports may be due to a misidentification of twaite shad (Duncker and Ladiges, 1960). Former reports of houting (Möller, 1984) were identified as powan (Dehus *et al.*, in preparation) that probably had escaped from aquaculture in the GDR (Morawa, 1982).

None of the six species mentioned above was found during the survey in 1981–82, except for two burbots caught in Hamburg harbour. During the survey in 1984–86, eight salmon of 15–80 cm were caught at different stations. Also the catches of trout have increased considerably. By weight, trout ranked fifteenth in the anchor net catches in 1984–86, and a single anchor net fishermen in 1986 landed up to 500 kg from the Elbe (Möller, 1988). The recovery of salmonid species was made possible by the restoration of several small tributary rivers. However, it is doubted whether the catches are from fish that had developed from naturally spawned eggs instead of artificially bred ones (Hartmann, 1987).

Changes in the Elbe fish fauna can hardly be quantified as only few reliable catch statistics exist and as today a number of special gears, designed to catch species like lamprey, sturgeon, salmon, shad, ruffe, and burbot are no longer in use. Before 1981, only one scientifically based analysis of catch data had been carried out. It had been done by Apstein (1895), who in summer 1894 studied the catches of anchor net fishermen in order to estimate the contribution of this fishing method to overfishing local resources.

The principles of anchor net fishing have not changed since the end of the last century. Catches from the same vessel as was chartered for the 1981–82 survey, had been analysed by Apstein (1895). However, the net openings have become larger after the introduction of mechanization and today the cotton nets have been replaced by synthetics. In accordance with different net openings used, the data presented in Table 4 were adjusted and calculated for a unit catch per hour of 80 m^2 net opening (1894: $\times 4$, 1981–82: $\times 1.4$). No possibility was seen to correct the effects of different mesh size used or of the possibly different hydraulic conditions.

Apstein's report is hand-written in Sütterlin script and has not been published. An analysis of his protocols revealed that 104 of his samples were comparable to 31 samples from 1982, 39 samples from 1985, and 34 samples from 1986. All those catches were made between stations 25 and 32 (Figure 1) from June to September, the majority of them between the water surface and approximately 2 m above the bottom on average.

From the comparison given in Table 4, no dramatic changes in the unit catches made since 1894 can be deducted. Nevertheless, the real catches per vessel and hour of eel were 11 times, and those of adult smelt 5 times, higher on average for the 1980s as compared with 1894. The larger total unit catch in 1894 obviously is due to higher catches of 0-group smelt which may be explained by smaller mesh size used during that year. A similar explanation may be valid in the case of stickleback. However, 0-group shad with a similar length occurred more frequently in the catches from the 1980s. The significant decrease in the catch of ruffe may indicate a decline in the stock, although good ruffe catches were still reported a few years ago (Wilkens and Köhler, 1977). Further investigations will have

TABLE 4. Comparison of catches made with anchor nets during different years in the lower Elbe between Freiburg and Hamburg. All data are averages of monthly mean catches made in June, July, August, and September (1986: except September and only one station with 16 samples in August) and are calculated for a net opening of 80 m^2

	1894	1982	1985	1986
Average net opening in m^2	20	56	80	80
Mesh size of cod end in mm	4.0	7.5	7.5	7.5
Number of hauls examined	104	31	39	34
Average duration per haul	3h50	3h37	4h20	4h15
Total unit catch in kg hr^{-1}	51	32	19	26
Number of species caught	21	23	22	19
Unit catch in n h^{-1}				
Alosa fallax, 0 group	712	5834	6639	2799
Osmerus eperlanus, 0 group	13050	2217	755	2122
Osmerus eperlanus, 1+ group	139	437	156	56
Anguilla anguilla	16	33	60	53
Gasterosteus aculeatus	118	1	3	1
Gymnocephalus cernuus	190	1	3	1
Platichthys flesus	48	14	46	29
Stizostedion lucioperca	0	0	2	113
Other species recorded				
Abramis ballerus		+	+	
Abramis brama		+	+	+
Alburnus alburnus	+	+		
Aspius aspius		+	+	+
Barbus barbus	+			
Blicca bjoerkna	+	+	+	+
Coregonus sp.	+	+		+
Cyprinus carpio				+
Esox lucius	+	+		
Lampetra fluviatilis	+	+	+	
Leuciscus idus	+	+		+
Lota lota	+			
Misgurnus fossilis			+	
Perca fluviatilis	+	+	+	+
Petromyzon marinus		+		
Pungitius pungitius	+			
Rutilus rutilus	+	+	+	+
Salmo trutta		+	+	+
Scardinius erythrophthalmus			+	+
Vimba vimba	+			
Clupea harengus	+	+	+	+
Gobiids	+	+	+	
Pleuronectes platessa			+	
Psetta maxima			+	
Scomber scombrus				+
Sprattus sprattus		+	+	

to show whether the high catches of pike-perch in 1986 are an exception or whether they indicate a long-term population growth. No clear tendencies are evident in the cases of adult smelt and flounder. Eel catches in the 1980s had increased by a factor of three. Interestingly, Apstein (1895) noted that the eel catches during summer of 1894 were classified as being exceptionally high for that time. Except for ide, all other species mentioned in Table 4 were caught at a rate of less than 0.1 individuals per hour in 1894. The assumptions made above were doubted by Peters *et al.*, (1986). Claiming to have operated an old museum net in the same way as used at the turn of the century, these authors made an interesting comparative fisheries in July and August 1985. Twelve hauls were made on 4 stations between Glückstadt and Hamburg (Figure 1), resulting in 7.0 kg total fish per hour, including 17 smelts of the 1+ group and 4 eels. Unfortunately, the number of flounders caught was not given. These catches represent 48% of the total catch, 75% of the eels and 31% of the 1+ smelts caught on average per hour by Apstein in July and August 1894. The statement of Peters *et al.*, however, that their results indicate a decline of fish stocks, is not convincing: while the catches examined by Apstein were made in a professional way by placing the nets close to the bottom (bottom distance 0–2 m in 59% of the hauls), Peters *et al.* during their fisheries kept a bottom distance of more than 2.5 m, thus avoiding the water layers preferably inhabited by flounder, eel, and smelt. The anchor net as used during these surveys is not the typical gear to catch lampreys, coregonids, cyprinids and bottom-dwelling species like the burbot. No serious statements about their present population sizes will be possible unless special gears like drift nets, shallow-water anchor nets and burbot traps are included in further surveys.

A number of species listed in Table 1, such as sterlet, rainbow trout and powan, do not belong to the original Elbe fauna and obviously have escaped from fish farms. Also carp is not indigenous to the lower Elbe, although its occurrence in sidewaters has been well documented for many decades (Duncker and Ladiges, 1960; Sterner, 1916a).

DEVELOPMENT OF FISHERIES

In 1894, 75 anchor net vessels were fishing on the lower Elbe River (Apstein, 1895) and 120 similar gears were operated from land during summer months (Voigt, 1870). Sterner (1916b) listed 90 motor vessels, 134 sailing vessels, 116 jollies, and 602 smaller boats which were used for fishing mainly in the lower Elbe or its estuary, offering work for 1650 men. Many of the smaller boats, however, were operated by part-time fishermen, for recreational fishing or by those 140 professional cast net fishermen who, at the turn of the century, were operating downstream from Hamburg harbour towards the mouth of the Lühe River (Schnakenbeck, 1928). The general decline of the Elbe fisheries began in the mid-1920s. Today, eight anchor net vessels are left, cast net fishing has been abandoned, and only 65 boats of part-time fishermen still are officially registered (Rönnau, 1987). Apart from this, a flourishing shrimp fishery has developed in the estuary downstream from Cuxhaven.

Fisheries for sturgeon ceased early this century. Until 1840, the formerly low valued sturgeon had been caught mainly by fishermen from the village of Glückstadt. A boom in sturgeon fishing started a few years later. Within half a century, overfishing depleted the stock. The annual landings from the Elbe and the Waddensea decreased from an estimated average of 7000 in the 1880s to 2800 in 1890 and to 19 in 1919 (Ehrenbaum, 1913; Schnakenbeck, 1928). In order to compensate for the decreasing catches, the mesh size of sturgeon nets was reduced from 20 to 12.5 cm, leading to an increase of small (< 150 cm) sturgeon in the catches from 3 per cent in 1898 to 12 per cent in 1901 and 88 per cent in 1911–13 (Ehrenbaum, 1916). Male sturgeon were known to mature at a size above 120 cm, females at a size above 160 cm. Mainly trawl fisheries in the Wadden Sea were accused of catching juvenile sturgeon, thus initiating the breakdown of the Elbe stock (Blankenburg, 1910). Ambitions to raise the minimum size of marketable sturgeon from 100 cm to 125 cm were delayed by a report of the 'Commission zur wissenschaftlichen

TABLE 5. Estimated catches from the lower Elbe (1886–1919) and landings by Elbe anchor net vessels at Cuxhaven fish market (1961–1986) and at the central marketing office for stock eel. All data in tons per year, except for sturgeon and salmon in numbers per year. See text for reliability of these data
Sources: 1886–1919: Calculated from 'Mitteilungen des Deutschen Seefischereivereins' (Rönnau, 1987)
1961–1975: Kühl (1976), Saint-Paul (1977)
1979–1986: Rönnau (1987)

Year	Sturgeon	Salmon	Shad	Smelt	Eel ≤35 cm	Eel > 35 cm	Ruffe	Cod	Flounder
1886–90	2800	320	37	?	175		150	0	122
1891–95	2737	1432	27	409	109		356	0	136
1896–00	1347	564	64	485	121		380	0	215
1901–05	982	1372	83	365	77		307	0	177
1906–10	412	209	78	367	112		217	0	226
1911–15	?	?	29	1230	205		150	0	406
1916–19	8	?	11	698	191		86	0	410
1961–65	0	?	4	175	66	4	11	19	12
1966–70	0	?	0	144	73	3	6	41	7
1971–75	0	?	0	152	55	2	5	47	6
1979	0	0	0	84	55	0	0	9	0
1980	0	0	0	125	65	0	0	33	0
1981	0	0	0	144	62	3	0	18	0
1982	0	0	0	156	70	4	0	55	1
1983	0	0	0	47	72	0	0	5	0
1984	0	0	0	11	70	1	0	4	0
1985	0	0	0	18	62	1	0	0	0
1986	0	0	0	21	59	0	0	8	0

Erforschung der Deutschen Meere' which failed in 1890 to accept the need for restrictions. The increase of the minimum size to 150 cm in the 1920s as well as efforts in artificial restocking (Ehrenbaum, 1913) were unsuccessful.

Early catch reports from the lower Elbe include salmon, but not trout (Table 5). It is supposed here, that both species had been marketed as salmon. The importance of 'salmon' fisheries clearly ranked behind that of sturgeon fisheries in the lower Elbe (Sterner, 1918e). The decrease of both stocks started during the same time, but still in the early 1930s some good salmon catches were made (Fick, 1932). In the case of salmon, the decline was mainly attributed to weir construction and the destruction of spawning grounds in upstream tributary rivers (Bauch, 1957). The same reason, in combination with water pollution, seems to have adversely affected attempts at restocking (Riedel-Lorjé and Gaumert, 1982). Even before 1800, salmon catches were known to be highly variable. Although salmon might have been more frequent in former centuries, reports of legendary mass occurrences have been doubted by Sterner (1918d) and Rambach (1801). The latter author argued that the name of the cheap and unpopular 'salmon' available around 1800 in Hamburg did not refer to fresh Elbe salmon, but to salted and dried 'nordic salmon'. Although two valuable species, neither sturgeon nor salmon ever have been the backbone of fisheries in the lower Elbe.

Due to its bony flesh, the twaite shad had never been accepted to any larger degree as a food fish in the Elbe region. Nevertheless it was caught during its spawning season in spring and mainly sold to the Rhine area, where the highly priced, morphologically similar allis shad had disappeared at the end of the 19th century (Sterner, 1918a). Today, shad is no more acceptable to consumers. Also the fishery for ruffe has ceased due to lack of demand. To a certain degree, both species today have been replaced by cod that during some winter seasons can be caught in profitable quantities by anchor net vessels in the estuary (Table 5).

By weight, smelt has always been the most important species on the Elbe. The pre-spawners, especially, caught from November to March, were a popular and cheap food. Surplus catches were sold to farmers as fertilizers and as chicken and cattle food (Schnakenbeck, 1928). During summer, a special anchor net fishery established to catch smelt fry which was used as bait for eel traps that were anchored in the vicinity of the vessels. This destructive fishing method has been accused of being detrimental to young year classes of different fish species (Voigt, 1870), although effects on local stocks could not be proved (Apstein, 1895). Highest smelt catches are reported for the period of World War I (Table 5), when prices for fish increased rapidly (Koos, 1922). Landings have constantly decreased thereafter. Although still available in huge quantities (Möller, 1984, 1988), nowadays only about 20 t can be sold annually, mainly for export to Belgium and France.

The traditional backbones of the Elbe fisheries have been flounder and eel. Their officially registered catches have sharply decreased during the past three decades (Table 5). At the beginning of this century, two thirds of the registered vessels and boats were mainly operating flounder gears (Sterner, 1916c). In 1880 trawl fishing for flounder was introduced into the Elbe estuary. It was mainly practised in winter, when storms made fishing along the open coast too dangerous. This fishing method was rather effective and soon became a serious competition to traditional drift and gill net methods. Within only four months from October to December 1923, 462 t of flounder were landed by trawlers from the Elbe estuary (Schnakenbeck, 1926). Trawling was restricted by order downstream from a line between Neufeld and Oste. Obviously, this restriction was not accepted by many fishermen (Heidrich, 1924). Sterner (1916c) estimated that at the beginning of this century professional flounder fishermen caught 400 t annually, while the amount caught by part-time fishermen remained unknown. The official landings of flounder as compiled in Table 5 were doubted by Sterner (1916c) who from more precise sources estimated that the real catch was about 2.5 times higher. Most flounders were sold fresh, others were dried. Being fatter and more tasty, the 'freshwater' flounder caught near Hamburg was preferred to the meager and frequently diseased flounder from the estuary (Schäperclaus, 1927; Sterner, 1916c). This difference in quality is still evident today (Möller, 1984; in preparation), although Schnakenbeck (1926) demonstrated by tagging experiments that both groups belong to the same species and mix during winter months in the estuary. Flounder is still abundant today but it is an unwanted catch, in shrimp and anchor net fisheries, due to low market price. Trawl fishing for flounder in the estuary as well as gill net fishing has ceased.

Elbe eel is caught either for human consumption (above 35 cm) or for stocking purposes (below 35 cm). Official statistics report annual average catches of approximately 140 t for the period 1886–1919. Based on investigations by Sterner (1916d) for the year 1908, it is estimated that the data in Table 5 have to be multiplied by a factor of 2 to obtain more realistic values. During this year, about 4 million eel, weighing 100 t were sold for stocking purposes from the village of Altenwerder, the traditional center of eel fisheries. Today, stock eels are marketed centrally through the 'Aalversandstelle' in Hamburg. Landings, which have been reported since 1965, fluctuate around an annual mean of 65 t (Table 5).

The high demand for large eel today enables the fishermen to sell their catches completely outside the fish market. Landings as recorded for the period since 1961, in Table 5, are not at all representative. From the known weight relation between stock eel and market eels in the catches, as well as from research catches made in 1985, it can be estimated that today 100–120 t eel are caught annually in the lower Elbe, about half for stocking purposes and half for direct human consumption. Due to its high price and the rather constant catches, today the eel maintains a financially healthy Elbe fishery.

Although incomplete data are available, it may be estimated that around the turn of the century (1896–1905) annually about 1000 sturgeons, 1000 salmons, 500 t flounder, 450 t smelt, 350 t ruffe, 200 t eel, and 75 t shad were caught in the lower Elbe. The annual landings in the 1980s are estimated to be 100 t eel, 75 t smelt, and 18 t cod with a strong decline in smelt landings and recently increasing catches of trout.

Reasons for change

The tonnage of ocean-going ships, as well as of river boat cargo landed at the port of Hamburg tripled from 1873 to 1893 and again from 1893 to 1913 (Sterner, 1916b). The rise of Hamburg to be the largest port in continental Europe led to a rapidly increasing population and, consequently, to increased demand for fresh fish. As most kinds of fish were available in sufficient quantities, fisheries flourished and fishing villages like Altenwerder increased in prosperity (Breckwoldt, 1894). Species like herring, ruffe, shad or smelt that, during former decades, often had been sold to farmers as fertilizers or chicken food (Duge, 1927; Schnakenbeck, 1928; Voigt, 1870), now could be sold more profitably for human consumption. Fishermen and fish processors learned to smoke sturgeon flesh that previously had little value, and sturgeon eggs, formerly used as bait in eel traps, were now used for preparing kaviar (Rübcke, 1910).

The same cause that gave rise to the boom in fisheries a short time later caused its biggest problems. Demand for deeper shipping lanes led to further stream regulation; ship traffic seriously interfered with fishing activities and the bow waves generated by the ships adversely affected shallow nursery areas; high prices offered for species like sturgeon led to overfishing; and the increasing human population produced more and more sewage which affected fishing grounds in Hamburg harbour. Until the end of last century, Hamburg was supplied all year round with live fish, which was transported in tanks from the fishing grounds to the city. During the following years, however, complaints grew that the fish were asphyxiating during transportation through Hamburg harbour (Niemann, 1928; Puls, 1912). Mortalities resulted from the effects of increased sewage discharges; the advantages and disadvantages of which were discussed vehemently (Bonne, 1905; Schiemenz, 1912; Volk, 1910).

The decline of fisheries in the mid-1920s was commented upon in a number of articles (Bornholdt, 1925; Duge, 1927; Sicks, 1928; Wegener, 1927). The authors, however, could only present speculations about its causes. Water pollution, stream regulation and overfishing were the favourite themes. During the same period, serious marketing problems were created for the majority of small fishermen by banning the street sale of fish in Hamburg (Anonymous, 1925). Elbe fish now had to be sold through auctions at the fish market where they had to compete with increasing amounts of fresh sea fish landed by the expanding ocean fleet.

The first report on fish landings after World War II was published by Koops (1959). For the period 1949–56 he noted a drastic decline in the landings of all species, except for lamprey, eel and burbot (data for smelt and flounder were not presented). The reason for this was seen in the generally increasing wealth of the local population who now rejected the 'low quality' freshwater species. Another problem that the Elbe fisheries had to face was that fish caught near Hamburg were difficult to sell as they frequently tasted of phenol

(Imam *et al.*, 1958; Mann, 1969). This observation can still be made today, although a reduction of the problem was noted in the mid-1960s (Mann, 1968).

In recent reviews, however, a mainly pollution- and stream regulation-induced decline of fish stocks has been identified as responsible for the decrease of fisheries in the lower Elbe (Kühl, 1976; Mann, 1968; Peters, 1978; Wilkens and Köhler, 1977). This assumption was based on the observations that landings declined and major fishing activities shifted towards the estuary while the industrial and commercial use of the river increased. This evaluation has to be considered a simplification which (like others) cannot be proved as both reliable catch and profit statistics from fisheries are lacking.

Flounder, eel and smelt which supported the flourishing fisheries early this century can still be caught in huge quantities in the lower Elbe today (Möller, 1984, 1988). The most aggravating change has been the changing behaviour of consumers. Today, ready-to-eat products and fillets are preferred to the relatively small, bony, and often heavily parasitized, species caught in the Elbe. Flounder can no more be marketed at reasonable prices. Smelt catches have reduced considerably, not because of the lack of fish, but because fishermen themselves have organized catch restrictions in order to keep the price at an acceptable level. Cod has become a profitable species also for anchor net fisheries, but its occurrence in the estuary is restricted to the winter months and there are large annual fluctuations. The Elbe fisheries today depend on eel and, despite annual fluctuations, there has been no dramatic decline in the catches neither for stock eel nor for large eel since the 1960s (Table 5).

PERSPECTIVES

The lower Elbe has suffered far less from stream regulation than most other large European rivers. Since the beginning of this century, the Wadden area between Cuxhaven and Hamburg has been reduced by 11 per cent and the shallow water area by 26 per cent (calculated from ARGE Elbe, 1984). However, serious alterations have taken place especially in the most important nursery grounds for Elbe fishes between Finkenwerder and the Schwinge estuary, where more than half the shallow water area has been lost. Despite these disadvantages, the present situation maintains a rich fish population which sustains a profitable business for the remaining eight anchor net vessels, although the most frequent species are of no, or only of low, commercial value. The greatest threat to fisheries today is not seen in any pollution-induced decline of fish stocks, but in the occurrence of pesticides and parasites in the fish, especially in eel.

The recent introduction of the swimbladder nematode *Anguillicola crassa* (Peters and Hartmann, 1986) is assumed to lead to a 100 per cent infection in the near future. First

TABLE 6. Percentage of Elbe eel exceeding the levels of toxicants tolerated by German food law (eel of various size and from various places between Cuxhaven and the boarder to the GDR) (Möller 1984, data provided by Veterinäruntersuchungsamt Itzehoe)

Substance	Tolerated level	Number of eels examined	Percentage exceeding tolerated level
Hg	1.0 ppm in muscle	194	53
$\alpha+\beta$HCH	0.5 ppm in fat	70	69
γHCH	2.0 ppm in fat	70	13
DDT	3.5 ppm in muscle	70	11
HCB	0.5 ppm in fat	70	96

observations indicate a severe pathogenicity of this parasite. Also the prevalence of the pathogenic nematode *Pseudoterranova decipiens*, which today can be found in the flesh of 10 per cent of Elbe eel and in the majority of adult smelt, may increase as the population size of the final host, the common seal, is increasing in the Wadden Sea (Möller and Klatt, 1988; Möller and Schröder, 1987). The most serious problem is the accumulation of various anthropogenic toxicants in the flesh of eel. According to investigations made in 1980–82, half of the specimens examined exceeded the accepted level of mercury and nearly all eel exceeded the level of at least one of the chlorinated hydrocarbons considered by German food legislation (Table 6). Similar results were presented by Krüger and Kruse (1982) and Kruse *et al.* (1983). This means a ban on marketing Elbe eel for human consumption. The situation is similar for bream (ARGE Elbe, 1982) and it seems to be similar for other species which spend all, or most, of their lifetime in the lower Elbe (Möller, 1984).

The conclusion can be drawn that today it is not fish stocks but fisheries that suffer chiefly from the introduction of toxicants into the lower Elbe. Encouraging progress, both from the view of the fish and the fisherman, has been achieved towards a more acceptable water quality by the restoration of several tributary rivers and by a considerable reduction of waste input from the city of Hamburg and several industrial areas. However, the key to further fundamental improvement of conditions in the lower Elbe is in the future cooperation in water protection by the three Elbe countries: Czechoslovakia, GDR and West Germany.

REFERENCES

Anonymous 1925. Die Lage der Kleinfischer an der Unterelbe. *Fischerbote*, **17**, 428.

Apstein, C. 1895. *Bericht über die im Auftrage des Deutschen Seefischerei-Vereins unternommene Untersuchung der Steerthamen in der Unterelbe*, Unpublished manuscript, Zoologisches Institut der Universität Kiel, 229 pp.

ARGE Elbe 1982. *Chlorierte Kohlenwasserstoffe—Daten der Elbe*, Hamburg, Wassergütestelle Elbe, 107 pp.

ARGE Elbe 1984. *Gewässerökologische Studie der Elbe*, Hamburg, Wassergütestelle Elbe, 98 pp.

Bauch, G. 1957. Der Elblachs (*Salmo salar* L.), seine Biologie und wirtschaftliche Bedeutung. *Zeitschrift für Fischerei*, **6**, 241–250.

Blankenburg, U. 1910. Von der Störfischerei in der Elbe. *Fischerbote*, **2**, 7–12.

Blankenburg, U. 1912. Die Fischerei in der Unterelbe. *Fischerbote*, **4**, 241–247.

Bonne, G. 1905. Die Vernichtung der deutschen Flussfischerei durch die Verunreinigung unserer Gewässer, mit besonderer Berücksichtigung der Verhältnisse auf der Unterelbe. *Zeitschrift für Fischerei*, **12**, 1–28.

Bornholdt, H. 1925. Von der Kleinfischerei auf der Unterelbe im Jahre 1924 und was uns davon unsere Fischer erzählen. *Fischerbote*, **17**, 27–28.

Breckwoldt, J. J. 1894. *Historische Nachrichten über die Elbinsel Altenwerder*, Harburg, Verlag Danckwert'sche Buchhandlung, 60 pp.

Dehus, P., Neuman, M. and Schubert, C. Daten zur limnischen Fischfauna Schleswig-Holsteins. *Berichte aus dem Institut für Meereskunde*, Kiel (in preparation).

Dieckwisch, B. 1987. *Die Verteilung der Fischbrut in der Unterelbe 1985*. Unpublished Diplomarbeit, Universität Kiel.

Diercking, R. personal communication. Institut für Hydrobiologie und Fischereiwissenschaft, Hamburg.

Duge, F. 1927. Befürchtungen für die Elbfischerei vor 150 Jahren. *Fischerbote*, **19**, 3–10.

Duncker, G. and Ladiges, W. 1960. *Die Fische der Nordmark*, Hamburg, Kommissionsverlag Cram, de Gruyter & Co., 432 pp.

Ehrenbaum, E. 1913. Über den Stör. *Fischerbote*, **5**, 142–149.

Ehrenbaum, E. 1916. Zum Antrage auf Erhöhung des gesetzlichen Mindestmaßes für den Stör. *Fischerbote*, **8**, 31–33.

Fick, H. 1932. Vom Lachsfang in der Niederelbe 1932. *Fischerbote*, **24**, 197–199.
Fiedler, M. 1988. *Die Bedeutung von Zoobenthos und Zooplankton der Unterelbe als Fischnahrung*, Unpublished Dissertation, Universität Kiel.
Flügge, H. 1988. Hydrological changes and pollution of the lower Elbe, Germany (personal communication).
Fric, A. 1859. Kritisches Verzeichnis der Fische Böhmens. *Lotos*, **9**, 199–205.
Gaumert, D. 1981. *Süßwasserfische in Niedersachsen. Arten und Verbreitung als Grundlage für den Fischartenschutz*, Hannover, Niedersächsisches Ministerium für Ernährung, Landwirtschaft und Forsten, 134 pp.
Hartmann, U. 1987. *Ökologische Untersuchungen zur Fortpflanzungsbiologie einiger Fischarten in der Stör*, Unpublished Diplomarbeit, Universität Hamburg.
Heidrich, H. 1924. Die Schleppnetzfischerei in der Unterelbe und die Wanderungen des Elbbutts. *Fischerbote*, **16**, 73–79.
Holst, J. 1931. *Das Leben und Fischen der Altenwerder Fischer in alter und neuer Zeit*, Altenwerder, Bielenberg, 49 pp.
Imam, A. K., Lühmann, H. and Mann, H. 1958. Über Neunaugen und Neunaugenfischerei in der Elbe. *Fischwirt*, **8**, 249–261.
Koops, H. 1959. *Der Quappenbestand der Elbe*. Untersuchungen über die Biologie und fischereiliche Bedeutung der Aalquappe im Hinblick auf die Auswirkungen des im Bau befindlichen Elbstaues bei Geesthacht. *Kurze Mitteilungen des Instituts für Fischereibiologie der Universität Hamburg*, **9**, 1–60.
Koos, H. 1922. Die Lage der Elbfischerei. *Fischerbote*, **14**, 43–44.
Krüger, K. E. and Kruse, R. 1982. Der Quecksilber-, Blei- und Cadmiumgehalt in Aalen aus der Elbe und ausgewählten Vergleichsgewässern. *Archiv für Lebensmittelhygiene*, **33**, 123–128.
Kruse, R., Boek, K. and Wolf, M. 1983. Der Gehalt an Organochlor-Pestiziden und polychlorierten Biphenylen in Elbaalen. *Archiv für Lebensmittelhygiene*, **34**, 81–86.
Kühl, H. 1976. Zum Stand der Hamenfischerei in der Unterelbe. *Neues Archiv für Niedersachsen*, **25**, 315–324.
Lillelund, K. 1961. Untersuchungen über die Biologie und Populationsdynamik des Stintes Osmerus eperlanus (Linnaeus 1758), der Elbe. *Archiv für Fischereiwissenschaft*, **12**, Beiheft, 1–128.
Lübbert, H. and Wiese, E. 1949. *Hamburger Fischerei in zehn Jahrhunderten*, Hamburg, Verlag Gesellschaft der Freunde des vaterländischen Schul- und Erziehungswesens, 122 pp.
Mann, H. 1968. Die Beeinflussung der Fischerei in der Unterelbe durch zivilisatorische Maßnahmen. *Helgoländer wissenschaftliche Meeresuntersuchungen*, **17**, 168–181.
Mann, H. 1969. Geschmacksbeeinflussung bei Fischen. *Ernährungswissenschaft*, **71**, 10–24.
Möller, H. 1984. *Daten zur Biologie der Elbfische*, Kiel, Möller, 217 pp.
Möller, H. 1988. *Fischbestände und Fischkrankheiten in der Unterelbe 1984–1986*, Kiel, Möller, 344 pp.
Möller, H. 1987. Spawning and feeding migrations of fishes in the Elbe estuary. In Tougaard, S. and Asbirk, S. (Eds), *Proceedings of the 5th International Wadden Sea Symposium*, Copenhagen, The National Forest and Nature Agency, 87–101.
Möller, H. Adverse hydrographical living conditions for fish in an estuary (in preparation).
Möller, H. and Klatt, S. (in press). The role of smelt as transmitter of sealworm Pseudoterranova decipiens in the Elbe estuary. *International Council for the Exploration of the Sea*, C.M./E:130.
Möller, H. and Schröder, S. 1987. Neue Aspekte der Anisakiasis in Deutschland. *Archiv für Lebensmittelhygiene*, **38**, 121–128.
Morawa, F. 1982. Süßwasserfische in Niedersachsen. *Fischer und Teichwirt*, **33**, 200–202.
Nellen, W. and Dehus, P. 1985. Jüngere Ergebnisse zur Situation der Süßwasserfischfauna in Schleswig-Holstein. *Schriften des naturwissenschaftlichen Vereins Schleswig-Holstein*, **55**, 63–80.
Niemann, H. 1928. Das Fischabsterben im Sommer 1928. *Fischerbote*, **20**, 164–165.
Peters, G. and Hartmann, F. 1986. Anguillicola, a parasitic nematode of the swim bladder spreading among eel populations in Europe. *Diseases of Aquatic Organisms*, **1**, 245–260.
Peters, N. 1978. Die Niederelbe—eine Landschaft im Umbruch. *Umschau*, **78**, 35–40.
Peters N., Schmidt, W., Gercken, J., Kranz, H. and Waterman, B. 1986. Die Fische der Unterelbe im Vergleich der Jahre 1894 und 1982—eine Richtigstellung, *Archiv. für Fischereiwissenschaft*, **36**, 287–304.
Puls, G. 1912. Die Eingabe der unterelbischen Fischer. *Fischerbote*, **4**, 15–16.

Rambach, J. J. 1801. *Versuch einer physisch-medicinischen Beschreibung von Hamburg*, Hamburg, Rambach.
Riedel-Lorjé, J. C. and Gaumert, T. 1982. 100 Jahre Elbe-Forschung. Hydrobiologische Situation und Fischbestand 1842–1943 unter dem Einfluß von Stromverbau und Sieleinleitungen. *Archiv für Hydrobiologie Supplement*, **61**, 317–376.
Röbcke, E. personal communication. *Groß Kampen*, Beidenfleth.
Rönnau, K. 1987. *Historische Entwicklung und gegenwärtiger Stand der Fischerei auf der Unterelbe*, Unpublished Diplomarbeit, Universität Kiel.
Rübcke, R. 1910. Die Finkenwärder Elbfischerei in früherer Zeit. *Fischerbote*, **2**, 148–149.
Rübcke, R. 1911. Die Fischerei in der Unterelbe. *Fischerbote*, **3**, 221–222, 262–263, 294, 325, 353.
Rübcke, R. 1912. Die Fischerei in der Unterelbe. *Fischerbote*, **4**, 255, 292, 321, 364–365.
Rübcke, R. 1914. Die Fischerei in der Unterelbe. *Fischerbote*, **6**, 264, 316, 419.
Saint-Paul, U. 1977. Einige Aspekte der deutschen Satzaalfischerei. *Archiv für Fischereiwissenschaft*, **28**, 123–133.
Schäperclaus, W. 1927. Lymphocystiserkrankungen bei Flundern und Schollen und ihre Bekämpfung. *Mitteilungen des deutschen Seefischerei-Vereins*, **43**, 96–99.
Schiemenz, P. 1912. Die Elbfischerei und die Abwässer von Hamburg-Altona. *Fischerbote*, **4**, 48–50.
Schnakenbeck, W. 1926. Biologische Untersuchungen über den Elbbutt und die Schleppnetzfischerei in der Unterelbe. *Fischerbote*, **18**, 232–241.
Schnakenbeck, W. 1928. Die Nordseefischerei. In Lübbert, H. and Ehrenbaum, E. (Eds), *Handbuch der See-Fischerei Nordeuropas*, Vol. 5(1), Stuttgart, Schweizerbart'sche Verlagsbuchhandlung, 229 pp.
Schnakenbeck, W. 1937. Wie Hamburg und Altona Fischereistädte wurden. In *Von Fischerei und Fischmärkten in Hamburg und Altona*, Altona, Alster Verlag, 11–172.
Sicks, H. 1928. Blüte und Niedergang der Altenwerder und Neuhofer Fischerei. *Fischerbote*, **20**, 186–188.
Sterner, E. 1916a. Alsterbefischung 1915. *Fischerbote*, **8**, 45–46.
Sterner, E. 1916b. Die niederelbische Küstenfischerei. *Fischerbote*, **8**, 69–76.
Sterner, E. 1916c. Die niederelbische Küstenfischerei. II. Die Fischereiarten. 1. Die Buttfischerei. *Fischerbote*, **8**, 117–132.
Sterner, E. 1916d. Die niederelbische Küstenfischerei. II. Die Fischereiarten. 2. Die Aalfischerei. *Fischerbote*, **8**, 190–202.
Sterner, E. 1916e. Die niederelbische Küstenfischerei. II. Die Fischereiarten. 3. Die Stintfischerei. *Fischerbote*, **8**, 240–244.
Sterner, E. 1916f. Ergänzungsbestimmungen zum Fischereigesetz im Gebiet der niederelbischen Küstenfischerei. *Fischerbote*, **8**, 253–257.
Sterner, E. 1916g. Die niederelbische Küstenfischerei. II. Die Fischereiarten. 4. Die Sturenfischerei. *Fischerbote*, **8**, 311–314.
Sterner, E. 1918a. Die niederelbische Küstenfischerei. II. Die Fischereiarten. 5. Der Maifischfang. *Fischerbote*, **10**, 13–16.
Sterner, E. 1918b. Die niederelbische Küstenfischerei. II. Die Fischereiarten. 6. Die Schnepel- und Neesenfischerei. *Fischerbote*, **10**, 17–19.
Sterner, E. 1918c. Die niederelbische Küstenfischerei. II. Die Fischereiarten. 7. Die Quappenfischerei. *Fischerbote*, **10**, 72–75.
Sterner, E. 1918d. Die niederelbische Küstenfischerei. II. Die Fischereiarten. 8. Die Lachsfischerei. *Fischerbote*, **10**, 76–83.
Sterner, E. 1918e. Die niederelbische Küstenfischerei. II. Die Fischereiarten. 9. Die Störfischerei. *Fischerbote*, **10**, 179–203.
Sterner, E. 1918f. Die niederelbische Küstenfischerei. II. Die Fischereiarten. 10. Die übrigen Fischereiarten im oberen Teil. *Fischerbote*, **10**, 245–253.
Voigt, J. F. 1870. *Ueber den Fischereibetrieb auf der Unter-Elbe*, Hamburg, Grüning, 27 pp.
Volk, R. 1910. Die Bedeutung der Sielabwässer von Hamburg-Altona für die Ernährung der Elbfische. *Fischerbote*, **2**, 57–60, 84–89.
Wegener, H. 1927. Die Kurrenfischerei in der Unterelbe und die Erhaltung der Existenz unserer Elbe-Berufsfischer. *Fischerbote*, **19**, 10–13.
Wilkens, H. and Köhler, A. 1977. Die Fischfauna der unteren und mittleren Elbe: die genutzten Arten, 1950–1975. *Abhandlungen und Verhandlungen des naturwissenschaftlichen Vereins zu Hamburg*, **20**, 185–222.

Historical Change of Large Alluvial Rivers: Western Europe
Edited by G. E. Petts
©1989 John Wiley & Sons Ltd

13

Historical Changes in the Fish Fauna of the River Duero Basin

JAVIER LOBON-CERVIA*, BENIGNO ELVIRA and PEDRO A. RINCON

National Museum of Natural Sciences, CSIC, José Gutierrez Abascal, 2, 28006 Madrid, Spain

INTRODUCTION

The River Duero rises in Los Picos de Urbion (Soria) at about 2080 m above sea level, and flows westwards through the flat lands of Old Castile, Leon and Portugal. After about 937 km the river flows into the Atlantic Ocean close to Porto (Figure 1). Its drainage basin of 98 375 km^2 is shared between Spain (80 per cent) and Portugal (20 per cent). The central area is high at about 800 m above sea level. Rain and snow combine to generate an average discharge of 570 $m^3 s^{-1}$ with a maximum recorded discharge of 20 000 $m^3 s^{-1}$ and a minimum of 1 $m^3 s^{-1}$. Other general physiographic and biological data are given in Sole-Sabaris (1978), Lobon-Cervia *et al.* (1986) Garcia de Jalon and Lopez Alvarez (1983) and Garcia de Jalon and Gonzalez del Tanago (1983). Although sharing a topographic divide, the fish fauna of the Duero and Ebro differ markedly (Sostoa and Lobon-Cervia, this volume). Moreover, unlike the Ebro, there is a valuable taxonomic record for assessing historical change.

THE FISH FAUNA

For the purposes of this study, the fish species of the River Duero can be classified into three groups; (1) freshwater species (*sensu strictu*), (2) migratory species, both catadromous and anadromous, and (3) eurihaline species. In this paper, the major concern is with the freshwater species because this group has been subjected to more important historical changes than any other group. Some migratory and eurihaline species have also been severely damaged and have declined in numbers dramatically.

Freshwater species

A list of fish species living in the River Duero catchment area together with their origin, is given in Table 1. Only 20 species comprise the freshwater group, i.e. 77 per cent of the

*Correspondence address to this author, Department of Organisms and Systems Biology, University of Oviedo, 33071 Oviedo, Spain

FIGURE 1. Geographical situation of the River Duero drainage basin within the Iberian Peninsula, Letters by the map denote the main tributaries: P is River Pisuerga; E, Esla; A, Adaja and T, Tormes, D is the main Duero corridor

TABLE 1. Native (N) and introduced (I) species living in the River Duero drainage basin and their present status. A and C refer to anadromous and catadromous species and F is freshwater species (*sensu strictu*). Ab is abundant; C, common; R, rare; V, vulnerable; E, extinct and ? is undocumented. Note that eel (*A. anguilla*) is extinct throughout most of the basin (Figure 2) but is abundant below the reservoirs of the central corridor. Notations are used as recommended by ICONA (1986)

Petromizontidae			
Petromizon marinus	N	A	R
Acipenseridae			
Acipenser sturio	N	A	E
Clupeidae			
Alosa alosa	N	A	V
Alosa fallax fallax	N	A	V
Anguillidae			
Anguilla anguilla	N	C	E & Ab
Salmonidae			
Hucho hucho	I	F	?
Salmo gairdneri	I	F	C
Salmo salar	N	A	E
Salmo trutta	N	F	Ab
Salvelinus fontinalis	I	F	R
Esocidae			
Esoz lucius	I	F	C
Cyprinidae			
Barbus bocagei	N	F	Ab
Chondrostoma polylepis	N	F	Ab
Carassius auratus	I	F	C
Cyprinus carpio	I	F	C
Gobio gobio	I	F	Ab
Leuciscus sp.	N	F	Ab
Phoxinus phoxinus	I	F	Ab
Rutilus arcasii	N	F	Ab
Rutilus arburnoides	N	F	Ab
Rutilus lemmingii	N	F	C
Rutilus macrolepidotus	N	F	C
Tinca tinca	N	F	Ab
Cobitidae			
Cobitis calderoni	N	F	C
Poecilidae			
Gambusia affinis	I	F	Ab
Centrarchidae			
Micropterus salmoides	I	F	C

total number of species, with 50 per cent of them being native whilst 50 per cent have been introduced.

The taxonomy of the native species has been recently reviewed and is rather well documented (Collares-Pereira, 1983; Doadrio, 1984; Elvira, 1987a, 1987b). The only exception seems to be the population of chub. For many years (Lozano-Rey, 1935; IFIE, 1952; Almaca, 1965; and others), their populations were considered as belonging to the taxonomic group of *Leuciscus cephalus cabeda* but recently they have been found to be a new species although its description has not been published yet (Doadrio, 1987); in this paper, we deal with the species as *Leuciscus* sp. Native freshwater species are the following: brown trout (*Salmo trutta*), Iberian barbel (*Barbus bocagei bocagei*), Iberian nase (*Chondrostoma polylepis polylepis*), duero's chub (*Leuciscus* sp.), four species of roach (*Rutilus alburnoides, R. arcasii, R. macrolepidotus* and *R. lemmingii*), tench (*Tinca tinca*) and calderon's loach (*Cobitis calderoni*).

All the native Cyprinidae and Cobitidae except tench are endemic to the Iberian Peninsula although none of them is restricted to the River Duero Basin. However, in spite of the large catchment area these families contain a low number of species in comparison to other European basins; a fact that applies to all the rivers of the Iberian Peninsula (Daget, 1968; Sostoa and Lobon-Cervia, this volume). Among many comparative examples in rivers flowing to the Mediterranean Sea Basin, Daget and Economidis (1975) recorded 17 species for the River Evros and 18 for the Strymon, both in the balkan Peninsula, while Gandolfi and Le Moli (1977) reported 13 species for the Italian River Po. A comparison with Central and Northern European rivers flowing to the Atlantic Ocean would increase the difference. In Poland, for much smaller drainage basins, Kolder *et al.* (1974) recorded 32 species for the River Raba and tributaries while Mahon (1984) reported 22 species for the River Nida.

The contemporary distribution of fishes within the Duero reflects a long history of stocking and introduction. The original distribution of brown trout (*S. trutta.*) in the River Duero has drastically changed in the last century mainly because of human impacts. The first known trout-farm in Spain was built in 1866 in La Granja (Segovia) taking water from the River Eresma, a southern tributary of the River Duero. As this farm succeeded in raising fish, other farms were established. The progressive development of this industry has had several consequences, including: (1) continuous escapes of individuals from these farms; (2) colonization of new reaches of streams by escaped individuals; and (3) mixing of their genetic pools with those of natural populations. Further, in the 1940s and later years, the Fisheries Services of the State Administration, in order to promote angling and due to its poor facilitates to manage fish, used trout from these farms to stock streams, thereby increasing their distribution and genetic mixing.

Throughout the River Duero and its tributaries, four species (*B. bocagei, C. polylepis, Leuciscus sp.* and *C. calderoni*) are common and widely distributed, while roach (*Rutilus* spp.) show more limited distribution. *R. arcasii* occupies all the central, northern and eastern areas of the basin while *R. macrolepidotus*, its closest relative (Elvira, 1987b), lives only in the westernmost streams. *R. lemmingii* is restricted to the River Huebra-Yeltes drainage basin (lefthand-side tributary), and *R. alburnoides* is known to occur in the southwestern waters (i.e. River Agueda).

Tench (*T. tinca*) has also been traditionally grown in ponds throughout the river basin, although not intensively, and it has been further subjected to river stocking. Today it is not possible to separate accurately its original distribution from that resulting from human manipulation.

Ten species belonging to five families have been introduced in historical times (Table 2). Carp (*C. carpio*) and goldfish (*C. auratus*) have been reported in this river since the 17th century and have been stocked up continuously to the present day. Gudgeon (*G. gobio*) is thought to have been introduced by the mid-19th century and it seems to be associated with the first stages of trout farming. Earliest references are due to Ugarte (1929) and Lozano-Rey (1935). Since these records, its distribution has increased significantly, nowadays occupying most, if not all, the river basin (Figure 2). A similar situation prevails

TABLE 2. Historical changes observed in the fish fauna of River Duero drainage basin in relation to dates of introduction and extinction (—) of species. The top total number of species in each particular period of time is given together with the theoretical number of species (extinctions are taken into consideration) which is given in parenthesis

	Dates							
	Native	17th	1865	1900	1925	1950	1975–present	Introduced in
P. marinus	*	*	*	*	*	*	*	
A. sturio	*	*	*	*	*	—	—	
A. alosa	*	*	*	*	*	*	*	
A. fallax	*	*	*	*	*	*	*	
A. anguilla	*	*	*	*	*	*	*	
S. salar	*	*	*	—	—	—	—	
S. trutta	*	*	*	*	*	*	*	
B. bocagei	*	*	*	*	*	*	*	
C. polylepis	*	*	*	*	*	*	*	
Leuciscus sp.	*	*	*	*	*	*	*	
R. arcasii	*	*	*	*	*	*	*	
R. alburnoides	*	*	*	*	*	*	*	
R. lemmingii	*	*	*	*	*	*	*	
T. tinca	*	*	*	*	*	*	*	
C. calderoni	*	*	*	*	*	*	*	
C. auratus		*	*	*	*	*	*	17th cent.
C. carpio		*	*	*	*	*	*	17th cent.
S. gairdneri				*	*	*	*	late 19th
S. fontinalis				*	*	*	*	late 19th
G. gobio					*	*	*	19th–20th
P. phoxinus					*	*	*	19th–20th
G. affinis						*	*	1926
E. lucius						*	*	1950
M. salmoides							*	1955
H. hucho							*	1968
Number of species	16	18	18	20 (19)	22 (21)	24 (22)	26 (24)	

for minnow (*P. phoxinus*). Some earlier reports (Lozano-Rey, 1935 and references therein) lead us to believe that minnow is native to the northwestern rivers of the Iberian Peninsula and its colonization in the Duero must be due to similar reasons to that of gudgeon. It is likely that its colonization of the River Duero occurred in the early 20th century because the same author (Lozano-Rey, 1919) reported 12 specimens from Vizcainos de la Sierra (Burgos) captured in 1918. Because minnow select very different habitats to gudgeon, their distribution is limited to the uppermost reaches of streams and rivers although they are also spreading rapidly.

Rainbow trout (*S. gairdneri*) was also introduced to Spain and specifically to the Duero Basin by the end of the 19th century. Early references were reported by Ugarte (1929) and Nobre (1932). Actually, their distribution is limited to places where they are frequently stocked and/or to rivers and streams flowing close to trout-farms from where they frequently escape. As with rainbow trout, char (*S. fontinalis*) was brought from North America by the end of the 19th century and was stocked in several places mainly in lakes and reservoirs

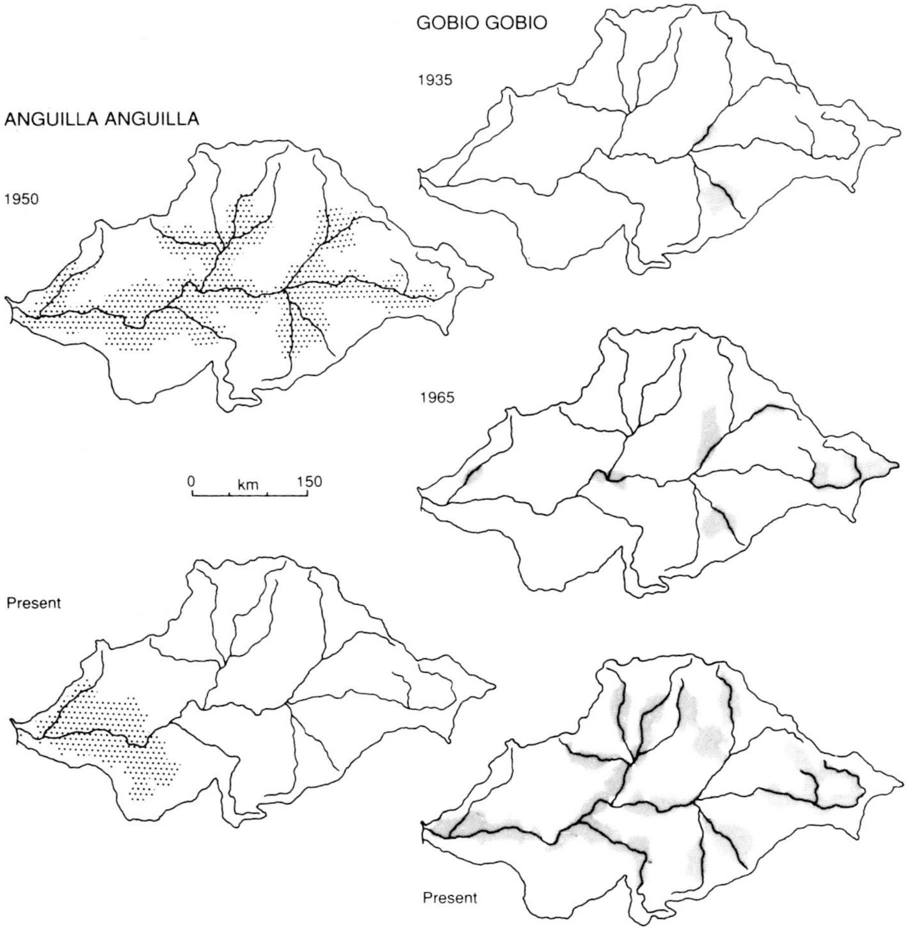

FIGURE 2. Historical changes in the distribution of gudgeon (*G. gobio*) and eel (*A. anguilla*) in River Duero basin

(Lozano-Rey, 1935). Because of recent introductions they occur, at least, in Lake Sanabria (Zamora), Laguna deBeila (Burgos) and Navacerrada (Segovia). One more salmonid, the Huchen (*H. hucho*) originally from Czechoslovakia, was stocked in 1968 in two tributaries of the River Duero; in the River Orbigo where the species rapidly disappeared and in the River Tormes in which the species thrived (Anonymous, 1969, 1973; Mondejar, 1981; Holcik *et al.*, 1984).

Pike (*E. lucius*) was brought from France in 1950 (Calderon, 1955) and spread throughout the basin. A few years later in 1955 the Fisheries Services introduced black-bass (*M. salmoides*). Both species acclimatized very rapidly and today, are common in lowland reaches.

The mosquito fish (*G. affinis*) is perhaps the only case of a fish introduced for purposes other than angling. It was brought from North America in 1926 by the Medical Services to assist in the struggle against paludism and was, subsequently, stocked in several rivers (Sostoa *et al.*, 1984). Although the effect of *G. affinis* on the mosquito population (host of the paludism virus) was never assessed, or otherwise studied in any way, paludism disappeared years later. However, it is interesting to note that some references on the biometrical features of these introduced populations were published by de Buen (1930). Like the previously mentioned species, mosquito fish spread throughout the lowland reaches of the River Duero and its tributaries where it is still common.

Migratory species

At least six migratory taxa have been reported in the River Duero in historical times, these are: *Anguilla anguilla* (catadromous) and *Petromyzon marinus*, *Acipenser sturio*, *Salmo salar*, *Alosa alosa* and *Alosa fallax fallax* (anadromous). The populations of all these species have sharply declined in this century to the extent that they are either threatened or extinct (ICONA, 1986). In the last 30 years, *A. anguilla* has been reduced both in distribution area and numbers (Figure 2). Before the 1960s eel was really the most common and widely distributed species in all the basin. Its decline is consequent upon the construction of a dam in the main channel of the River Duero.

P. marinus seldom appears in the literature but has experienced similar changes to the eel. Today this species seldom reaches the Carrapatello Reservoir (Portugal) situated 60 km from the river mouth (Eiras, 1977). This reservoir is also the upstream limit of both shads (*A. alosa* and *A. fallax*) although their populations are still abundant enough to support some professional fisheries (Eiras, 1980). Sturgeon (*A. sturio*) were recorded in the River Duero upstream to Barca d'Alva (Portugal) at about 200 km from the Duero mouth. In the last decades, and presumably for the same reasons, no new records have been reported. Salmon (*S. salar*) were once common, but none has been reported in this century. There are also historical records of, at least, *Gasterosteus aculeatus*, *Chelon labrosus*, *Liza ramada*, *Liza aurata*, *Mugil cephalus cephalus* and *Platichthys flesus*. All of these have disappeared from the Duero.

HISTORICAL CHANGES: AN OVERVIEW

Although the construction of main-stream reservoirs since 1940's has confined the distribution of eel and shads, and contributed to the disappearance of Sturgeon and Atlantic salmon, the greatest impact of Man upon the fish fauna of the Duero, over the historic timescale has been stocking with exotic species. The first study of the icthyology of the River Duero was undertaken by Steindachner (1866a, 1866b) who visited the river between 1864 and 1865. For those years, at least 17 taxa, 11 freshwater (*sensu strictu*) and six migratory species were present in the Duero (Table 2). However, Steindachner failed to record three native (*R. alburnoides*, *R. lemmingii* and *R. macrolepidotus*), two migratory (*P. marinus* and *A. fallax*) and one introduced (*C. auratus*) species. Their absence from Steindachner's work could be due simply to inadequate sampling techniques (fishing with nets, etc). Other studies made during the late 19th century by several Spanish and Portuguese workers (Capello, 1881; Osorio, 1888, 1894, 1895, 1896; Vieira, 1894, 1897) do not offer any new contributions to the statements of Steindachner. Introductions, adaptations and fish farming of foreign species began in these years as Maximiliano de la Paz Graells, a well-known naturalist of the time, was appointed as advisor to Queen Elizabeth II. His trips to France strongly influenced the development of natural sciences and fish farming (Graells, 1867).

Notable contributions to the knowledge of the Duero fish fauna were made during the early 20th century. Motivated by the general lack of information, Seabra (1911), Ugarte (1929), Velaz de Medrano and Ugarte (1930). Nobre (1932), de Buen (1930, 1935) and Lozano-Rey (1935) built the base of freshwater ichthyology. Furthermore, from these papers we have been able to infer, among many other things, that gudgeon (*G. gobio*) and minnow (*P. phoxinus*) were introduced and established in this period (Table 2) and that the first specimens of both species probably escaped from trout-farms where they were grown as food for trout. Still earlier, these specimens could originally have come from France as this country was the main supplier of trout for farms and, later on, of pike and black-bass.

Prior to the mid-20th century exotic species generally appeared in rivers as a result of accidents. Since 1950 the colonization of the Duero system by foreign species has been the result of administrative decision. In this way, pike (*E. lucius*), black-bass (*M. salmoides*) and huchen (*H. huicho*) were successively introduced in 1950, 1955 and 1968 respectively by administrative agencies lead by forest engineers. Furthermore, over the same period, these agencies increased and supported intensive stocking of those exotic species formerly introduced: *S. gairdneri*, *S. Fontinalis* and *C. carpio*. Trout-farming used foreign specimens imported from Germany, Italy and France.

No fisheries research at all was undertaken during the mid-20th century partly due to the civil war (1936–1939). Scientists, such as Dr. F. de Buen, left Spain and at this time Institutions devoted to research in these fields, such as the National Museum of Natural Sciences (CSIC, Madrid) were neglected, or were otherwise forced to abandon these branches of science (i.e. Instituto Forestal de Investigaciones y Experiencias, Ministry of Agriculture, Madrid). As a result, in Spain no more than six references on freshwater fish were published between 1940 and 1978 and none of them deal with the Duero fish!

Present and future

Although, since 1970's, the administration has reduced the intensity of stocking, particularly of *E. lucius* and *H. hucho* and, since 1981, of *S. fontinalis* (Figure 3), they are stocking four species (*S. trutta, S. gairdneri*, *M. salmoides* and *T. tinca*) in increasing numbers. As an example, data concerning stocking of individuals released in the streams and rivers of the River Duero during the period 1973–1983 are shown in Figures 3 and 4. Numbers of brown trout, rainbow trout and black-bass released in these waters in 1982 were 3.5, 5 and 10 times higher respectively, than in 1973 and 1974; totalling 3.5 millions in 1982.

It is worthwhile mentioning that if biological interactions have occurred after the introduction of these foreign species, they have not occurred to the extent of replacing native species by foreign species by competition or to the extinction of a native-prey by a foreign predator. Although these biological interactions have never been assessed in the River Duero, it appears that native species of the freshwater group (Cyprinidae and Cobitidae) have been, historically, rather stable in their distribution.

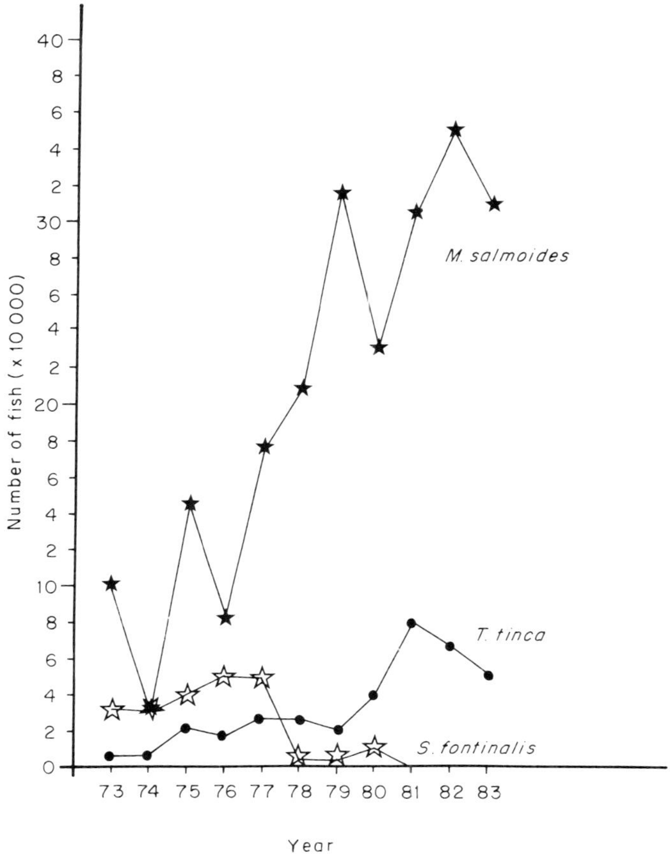

FIGURE 3. Number of black-bass (*Micropterus salmoides*), tench (*T. tinca*) and char (*Salvelinus fontinalis*) released per year in River Duero and tributaries for the period 1973–1983. Note that char (*S. fontinalis*) has not been stocked since 1981

In the 1980s, the problem remains that fish management is not based on scientific arguments but on the stocking of any river or stream without considering the status of the existing population. Moreover, stocking is made from farms, and farms still grow foreign specimens. This is a little surprising because in the last 10 years an important increase in the number of professionals working on fish has resulted in a huge volume of publications, especially on biology, ecology and culture.

This decade has witnessed a new problem: the last six new exotic species recorded in the Iberian Peninsula (Sostoa *et al.*, 1984) are not the result of fish escaping from farms (second half of 19th century) or administrative decisions (first half of 20th century) but of casual introductions. The two most important groups of people acting in this way are anglers and aquarium lovers. Anglers transport fish from one river to another as they use small, live cyprinids, as bait to fish pike, black-bass and trout; they press the administration to improve fisheries by increasing stocking; and they take their own decisions on which

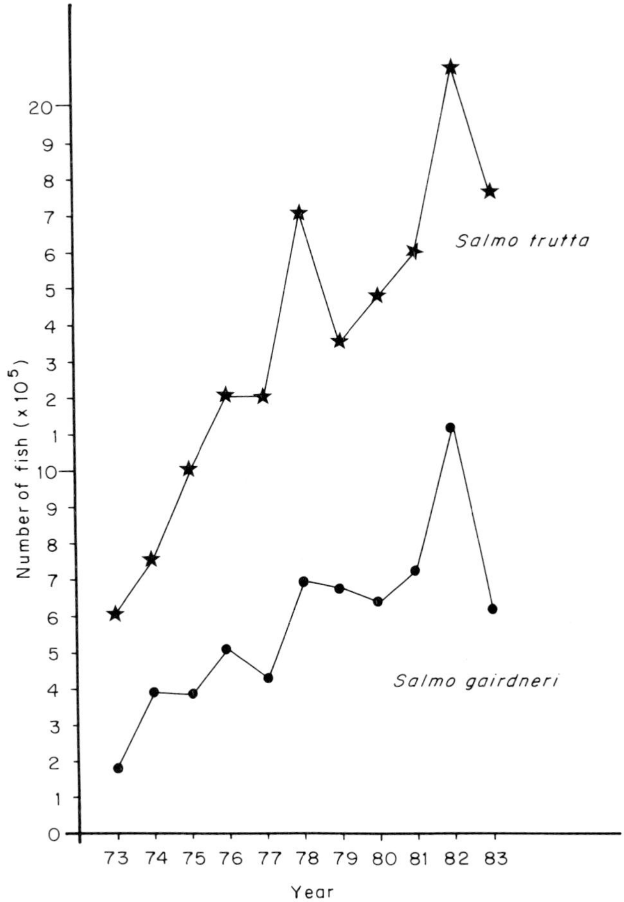

FIGURE 4. Number of individuals of brown trout (*Salmo trutta*) and rainbow trout (*S. gairdneri*) released per year in River Duero and tributaries for the period 1973–1983

species must be introduced to any particular water. Aquarium lovers release fish in any water where they assume fish can survive. Until recently, the River Duero has not been affected by these problems but several species are spreading rapidly in neighbouring rivers: *Lepomis gibbosus* in the River Tajo (Sostoa *et al.*, 1987), and *Ictalurus melas* in the River Ebro (Elvira, 1984). Clearly, such casual introductions are a potential threat in the near future.

ACKNOWLEDGEMENTS

Sampling in a large river basin like the one of the River Duero takes effort and time. Data presented herein is, therefore, the result of several years and were taken during different research projects. Sampling on River Adaja and other streams were supported by

CSIC–CAYCIT project n, 211. We acknowledge to all those people who assisted us in the field, especially to Cecilio Montañes, Miguel Guirado and Juan Cubo. Data on River Duero comes from samples of one of the authors (B.E.) and were taken during his Ph.D. Thesis research. All the samples were taken thanks to the facilities given by ICONA and the Dirección General de Pesca, Caza y Montes of the Communidad Autónoma de Castilla-León. The enthusiasm and kindness shown by Jesús Rueda (Director General) and R. Palomino (hunting and fishing staff) are especially and strongly acknowledged. We expect this and other studies actually in course to improve in some way conservation and management of one of the most interesting, historical and beautiful rivers of the Iberian Peninsula, the River Duero.

REFERENCES

Almaca, C. 1965. Contribution à la connaissance des poissons des eaux intérieures du Portugal. *Arq. Mus. Boc.*, **1**, 9–39.

Anonymous 1969. Recent introductions of fish, shrimps and oysters. *FAO Fish. Cult. Bull.*, **2**, 15.

Anonymous 1973. Sulec v Spaniji. *Ribic*, **32**, 255.

Capello, F. 1881. Catalogo dos peixes de Portugal. *Mem. Acad. R. Cienc. Lisboa*, **6**, 1–73.

Calderon, E. G. 1955. Acclimatation du brochet en Espagne. *Proc. Int. Assoc. Theor. Appl. Limnol.*, **12**, 536–542.

Collares-Pereira, M. J. 1983. *Estudo sistemático e citogenético dos pequenos ciprinideos ibéricos pertenecientes aos géneros Chondrostoma* Agassiz, 1835, *Rutilus* Rafinesque, 1820 e *Anaecypris* Collares-Pereira,1983. Ph.D. Thesis, Museu Bocage, University of Lisbon, 511 pp.

Daget, J. 1968. Diversité des faunes de poissons dans les cours déau du Portugal. *Arq. Mus. Boc. (2ª Ser.)*, 11, 21–26.

Daget, J. and Economidis, P. S. 1975. Richese spècifique de l'ichthyofaune de Macèdonie orientale et de Thrace occidentale (Grèce). *Bull. Mus. Nat. Hist. Nat.*, **346**, 82–84.

De Buen, F. 1930. Notas sobre la fauna ictiológica de nuestras aguas dulces. *Inst. Esp. Ocean. Not. y Res. Ser. II*, **46**.

De Buen, F. 1935. Fauna ictológica. Catálogo de los peces ibéricos, de la planicie continental, aguas dulces, pelágicos y de los abismos próximos. *Inst. Esp. Ocean. Not. y Res., Sr. II.*, **88**.

Doadrio, I. 1984. *Relaciones filogenéticas y biogeográficas de los barbos (Barbus, Cyprinidae) de la Peninsula Ibérica y aportes corológicos y biogeográficos a su ictiofauna continental*, Ph.D. Thesis, Dept. of Vert. Zoology, Complutense University of Madrid, 1119 pp.

Doadrio, I. 1987. *Leuciscus carolitertis* n. sp. from the Iberian Peninsula (Pisces: Cyprinidae). *Senkenb. biol.*, **68** (4/6) (in press).

Elvira, B. 1984. First records of North American catfish *Ictalurus melas* (Rafinesque, 1820) (Pisces, Ictaluridae) in Spanish waters. *Cybium*, **8**, 96–98.

Elvira, B. 1987a. *Revisión taxonómica y distribución geográfica del género Chondrostoma* Agassiz, 1835 (Pisces, Cyprinidae), Ph.D. Thesis, Dept. of Vert. Zoology, Complutense University of Madrid, 530 pp.

Elvira, B. 1987b. Taxonomic revision of the genus *Chondrostoma* Agassiz, 1835 (Pisces, Cyprinidae). *Cybium*, **11**, 111–140.

Eiras, J. C. 1977. Algumas medidas de protecçao ao sàvel (*Alosa alosa* L.) do Rio Douro. *Publ. Inst. Zool. 'Dr. A. Nobre'*, **134**, 3–8.

Eiras, J. C. 1980. *Alosa fallax* from North and South Portugal: study of its numerical characters. *Publ. Inst. Zool. 'Dr. A. Nobre'*, **155**, 7–15.

Gandolfi, G. and Le Moli, F. 1977. A preliminary report on fish distribution in the Po River. *Boll. Zool.*, **44**, 149–154.

Garcia De Jalon, D. and Gonzalez Del Tanago, M. 1983. Estudio biotipológico de las comunidades piscícolas de la cuenca del Duero. *Bol. Est. Cent. Ecol.*, **12**, 57–66.

Garcia De Jalon, D. and Lopez Alvarez, J. V. 1983. Distribución geográfica y mesológica de las especies piscícolas de la cuenca del Duero en el verano de 1981. *Actas Ier Congr. Esp. Limnol.*, 227–235.

Graells, M. de P. 1867. *Aqüicultura*, Madrid, 223 pp.

Holcik, J., Hensel, K., Nieslanik, J. and Skacel, S. 1984. *Hlavátka, Hucho hucho* (Linnaeus, 1758), Vyd. Slov. Akad. Vied, Bratislava, 307 pp.

Icona 1986. *Lista Roja de los Vertebrados de España*, Minist. Agricul. Pesc. & Alim., Secret. Gen. Tecn., 400 pp.

IFIE. 1952. Las colecciones de peces de la seccion de biologia de las aguas continentales. *Inst. For. Inv. Exp.*, **63**, 1–140.

Kolder, W., Skora, S. and Wlodek, J. M. 1974. Ichthyofauna of the River Raba and of its tributaries. *Acta Hydrobiol.*, **16**, 65–99.

Lobon-Cervia, J., De Sostoa, A. and Montanes, C. 1986. Fish production and its relation with the community structure in an aquifer-fed stream of Old Castile (Spain). *Pol. Arch. Hydrobiol.*, **33**, 333–343.

Lozano-Rey, L. 1919. Los peces de la fauna ibérica en la colección del museo en 1 de enero de 1919. *Trab. Mus. Nac. Cienc. Nat.*, **39**, 1–112.

Lozano-Rey, L. 1935. Los peces fluviales de España. *Mem. Real Acad. Cienc. Exact. Fis & Nat.*, 1–390.

Mahon, R. 1984. Divergent structure in fish taxocene of north temperate streams. *Can J. Fish. Aquat. Sci.*, **41**, 330–350.

Mondejar, C. 1981. Aclimatación del Salmón del Danubio (*Hucho hucho* L.) en el río Tormes. Escalimetría y crecimiento. *Abstracts of the Iª Jorn. Ict. Iber.*, León. Spain.

Nobre, A. 1932. Peixes das aguas douces de Portugal. *Bol. Minist. Agric. Portugal.*, **2**, 5–44.

Osorio, B. 1888. Additamento ao catalogo dos peixes de Portugal. *J. Sci. Mat. Phys. e Nat.*, **12**, 167–185.

Osorio, B. 1894. Algumas especies a juntar ao 'Catalogo dos peixes de Portugal' de F. Capello. *J. Sci. Mat. Phys. e Nat.*, **3**, 186–195.

Osorio, B. 1895. Segundo appendice ao 'Catalogo dos peixes de Portugal' de F. Capello. *J. Sci. Mat. Phys. e Nat.*, **3**, 254–275.

Osorio, B. 1896. Peixes de Matosinhos (terceiro appendice ao 'Catalogo dos peixes de Portugal' de F. Capello. *J. Sci. Mat. Phys. e Nat.*, **4**, 131–159.

Seabra, A. de F. 1911. Cataloque systematique des Vertebrés du Portugal. *Bull. Soc. Port. Sc. Nat.*, **5**, 129–224.

Sole-Sabaris, L. 1978. Las aguas, rios y lagos. In Teran, M. and Sole-Sabaris, L. (Eds), *Geografia general de Espua*, Ariel Publishing Co., Barcelona.

Sostoa, A., Sostoa, F. J., Lobon-Cervia, J., Elvira, B., Hernando, J. A., Doadrio, I. and Avila, M. 1984. Atlas y distribución de los peces de agua dulce: E proyecto, métodos y resultados preliminares. *Bol. Est. Cent. Ecol.*, **13**, 75–81.

Sostoa, A., Lobon-Cervia, J., Fernandez-Colome, V. and Sostoa, F. J. 1987. La distribución del Pez-Sol (*Lepomis gibbosus* L.) en la Peninsula Ibérica. *Doñana, Acta Vert.*, **14**, 121–123.

Steindachner, F. 1866a. Ichthyologischer bericht über eine nach Spanien und Portugal unternomene Reise. I. über die Fische des Tajo (portug. Tejo), Duero (portug. Douro), Miño (portug. Minho), deren Nebenflüssen und dem Jucar bei Cuenca. *Sitzungsb. Akad. Wiss. Wien.*, **54**, 1–27.

Steindachner, F. 1866b. Allgemeine Bemerkungen über die Süsswasswerfischce Spaniens und Portugal und Revision der einzelnen Arten. *Selbstverlag des Verfassers, Wien.*, 1–15.

Ugarte, J. 1929. Peces de agua dulce. Contribución a la formación de su catálogo. *Rev. Biol. Forest. y Limnol.*, **1**, 54–70.

Velaz de Medrano, L. and Ugarte, J. 1930. Contribución a la formación del Catálogo ictiológico de nuestras aguas continentales. *Rev. Biol. Forest. y Limnol.*, **3**, 25–52.

Vieira, L. 1894. Contribution à l'étude des poissons d'eau douce du Portugal d'après la collection du Musée de Zoologie de l'Université de coimbra. *Ann. Sci. Nat.*, **2**, 1–18.

Vieira, L. 1897. Catalogo dos peixes de Portugal em collecao no Museu de Zoologia de Universidades de Coimbra. *Ann. Sci. Nat.*, **4**, 1–159.

Historical Change of Large Alluvial Rivers: Western Europe
Edited by G. E. Petts
©1989 John Wiley & Sons Ltd

14

Fish and Fisheries of the River Ebro: Actual State and Recent History

ADOLFO DE SOSTOA

Dept. of Animal Biology, School of Biological Sciences, University of Barcelona, c/., Diagonal 645–647, 08028 Barcelona, Spain

and

JAVIER LOBON-CERVIA*

National Museum of Natural Sciences, CSIC, c/., José Gutierrez Abascal, 2, 28006 Madrid, Spain

INTRODUCTION

The 83 000 km^2 Ebro basin forms a triangle draining the eastern area of the Iberian Peninsula. The basin is delimited by the Pyrenees in the north and the Iberian and Catalan mountain ranges in the south (Figure 1). The basin formed during the Tertiary and was filled by marine sediments. Later on, it became a closed lake. The fluvial period started when the Ebro cut a route to the sea through the Catalan littoral mountains. In contrast to other Iberian river basins (not least the Duero, *cf.* Lobon-Cervia *et al.*, this volume), the River Ebro is a low altitude basin, its central area (Zaragoza) is about 200 m above sea level; the surrounding mountain ranges are responsible for its local climate, especially for the lack of marine influence—it is an arid region with a continental climate.

Rainfall is poor and decreases with altitude towards the centre of the basin. In the surrounding areas, rainfall reaches up to 600 mm annually. In the 'piedmont' regions annual rainfall is about 400 mm, and declines to 300 mm or less in the central areas (i.e. in Los Monegros). Both summer and winter are dry. Temperature varies from 42°C in the summer to a winter average of 3°C, with a minimum recorded at -15°C.

According to Sole-Sabaris (1978) the water capture system is mixed between snow, snow plus rain, and rain components, producing, as a consequence, several different runoff peaks in different months of the year (Figure 2). Water flow increases gradually from Reinosa to Miranda de Ebro (81 $m^3 s^{-1}$), where the river receives runoff from the Pyrenees. As

*Address for correspondence to this author: Department of Organisms and Systems Biology, University of Oviedo, 33071 Oviedo, Spain

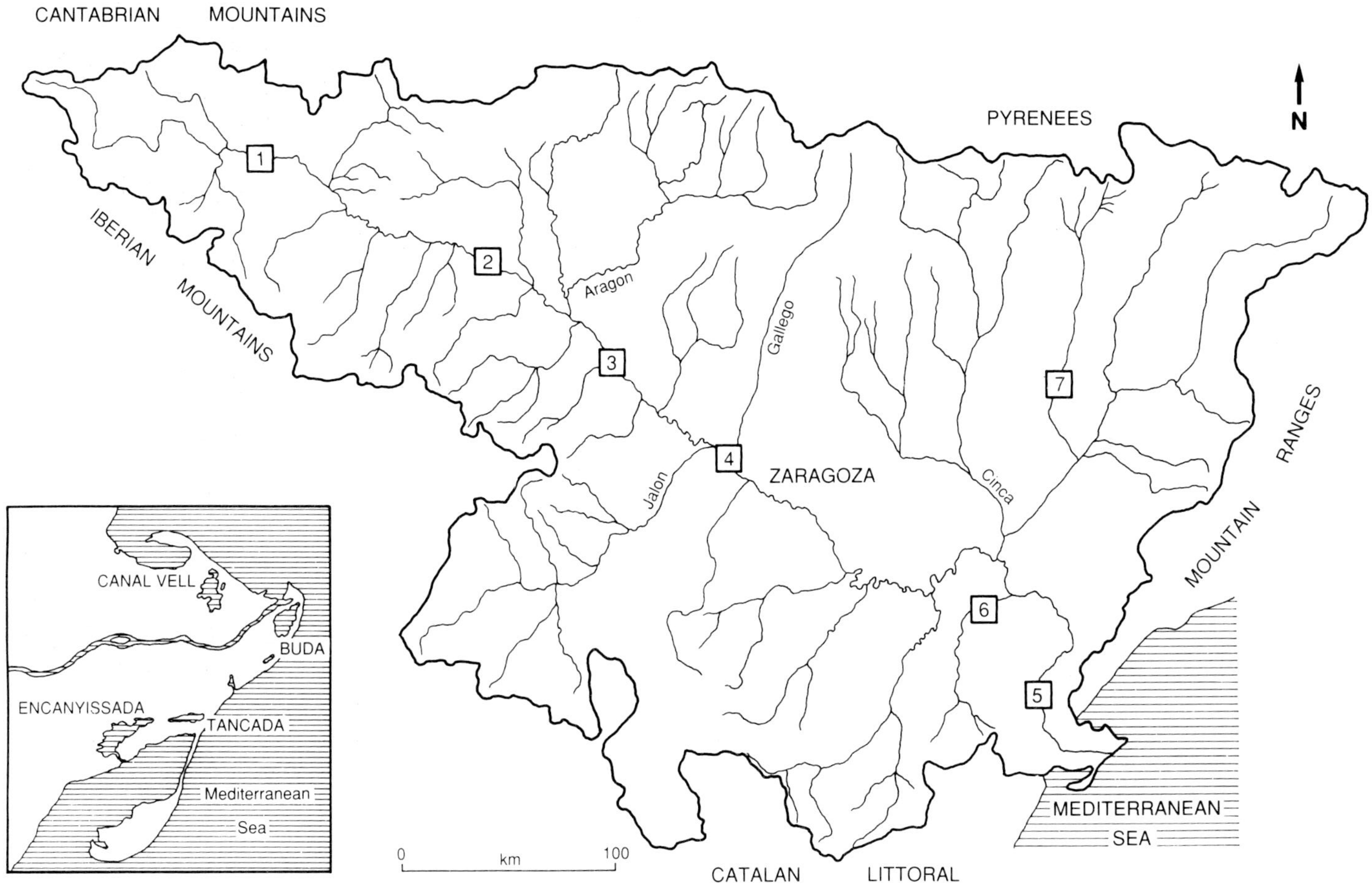

FIGURE 1. Map of the River Ebro basin showing the locations where water analysis was performed: 1. Palazuelos, 2. Mendaña, 3. Pignatelli, 4. Zaragoza and 5. Tortosa; numbers 6 and 7 refer to the Rivers Matarraña and Noguera Ribagorzana respectively. Insert shows a map of the Ebro Delta locating the four main coastal lagoons

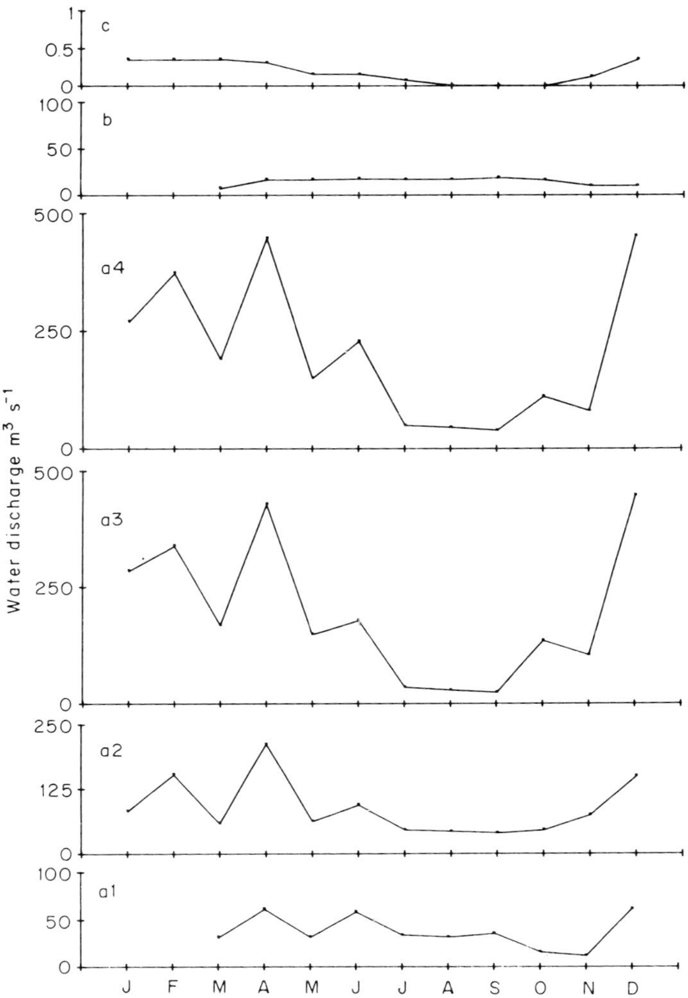

FIGURE 2. Monthly variation in the water discharge. a1, a2, a3, a4 and a5 refer to the same locations mentioned in Figure 1, b refers to the river Noguera Ribagorzana and c to the River Matarraña

the Ebro reaches Tortosa, its discharge is more than 600 $m^3 s^{-1}$. Over an historic timescale many catastrophic floods have been reported. The most important were those of 1582, 1617, 1787, 1866, 1871, 1884, 1907 (discharge of 23 500 $m^3 s^{-1}$) and 1937. The contrasting regimes of the major tributaries are shown in Figure 2b,c. The two tributaries, for which data are given are representative of all the tributaries of the Ebro basin. The Noguera Ribagorzana which drains the Pyrenes has a late spring–summer runoff maximum whereas the Matarraña, draining the limestone Catalan littoral range has a winter–spring maximum. By August and September the Matarraña is reduced to a succession of unconnected ponds.

Because of the continental climate, water temperature shows strong seasonal variations. In the main water corridor from 521 m above sea level in Palazuelos to 8 m above sea level

(Tortosa), minimum water temperature ranges between 2.5°C and 8.5°C in August. Stronger variations are evident in the tributaries as they are influenced by particular microclimates.

About 10 per cent of the total surface of the basin is cultivated, mainly for grapes and olive trees (wine and oil production). Wheat farms are also abundant and are distributed throughout. The traditional livestock are sheep and the main industry is devoted to agriculture and closely related products. Relatively recently a succession of reservoirs have been built to control floods and even more recently hydroelectric power dams have been introduced. The most important reservoirs in the River Ebro are: Flix Reservoir, built in 1945, Mequinenza (1964) and Ribarroja (1967). These reservoirs, together with those of the Rivers Segre and Cinca, produce 1800 MW of electric power a year. In the tributaries a few reservoirs and hydroelectric dams have been built along the course of the Noguera Ribagorzana. The three most important are Escales, Canellas and St. Anna, producing about half a million kilowatts per year.

The Ebro Delta is rather small (350 km^2), and is divided by the main river. The altitude is generally lower than 0.5 m above sea level although the levees rise up to 4 m above sea level. The Ebro Delta is young geologically, post dating the last glaciation (less than 15 000 years ago). Maps dating from the 16th century clearly show changes in its physical structure throughout the historic time period (Maldonado, 1972, 1977). Today, the landscape of the delta in the central area, parallel to the main river course, is characterized by a complicated system of man-made channels to irrigate fields and rice farms. Along the coastline, lagoons and marshy areas surround the delta, the largest being the Encanyisada (555 ha), Canal Vell (269 ha), calaix de d'Illa de Buda (408 ha) and Tancada (195 ha): the whole water surface accounts for 7.2 per cent of the delta (Figure 1a). Because of river regulation the seasonal flow variation is not so pronounced as it was historically. This regulation, however, allows a salt-water wedge of marine origin to move up to 40 km upstream along the main water course from the mouth. Mean river depth is about 6 m and mean water discharge 580 $m^3 s^{-1}$ (Maldonado, 1972). The incursion of the salt-water wedge leaves only about 2 m depth of freshwater along the lowest 15 km of channel; the conductivity rises from 960 $\mu S\,cm^{-1}$ to 4210 $\mu S\,cm^{-1}$ at 6 m. Further information is given by Sostoa and Sostoa (1985).

THE FISH FAUNA OF THE RIVER EBRO

In this section we discuss the fish fauna of the freshwater River Ebro, including anadromous, catadromous and euryhaline species. We exclude, therefore, all those marine species occurring in the Ebro Delta with stenohaline behaviour as these are discussed later.

In comparison with other Iberian river systems, the River Ebro is richest in number of species. This is due mainly to its very particular geographical situation which has contributed to the Ebro: (1) sharing the fish fauna of the Iberian Peninsula in geological times; (2) being a refuge for fish species during the Pleistocene glaciations, as also did the rest of the Iberian Peninsula; and (3) being, more recently, colonized by fish species which have not been able to migrate into the southern river basins of the peninsula. As a consequence, the River Ebro shares taxonomic groups of fish with central and northern European rivers as much as with those of the Mediterranean basin. Thirty-eight taxa belonging to 21 families are present in the drainage bsin (Table 1). Of these, 27 are indigenous and 11 have been introduced. Their taxonomic position has recently been studied and, in general, it is well documented (Demestre *et al.*, 1977; Doadrio, 1984; Elvira, 1986).

According to their origin, the fish fauna can be divided into five groups. The first group is composed of species with Mediterranean origin (freshwater and marine) and a natural distribution on the southern slopes of the Euromediterranean mountain ranges (Pyrenees, Alps, Balkans): *Rutilus arcasii*, *Chondrostoma toxostoma miegii*, *Barbus haasi*, *Aphanius iberus*, *Valencia hispanica* and *Blennius fluviatilis*. *A. iberus* has recently been sighted by Villwock (1982) in an oasis pond of the Algerian deserts but its taxonomic position requires further research.

The second group is composed of one species only, *Barbus graellsii*. According to Almaca (1978) this taxon is originally Iberian but according to Doadrio (1984) is more closely related to the North-African species (i.e. *B. callensis*).

The third group is formed by the 'younger species', i.e. those which have colonized the Ebro basin from the mid-northern European region, supposedly within, or after, the glaciations. This includes: *Noemacheilus barbatulus*, *Leuciscus cephalus cephalus*, *Phoxinus phoxinus* and *Cobitis paludicola*. None of the species included in the above groups is strictly endemic to the River Ebro. Nevertheless, *B. graellsii*, *B. haasi*, *Ch. toxostoma*, *N. barbatulus*, *Ph. phoxinus* and *V. hispanica* seldom occur in areas outside the River Ebro basin and these restricted areas are directly related to their zoogeographical origin. The first three species live in the northern Mediterranean flowing streams with a northern limit in the River Besos (Barcelona province). The fourth and fifth species inhabit streams of northern Spain flowing on the northern slope of the Cantabrian mountains towards the Cantabrian Sea, in the opposite direction to the Ebro River and *V. hispanica* is present in only a few, scattered, marshlands of the Mediterranean coast.

A still undecided problem is to assess whether or not *Ph. phoxinus* is a native species in the Ebro basin. It appears to be endemic in the previously mentioned rivers flowing to the Cantabrian Sea in Santander, Navarra and the Vasque country, very close to the sources of the River Ebro, and in the uppermost, eastern tributaries of the River Duero (Lobon-Cervia *et al.*, this volume).

Two more groups can be considered. One is composed by all the anadromous and catadromous species: *Acipenser sturio*, *Alosa alosa*, *Alosa fallax nilotica*, *Anguilla anguilla* and *Petromyzon marinus*. Perhaps brown trout (*S. trutta*) could be included in this group as some authors accept their colonization into the River Ebro through the Mediterranean Sea (see Hsu *et al.*, 1977 for the Tethys Sea movements) although earlier, or alternative, colonizations across the Pyrenean waterways can not be refuted).

The final group includes all the introduced species: *Salmo gairdneri*, *Salvelinus fontinalis*, *Esox lucius*, *Carassius auratus*, *Cyprinus carpio*, *Gobio gobio*, *Scardinius erythrophthalmus*, *Ictalurus melas*, *Silurus glanis*, *Gambusia affinis* and *Micropterus salmoides*. All these species have been stocked in different Spanish river basins and most of them are actually common. The exception is *S. glanis*, which has been recorded in the River Ebro only. Regarding the basin as a whole, the actual status of the species is also given in Table 1. Note that because of the Ribarroja and Mequinenza reservoirs in the lower reaches of the Ebro, that block the upstream migration of anadromous and catadromous species, eel (*A. anguilla*), sturgeon (*A. sturio*), shads (*A. alosa* and *A. fallax*) and *P. marinus* are now extinct upstream from these reservoirs. Downstream, in the lowermost reaches, eel is still very abundant. Sturgeon is totally extinct—the last specimens were captured in the

TABLE 1. Native (N) and introduced (I) species living in the River Ebro drainage basin and their current status. A and C refer to anadromous and catadromous species and F to freshwater species (*strictu sensu*). F and B refer to those species living in both, fresh and brackish waters in the Ebro Delta. Ab is abundant; C is common; R, rare; V, vulnerable and E, extinct (no records in the last 10 years). *refers to the status of a particular species in the Ebro Delta; for instance, eel (*A. anguilla*) is extinct (E) in practically all the basin but is still abundant in the Ebro Delta. **, Pyrenean trout populations are supposed to be morpha *lacustris*. ***, concerns those species considered to be threatened in Spain (ICONA, 1986), independently of their status in the River Ebro

Petromyzontidae			
Petromyzon marinus	N	A	V
Acipenseridae			
Acipenser sturio	N	A	E
Clupeidae			
Alosa alosa	N	A	V*
Alosa fallax nilotica	N	A	V*
Anguillidae			
Anguilla anguilla	N	C	Ab*, E
Salmonidae			
Salmo trutta m. fario	N	F	Ab
Salmo trutta m. lacustris	N	F	R**
Salmo gairdneri	I	F	C
Salvelinus fontinalis	I	F	R
Esocidae			
Esox lucius	I	F	C
Cyprinidae			
Barbus graellsii	N	F	Ab
Barbus haasi	N	F	Ab
Carassius auratus	I	F	R
Cyprinus carpio	I	F	C
Chondrostoma toxostoma	N	F	Ab
Gobio gobio	I	F	Ab
Leuciscus cephalus	N	F	Ab
Phoxinus phoxinus	N	F	C
Rutilus arcasii	N	F	Ab
Scardinius erythophthalmus	I	F	R
Cobitidae			
Cobitis paludicola	N	F	C
Homalopteridae			
Noemacheilus barbatulus	N	F	C
Centrarchidae			
Micropterus salmoides	I	F	R
Ictaluridae			
Ictalurus melas	I	F	R
Siluridae			
Silurus glanis	I	F	R
Cyprinodontidae			
Aphanius iberus	N	F, B	V***
Valencia hispanica	N	F, B	V***
Poecilidae			
Gambusia affinis	I	F	R

TABLE 1. (*continued*)

Blenniidae			
Blennius fluviatilis	N	F	C***
Atherinidae			
Atherina boyeri	N	F, B	Ab*
Gasterosteidae			
Gasterosteus aculeatus	N	F, B	V***
Syngnathidae			
Syngnathus abaster	N	F, B	C*
Mugilidae			
Chelon labrosus	N	F, B	Ab*
Liza ramada	N	F, B	Ab*
Mugil cephalus	N	F, B	Ab*
Gobiidae			
Pomatoschistus microps	N	F, B	Ab*
Moronidae			
Dicentrarchus labrax	N	F, B	C*
Pleuronectidae			
Platichthys flesus	N	F, B	C*

1940s—and shads and *P. marinus* have dramatically declined in numbers recently. Overall, however, reservoirs have markedly increased the number of species in headwater tributaries. For example, along the Noguera R. seven species mainly cyprinids, pike and black-bass (Table 2), have been stocked in reservoirs, and/or in the lower reaches of the rivers, close to the reservoirs. Short migrations upstream make them appear along the river courses.

Recent history of fisheries in the River Ebro

Except *C. carpio* which was supposedly introduced by the Romans, perhaps two thousand years ago, together with some unsuccessful stocking at the end of the 19th century (Darder, 1913), all the other species have been brought to these rivers since 1900. In general, no scientific arguments have been used to properly introduce fish species, although exceptions exist. Five species (*S. gairdneri*, *S. fontinalis*, *E. lucius*, *M. salmoides* and *G. affinis*) were stocked as a result of administrative and/or political decisions: the first four species to promote angling and the last one as a mosquito-eater to prevent paludism. All the other species have been introduced 'accidentally'; *S. erythophthalmus*, *I. melas* and *S. glanis*.

Professional freshwater fisheries in the River Ebro were common until the 1960s. In 1935, about 40 fishermen moved their boats between Tortosa and Amposta. In Faion another small group was settled, and upstream to Zaragoza several fishing boats were scattered along the river course. In general, fisheries were aimed at migrant species and cyprinids. Sturgeon (*A. sturio*) was exploited until the mid 1960s but catches declined following the construction of reservoirs. At the same time fisheries for cyprinids also decreased as these species became unmarketable due to changes in consumption habits. Today, the fish populations are exploited only by anglers.

Basically four fish species are stocked every year by the administration for angling purposes: *Salmo trutta*, *S. gairdneri*, *Cyprinus carpio*, and *Micropterus salmoides*. The

TABLE 2. Fish species captured along the course of the Rivers Noguera Ribagorzana and Matarraña (1 = presence)

River and site	Metres above sea level	S. trutta	S. gairdneri	Sa. fontinalis	A. anguilla	B. haasi	B. graellsii	Ch. toxostoma	L. cephalus	G. gobio	Ph. phoxinus	R. arcasii	Sc. erythrophthalmus	C. carpio	Ca. auratus	N. barbatulus	Co. paludicola	B. fluviatilis	E. lucius	Ga. affinis	I. melas	M. salmoides
River Matarraña																						
Parrisal	650	1	1			1		1														
Beceite	520	1	1			1		1														
Bec-Uldemó*	680	1				1	1	1														
Pes-Uldemó*		1				1	1	1														
Valderobles	500	1	1			1	1	1				1				1						
Val-Tasta*						1	1	1								1						
Tastavins*	400					1	1	1				1										
Tor-Compte						1	1	1				1				1						
Ejerique	360					1	1	1				1				1	1					
Massalió							1	1				1				1	1					
Maella	300						1	1				1				1						
Fabara	220						1	1	1	1									1			
Nonaspe	170						1	1	1	1		1		1	1	1	1	1	1		1	
Non-buitrera	130						1	1	1	1				1		1	1	1	1		1	
Faió	95						1	1	1	1	1			1		1	1	1	1	1	1	1
Non-Algas*	150						1	1	1										1			
Mas-Estudiant*							1	1	1												1	
Lledó*							1	1	1													

River Noguera Ribagorzana

Hosp.-Viella	1700	1																		
Bessiberris	1700	1		1																
Senet	1700	1																		
Pal. Passos	1700	1																		
LesBordes*	1600	1																		
A. Tortes†	1600	1																		
Bono	1500	1																		
Erta‡	1500	1																		
Adoms§	1350					1														
Vilaller	1200	1				1														
C. de Bohi	1200	1																		
Barruera	1200										1									
Durró	1200	1																		
Taüll	1200	1																		
Noales*	1100	1				1														
M. Malpas	1200	1																		
Pellarel	1200					1														
Mas Jaumet	900					1		1												
Ms. Tamurcia	1100					1		1												
Escales &	1000	1	1		1	1	1	1			1		1	1						
Escarlá	700						1	1												
Montanyana	550						1	1												
S. Ana &&	500	1				1	1	1	1	1						1			1	1
Almenar	350	1					1	1	1										1	

*Sites in River La Baliera (tributary of River Noguera Ribagorzana)
†Sites in River Sant Nicolau (tributary)
‡Sites in River D'Erta (tributary)
& Reservoirs

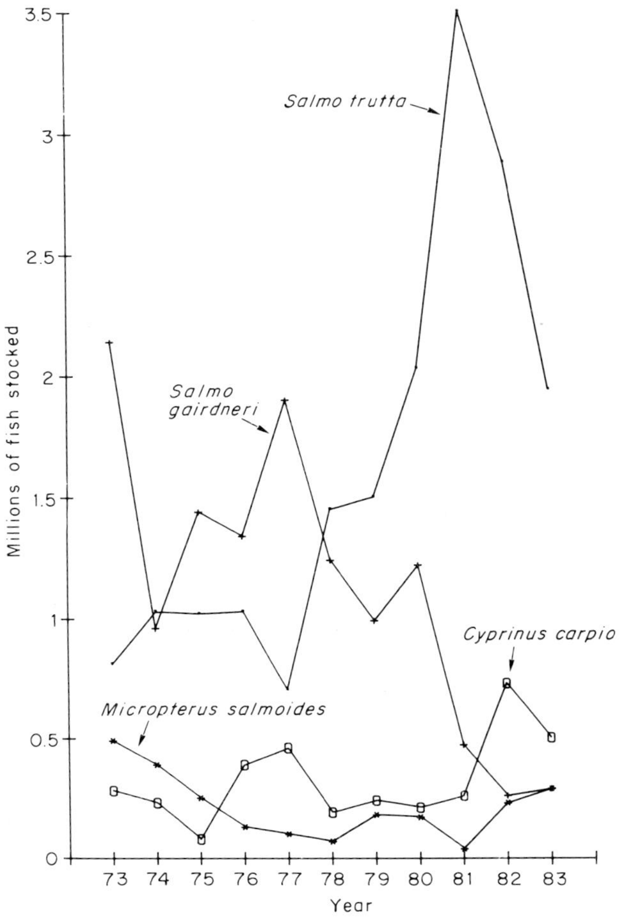

FIGURE 3. Numbers of individual fish of the following species: *Salmo trutta*, *Salmo gairneri*, *Micropterus salmoides* and *Cyprinus carpio*, stocked in the River Ebro basin in the period 1973–1983

first two species are stocked in all the provinces with streams and rivers flowing into the Ebro basin (Alava, Castellon, Huesca, Lerida, Logrono, Navarra, Santander, Tarragona, Teruel and Zaragoza), while the other two are mainly stocked in reservoirs (Figure 3). One more species, pike (*E. lucius*), which was introduced in Spain in the 1950s and whose populations appear in increasing numbers throughout the Ebro basin, is not included in the official statistics, at least, since 1973. A similar situation prevails for *S. glanis*. Besides, in 1973 about one million char (*Salvelinus fontinalis*) were stocked in lakes and streams of the Pyrenean Mountains in Huesca and Lerida. As far as we know this species has not been restocked but they still occur in several places.

In general, brown trout management in Spain and, of course in the Ebro basin, is very peculiar. In spite of huge natural populations throughout the country, the only form of management is stocking, i.e. no freshwater fisheries biologist work for the fisheries and game services and, therefore, management is not based on fishery science. Still more peculiar is that all individuals stocked in rivers are not grown from native river populations as they should be in order to prevent genetic contamination and related problems. Instead, since the early 1940s they have been imported especially from German and Italian farms. As there are no data on the actual stocks, trout are introduced in practically all the streams even in those where natural densities by the end of the angling season reach up to 15 000–20 000 individuals per hectare, or where natural trout production is in the order of 125–250 kg $ha^{-1}a^{-1}$, typical of most of the streams within its natural distribution area (Lobon-Cervia *et al.*, 1985, Lobon-Cervia, unpublished).

The consequences of such stocking are obvious. In the period 1973–1983 the number of brown trout released in the Ebro basin has increased by three times, from a little less than one million in 1974 to 3.5 million in 1981 (Figure 3). This tremendous increase in numbers is the result of several factors, among them:

1. A noticeable increase in the number of anglers from something like 300 000 in the early 1950s to nearly one million in 1984. These numbers refer to the whole country. A rough estimation for the Ebro basin is about 175 000;
2. An important reduction in the area of trout streams because of the construction of about 1000 reservoirs in the last 80 years, and
3. A change in the management policy which occurred in 1977–1978. In these particular years the administration decided to replace rainbow trout by brown trout (Figure 3).

No records of rainbow trout spawning in Spanish streams are available. So far, only one population has been recorded successfully spawning in spring in the River Matarraña in 1985. However, brown trout successfully spawn and mix with native brown trout and it is unlikely that 'genetic pools' of native brown trout are preserved, although this remains to be tested.

Recently, some interest by anglers has been focussed on black-bass (*M. salmoides*). The administration had introduced this species by the end of the 1960s, mainly in reservoirs. The species is spreading rapidly and, our recent observations suggest that it is colonizing streams and rivers. Carp (*C. carpio*) is stocked in rather large numbers (Figure 3) although angling for this species is poor.

As a whole, four million trout (rainbow + brown trout) were stocked in 1981. Together with black-bass and carp, 4.7 million fish were released into the rivers and streams of the Ebro basin in that year alone. For a maximum of 200 000 anglers this represents 23.5 fish per angler per year. Statistics for the last few years are not available but we suspect that this figure is small if we compare it with the actual situation in 1986–1988.

FISH FAUNA OF THE DELTA

From a zoogeographical point of view, the fish fauna of the Ebro Delta, except those species previously discussed of riverine origin, is composed of mediterranean–boreal taxa (44 per cent of the total number of species) and mediterranean–lusitanian taxa (33 per cent). In addition, some species of lusitanian and mauritanian origin account for 22 per cent.

Forty-eight species belonging to 29 families colonize seasonally or permanently the waters of the Ebro Delta. According to the degree of permanence in these waters and to their marine or limnotic origin, these species can be classified into five groups (Table 3). Distribution patterns are very variable because of the very strong environmental fluctuations.

TABLE 3. Classification of fish species inhabiting the Ebro Delta, indicating habitat preference

Accidental species 39.6 per cent	*Sardina pilchardus*	3	
	Engraulis encrasicolus	3	
	Brama brama	1	
	Argyrosomus regius	1	
	Umbrina cirrosa	1	
	Lithognathus mormyrus	1	
	Gobius cruentatus	4	
	Pomatoschistus minutus	4	Marine
	Blennius pavo	4	species
	Oedalechilus labeo		
	Mullus surmuletus	1	
	Pomatomus saltator	1	
	Lichia amia	1	
	Trachinus vipera	4	
	Esox lucius	1	
	Ictalurus melas	1	Limnetic
	Leuciscus cephalus	1	species
	Micropterus salmoides	1	
	Silurus glanis		
[Catadromous species 2.1 per cent]	*Anguilla anguilla*	5	
Migratory species 8.4 per cent	*Alosa alosa*	1	Migratory
[Anadromous species 6.3 per cent]	*Alosa fallax nilotica*	1	species
	Petromyzon marinus		
Seasonal species 22.9 per cent	*Platichthys flesus flesus*		
	Solea lascaris		
	Belone belone gracilis	3	
	Solea vulgaris vulgaris		
	Sparus aurata	3	
	Chelon labrosus	2	Marine
	Liza saliens	2	species
	Liza aurata	2	
	Liza ramada	2	
	Mugil cephalus cephalus	2	
	Dicentrarchus labrax	2	
Seasonal–sedentary species 6.3 per cent	*Pomatoschistus microps*	2	
	Syngnathus abaster	2	
	Atherina boyeri	2	
Sedentary species 22.9 per cent	*Aphanius iberus*	2	
	Gambusia affinis	5	
	Cyprinus carpio	5	
	Carassius auratus	5	
	Blennius fluviatilis	1	
	Barbus graellsii	1	
	Chondrostoma toxostoma	1	Limnetic
	Gobio gobio	1	species
	Gasterosteus aculeatus	6	
	Cobitis paludicola	6	
	Valencia hispanica	6	

1 riverine reaches in contact with marine waters
2 marshland
3 coastal lagoons
4 saltmarsh
5 cosmopolitan
6 dispersed and threatened

However, some generalizations can be made (Table 3) and patterns of distribution described (Sostoa and Sostoa, 1985).

In riverine environments fish assemblages are characterized by pronounced seasonal variations. In the first (freshwater) period, the assemblage is composed of riverine taxa. Only some euryhaline marine components appear, among them: *M. cephalus*, *L. ramada*, *L. saliens*, *D. labrax*, *S. abaster*, *P. microps*, and *A. boyeri*. Two more species *Ch. labrosus* and *L. ramada* with lesser osmoregulating ability may stay in the lowest, 12 km of the river.

During the dry season, brackish water moves upstream. This period is dominated by stenohaline marine species like *B. brama*, *U. cirrosa*, *A. regius*, *L. amia*, *L. mormyrus*, *M. surmuletus*, *P. saltator*, and *O. labeo* and, of course, by those eurihaline marine species like *S. vulgaris*, *S. lascaris* and *P. flesus*.

The coastal lagoons are influenced to a large degree by freshwaters coming from man-made irrigation channels. These systems, have two seasons with fresh and brackish waters but they follow the opposite cycle to that of natural coastal lagoons (Comin, 1981). In the longer freshwater period, 21 species have been recorded while in the shorter brackish period, this figure is reduced to 16 species only. Some species which are present at least in one of these environments, are: *B. belone*, *P. flesus*, *S. vulgaris* and *S. lascaris*. The first species migrates from the sea for spawning between February and June, while the other three spawn in the sea and colonize these lagoons where they stay until they complete their larval stages.

The 'Ullals' are dominated by small-sized native taxa, although in recent years some new, introduced species, have become common (*G. affinis*, *C. auratus*, *C. carpio* and *M. salmoides*). These assemblages are structurally more constant throughout the year. The native fish fauna of the 'Ullas' is extremely threatened (ICONA, 1986) and their total extinction may occur sooner than predicted. Moreover, *V. hispanica* has survived in these 'Ullals' and together with a few more, small populations, living along the Mediterranean coast, represent the last known individuals of the species. The same can be said of *A. iberus*. Therefore, protection of these environments and removal of allochthonous species, especially predatory fish like *M. salmoides*, is strongly recommended.

Recent history of fisheries in the Ebro Delta

Professional fisheries have been reported in the Ebro Delta, at least, since the 17th century. However, the fishermen became organized only in 1879. In this year, the foundation of the 'Societat de Sant Pere', granted by the State Administration to control fisheries in the coastal lagoons (Encanyissada, Tancada, Canal Vell and Les Olles), led to the regulation of fish yield, amount and season. Today, this association has over 1000 members and every year, by a peculiar lottery system, 78 particular sites—accounting for 1179 ha of water within the lagoons—are shared among the fishermen.

Basically, two fishing methods are used in the coastal lagoons. The first category includes all those using nets or line and hooks. These methods are employed within the lagoons and, in general, they are not very selective, although the 'ganguils' are specifically used for eel (*A. anguilla*) and the 'saltall', a sort of trammel-net, is devoted to mullet. Secondly, there is a group based on traps, fixed in the drainage channels which connect the inland lagoons and the sea. The channels are used by migrating fish and the methods are selective, for eel, mullet, flatfish, bass, pipefish etc.

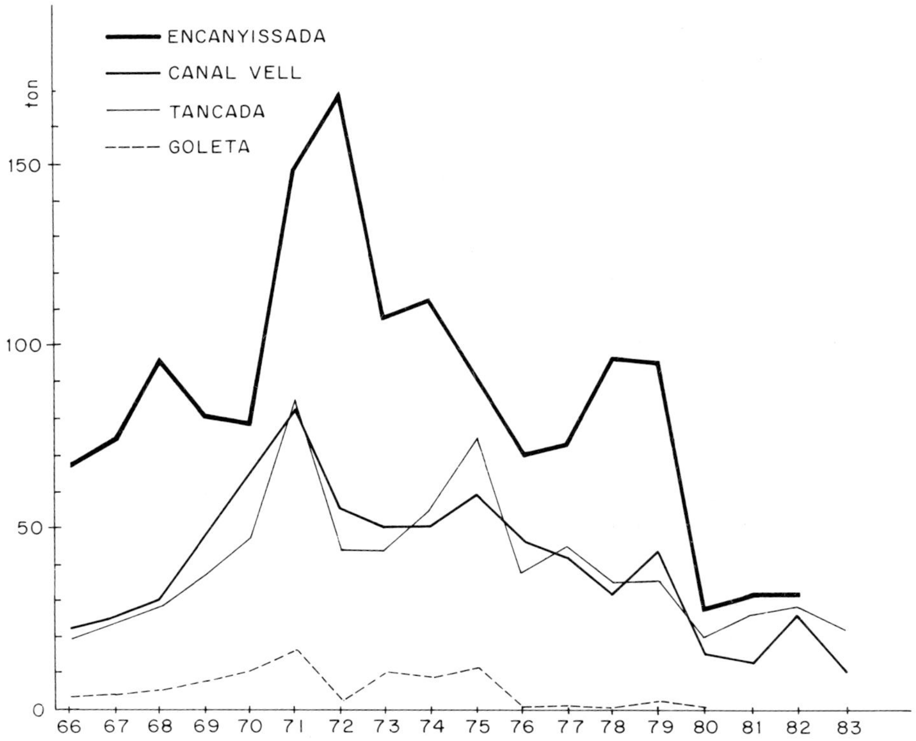

FIGURE 4. Total fish catches (tons per year) of dominant species in four main coastal lagoons of the Ebro Delta

For both types of method used in all the mentioned lagoons, the average yield for the period 1965–1983 was 144 kg. The total fish biomass captured during the same period reached 3 million kg wet weight, comprising: eel and elver (910 000 kg), carp and goldfish (912 000 kg), mullet (1059 000 kg), bass (120 000 kg) and *S. aurata* (7000 kg). The average contribution made by the main lagoons including their channels are: Encanyissada, 40 per cent; Tancada, 30 per cent; Canal Vell, 25 per cent and Les Olles, 5 per cent.

Year-to-year variations in the total yield obtained from these lagoons are shown in Figure 4. Between 1965 (probably earlier) and 1972, the annual yield increased progressively to reach 273 000 kg. However, from 1972 to 1980 the yield decreased sharply reaching a stable low of about 31 000 kg. Three main reasons have contributed to this change. First, traditionally after the fishing season, between March and October, all the fishing gear fixed in the channels connecting the sea with the lagoons, was removed. In this way, fish could move freely between the sea and the lagoons. However, in 1972, fishermen decided to leave the fishing gear fixed in their fishing sites until the next season, although not in use. This restricted fish migrations to only very small fish. The populations of the lagoons are, therefore, composed of fish growing up from these young stages.

Secondly, overfishing has become a significant problem. Although the number of fishermen has not increased in the last 20 years, there has been a marked rise in the fishing effort. For example, during 1975–1980 the number of 'ganguils' fixed in the four main lagoons to catch eel, increased from 400 to 1000. Thirdly, as mentioned for the freshwater fisheries, a general lack of science in management policy, together with the impact of increasing pollution, especially from fertilizers and herbicides, from the rice fields must have contributed to changing fish stocks.

ACKNOWLEDGEMENTS

Data presented herein, is the result of a few years sampling carried out by several people. Special thanks are due to our wives, Maria Pomes and Paloma Rovira, who patiently supported our long periods in the field, and to all those who worked out in the River Matarraña project, among them to Dolors Viñolas, Fredy Casals, Francesc Sostoa, Angels Puig, Gary Grossman and Mary Freeman. Most of the information on fish corology comes from Benigno Elvira and his ever-fruitful discussions on fish systematics and zoogeography. The professional fishermen of the Delta contributed with their effort and the ICONA supplied us with data on fish stocking. We expect both of them to receive our criticisms as a way to improve nature conservancy and fisheries management. Along the sampling period, the authors were partially supported by a CAYCIT project n. 211, and granted by the cooperation program USA–SPAIN through the River Matarraña project.

REFERENCES

Almaca, C. 1978. Repartition geographique des Cyprinidae Ibériques et secteurs ichthyogeographiques de la Peninsule Ibérique. *Vest. Cesk. Spol. Zool.*, **42**(4), 241–248.

Comin, F. A. 1981. *Limnología de las lagunas costeras del Delta del Ebro, características fisico-quçimicas y fitoplancton de La encañizada, Tancada y Buda*, Ph.D. Thesis. University of Barcelona.

Darder, F. A. 1913. *Cronica Piscatoria*, Publ. Aquar. Barcl., Barcelona.

Demestre, M., Roig, A., Sostoa, A. and Sostoa, F. J. 1977. Contributió a l'estudi de l'ictiofauna continental del delta de l'Ebre. *Treb. Inst. Cat. Hist. Nat.*, **8**, 145–226.

Doadrio, I. 1984. *Relaciones filogenéticas y biogeográficas de los Barbos (Barbus, Cyprinidae) de la Peninsula Ibérica y aportes corológicos y biogeográficos a su ictiofauna continental*, Ph.D. Thesis. Central University of Madrid, 1946 pp.

Elvira, B. 1986. *Revisión taxonómica y distribución geogrçafica del género Chondrostoma Agassiz, 1835 (Pisces, Cyprinidae)*, Ph.D. Thesis. *INYA (Ministry of Agriculture), Madrid*, **62**, 1–530.

Hsu, K. J., Montadert, L., Bernoulli, D., Cita, M. B., Garrison, E., Kidd, R. B., Melie Res, F., Muller, C. and Wright, R. 1977. History of the Mediterranean salinity crisis. *Nature*, **267**, 399–403.

ICONA. 1986. *Lista Roja de los Vertebrados de España*. Minist. Agricul. Pesca y Alim., Secret. Gen Tec., 400 pp.

Lobon-Cervia, J., Montañés, C. and Sostoa, A. 1985. Production and food consumption of a population of brown trout. (*Salmo trutta* c.) in an aquifer-fed stream of old Castile (Spain), *Proc. 4th Brit. Freshw. Fish Conf.*, 41–51.

Maldonado, A. 1972. *El Delta del Ebro. Estudio sedimentológico y estratigráfico*, Ph.D. Thesis. University of Barcelona.

Maldonado, A. 1977. Introducción geológica al Delta del Ebro. *Treb. Inst. Cat. Hist. Nat.*, **8**, 47–54.

Sostoa, A. and Sostoa, F. J. 1979. Sobre la ictiogauna del Delta del Ebro (NE de la Peninsula Iberica). *Misc. Zool.*, **5**, 178–179.

Sostoa, A. and Sostoa, F. J. 1985. The fish community of the Ebro Delta (Northeast Spain): a model of Mediterranean estuarine ecosystem. In Yaņex-Arancibia (Ed.), *Fish Community Ecology in Estuaries and Coastal Lagoons: Towards an Ecosystem Integration*, UNAM Press Mexico. Mexico, 654–694.

Sole-Sabaris, L. 1978. Las aguas, ríos y lagos. In Terán, M. and Solé-Sabaris, L. (Eds), *Geografía general de Espaņa*, Ariel Publis. Co., Barcelona.

Villwock, W. 1982. Ergzende mittelungen ber *Aphanius* ans der Oase Azraq/Jordenien sowie Betrachtungen zum taxonomischen status eines neuen *A. iberus* (Cyprinodontidae: pisces) ans dem oued zousfana, Igli/Norwest-Algerien. *Mitt. Hamb. Zool. Mus. Inst.*, **79**, 267–271.

Historical Change of Large Alluvial Rivers: Western Europe
Edited by G. E. Petts
©1989 John Wiley & Sons Ltd

15

Historical Changes of the Garonne River, Southern France

H. DECAMPS, M. FORTUNE and F. GAZELLE

Centre d'Ecologie des Ressources Renouvelables, CNRS, 29, rue Jeanne Marvig, 31055 Toulouse, France

INTRODUCTION

The Garonne is the third and the most southern of the large French atlantic rivers, following the Loire and the Seine. It is 580 km long and drains a basin of 57 000 km^2 and is characterized by high flow variability 70 to 8000 $m^3 s^{-1}$ downstream of the Tarn tributary. Man's impact on the Garonne has been progressive through the ages and has greatly changed the river over the last 300 years. Three stages may be distinguished: (1) changes to protect against flooding; (2) changes to aid navigation; and (3) developments in agriculture. These have all gradually modified the fluvial landscape and the changes have not been obscured by more recent industrial developments. Thus, this valley offers a useful model to study historical influences on the fluvial landscape in urban as well as in rural areas. The purpose of this chapter is to discuss the ecological consequences of the interactions between river dynamics and Man's activities related to the regional economic development of this essentially agriculture area.

HYDROLOGY AND GEOMORPHOLOGY

The Garonne river receives water draining from three contrasting regions: from the Pyrenees in the south of the drainage basis; from the Southern Central Massif in the East; and from the Aquitaine basin along its course. Connected with this diversity in the drainage basin the river can be divided into several longitudinal zones from its source in the Pyrenees to its mouth at the Atlantic (Lambert, 1982). Six natural sections may be distinguished (Figure 1):

—Section 1 (75 km long) crosses the Pyrenees, on a pebble bed in a glacial valley to the confluence with the Neste d'Aure;
—Section 2 (45 km long) is an alluvial valley parallel with the Pyrenean chain, from the Neste d'Aure to the Salat;
—Section 3 (65 km long) is a piedmont river flowing to the Ariege, which is the last Pyrenean tributary;

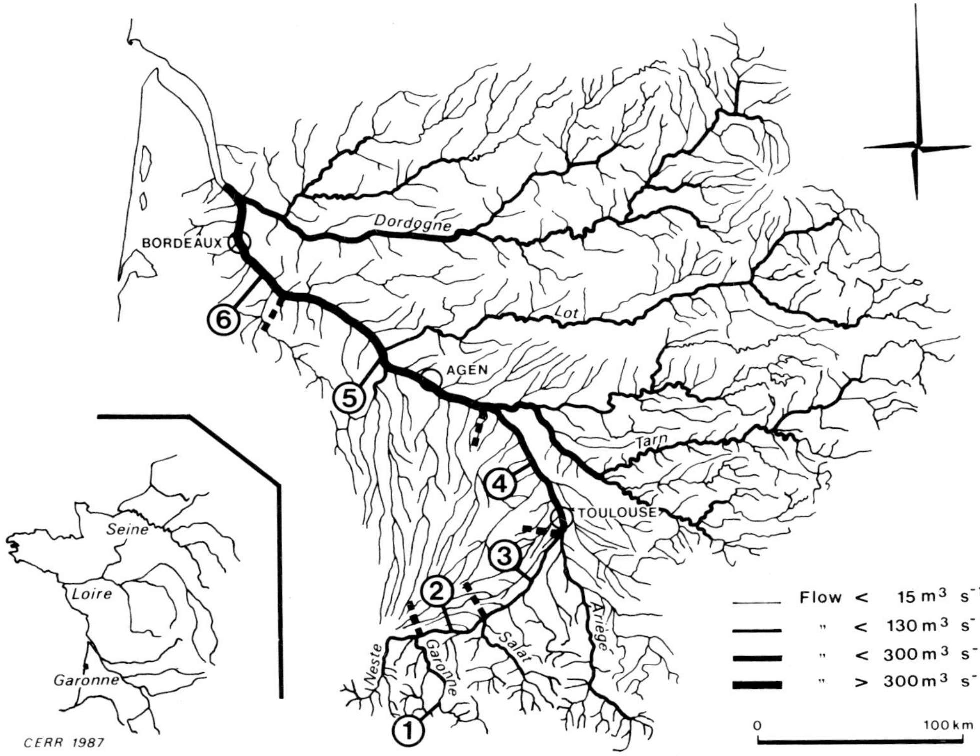

FIGURE 1. The River Garonne and its tributaries, with an indication of the mean annual flow, defining six sections (1 to 6) along the main course

—Section 4 (95 km long), between the Ariege and the Tarn it is still a piedmont river from a hydrological point of view but the valley becomes much larger and meanders are present;
—Section 5 (160 km long), often called the median valley of the Garonne, from the confluence with the Tarn to where the river becomes tidal—throughout this section the river bed was stabilized and regulated for navigational purposes between 1825–1835; and
—Section 6 (75 km long) the maritime river.

The physical characteristics of the river are summarized in Table 1. Most of the present study relates to sections 4 and 5 —the lower river below the Ariege confluence— which have been well documented.

The basin of the Garonne is influenced by two main climatic regimes: an oceanic one which may be subdivided into two types and a mediterranean one. Different types of flood events originate under each regime. Oceanic floods may be of a pyrenean or a classic type. Pyrenean floods occur in spring (May–June) and are most dangerous in sections up to section 4. Classic oceanic floods occur in winter (from December to March) and originate from the tributaries of the Massif Central. Section 5 is particularly affected by these events. Mediterranean floods occur in autumn and winter (from November to March) and originate in the southern part of the Massif Central. These affect the Tarn and section 5 of the Garonne.

TABLE 1. Some characteristics of the River Garonne

	Downstream to the Ariege tributary	Downstream to the Tarn tributary	Downstream to the Lot tributary
Distance from spring (km)	190	300	380
Drainage basin (km^2)	9980	32350	52000
Mean annual flow ($m^3 s^{-1}$)	200	460	630
Exceptional mean daily flow			
1875	8000	8000	8000
1952	4300	6200	5300
Mean flow in August (every 10 years)	51	70	92

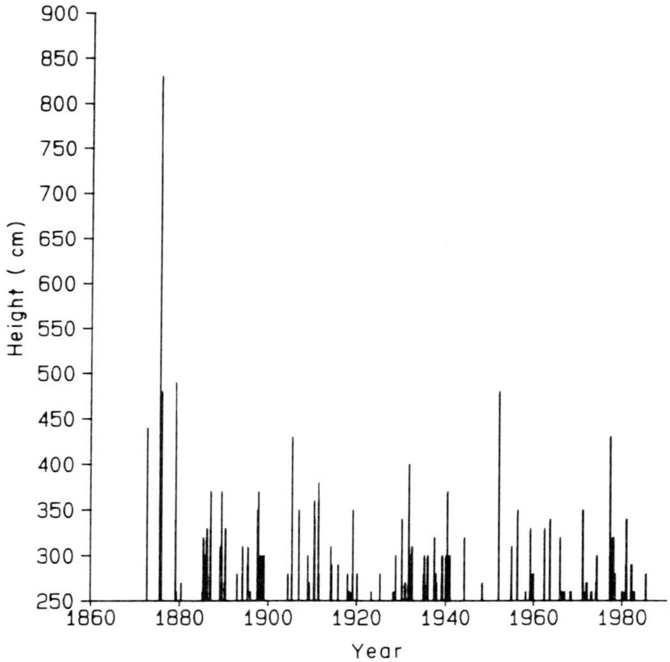

FIGURE 2. Floods above 2.5 m depth at Toulouse since 1872

The frequency and height of floods above the 2.5 m level at Toulouse since 1872 are shown in Figure 2. Flash floods on the Garonne are characterized by both their suddenness and their violence, and those with long duration have had catastrophic effects. Large areas can be affected by such floods, particularly in section 5 of the river. However, most of the time spates are short-lived: and the water level rises and falls very rapidly.

The intensity of interactions between the river and its terrestrial environment provides the basis for a typology of the floodplain (Décamps *et al.*, 1988). The highest parts of the floodplain which are subjected to fewer and shorter periods of submersion differ as

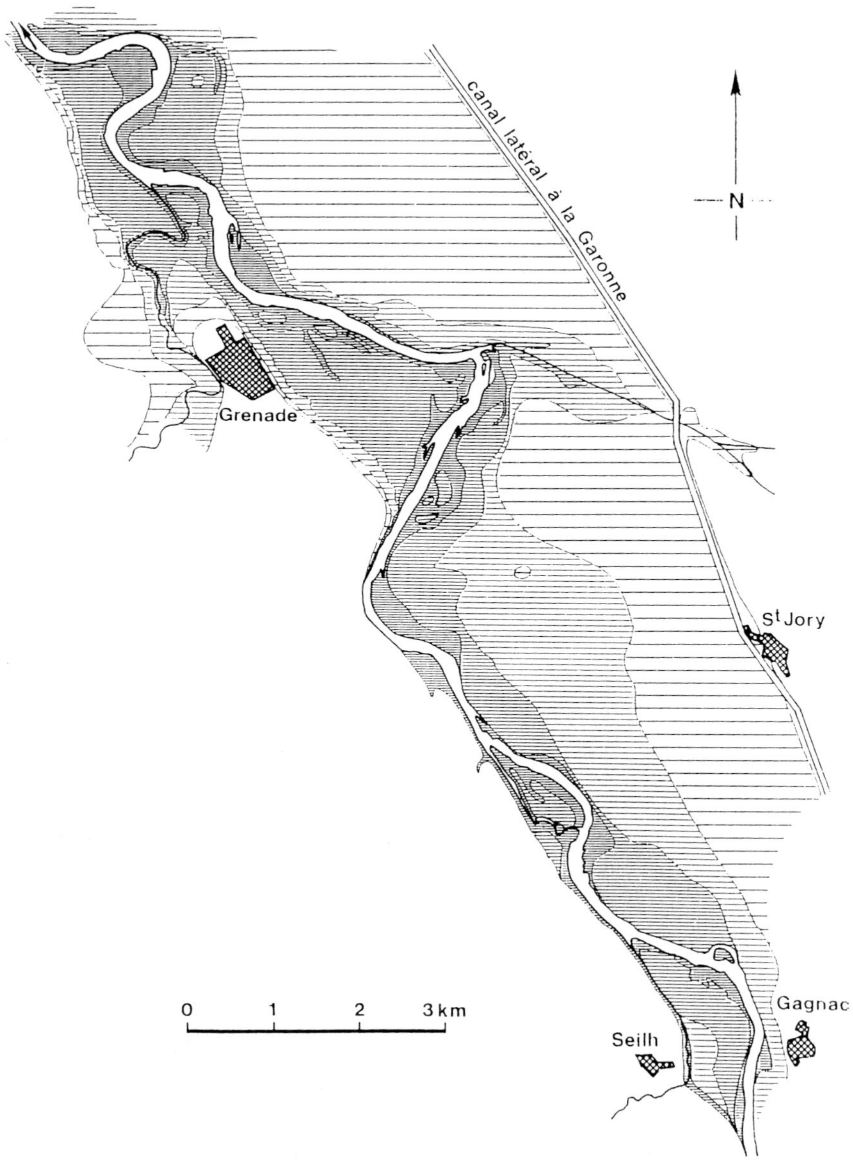

FIGURE 3. Typology of the floodplain of the River Garonne in section 5, based on flooding since 1875 with gradation from the lowest areas of the floodplain to the exceptional flood of June 1875

regards their quality of soil, the influence of the water table, and their land use. Three zones may be distinguished. These are best seen in section 4 downstream of Toulouse (Figure 3).

The lowest areas of the floodplain are submerged for about 80 days a year. This corresponds to a flow of 650 to 750 $m^3 s^{-1}$, i.e. a level of 1.8 to 2 m at Toulouse. Such areas are characterized by several factors including the height of banks with gentle slopes, distribution of recent alluvial deposits, depressed meander convexities, islands, dead arms which may occasionally function, and abandoned gravel-pits. This zone has never been used for habitation or agriculture; it is the area of natural riparian woods, the soil is frequently disturbed, and the surface is very rough.

Floodplain areas inundated by major floods corresponds to events with a frequency of between one in 2 to one in 20 years. The flow of the river is between 1500 and 2700 $m^3 s^{-1}$, i.e. a level of 3 to 4 m at Toulouse. Low parts of the floodplain, comprising ancient meanders and agricultural land, are submerged during these events. The land is affected by strong currents which modify the local morphology, eroding in some places and depositing sediments in others.

High areas of the floodplain are inundated only by exceptional floods occurring 3–4 times a century. The flow is above 4500 $m^3 s^{-1}$ and this corresponds to catastrophic events such as these of April 1770, September 1772, May 1835, June 1855, June 1875, February 1952 and May 1977. These areas are used as agricultural land, they are vast, flat and easy to work. Above these areas are a sequence of alluvial terraces: a low terrace at about 30 m above the river bed, a medium terrace at 50–80 m, a high terrace at 50 m, and a high-level 'flat' at 150 m (Hubschman, 1975).

HISTORICAL CHANGES

An historical account of the River Garonne must rely on cartographic documents, the history of which in itself is interesting. It was necessary to represent France on standard documents for government and economic purposes as early as Louis XIV and this was implemented by Colbert. Three generations of the Cassinis family carried out the work. A geometric outline of France was first drawn between 1682 and 1746. The next stage was a geometric description of the Kingdom at 1/86 400 scale. The area of the River Garonne was surveyed in 1769–1777. Other local surveys completed this work. In all, 179 sheets were edited by 1815, covering the whole country. This first map of France was used until the middle of the 19th century.

During the Revolution, France was divided into administrative departments, and a register was drawn up for tax purposes. A new map was required at a 1/80 000 scale. The survey began in 1818 and sheets were edited from 1832 to 1881. Several corrections and improvements were made later—using colour, contours and aerial photographs—mainly to suit the requirements of the army. However, the need of more precise mapping led to a tentative coverage at 1/50 000 and 1/20 000 scale in 1969. However, finally, it was decided to cover the country at 1/25 000 scale, with a revision every eight years.

Various other ancient documents have proved useful: drawings, plans, maps from Record Offices and Libraries. The older documents have often only simple drawings, but rich keys. These documents include requests for damages following flooding, law suits, and plans for the construction of bridges, quays or banks addressed to magistrates or even the King. All these documents provide precise information on the changes to the River Garonne during the past centuries as well as on the relationship between Man and the River.

Figure 4 summarizes the list of documents used in this study. Reliable cartographic documents dealing with the river in Toulouse have existed since the 17th century. These are the Tavernier map (1631) and the Jouvin de Rochefort map (1680), the latter being

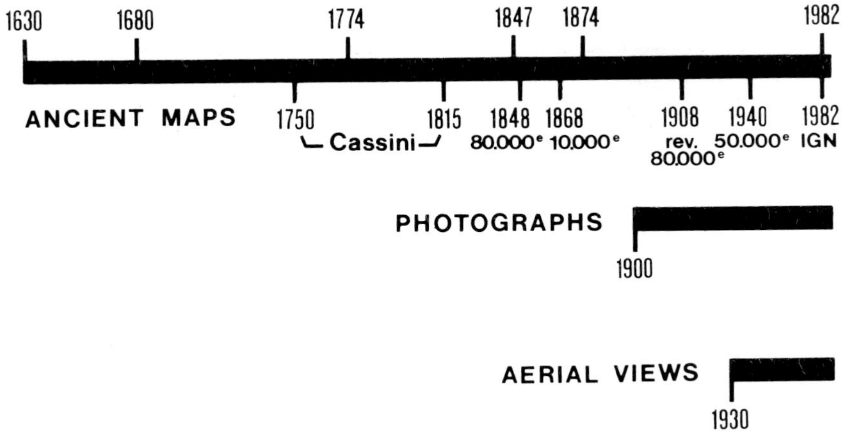

FIGURE 4. Documents available for a study of historical change in the Garonne

used in this study. Other maps which have been used are the Chalmandrier map (1774) and the first precise map of the City (Bellot, 1847). These maps have been compared with a 1982 map published by the French National Geographic Institute (IGN). Other maps of the River are those of Cassini (1753–1815), and maps edited in 1744, 1772 (plans of the River Garonne), 1859 (Carte d'Etat-Major), and 1868 (map at 1/10 000).

THE RIVER IN TOULOUSE

The morphology of the Garonne has changed, primarily since 1680, as a result of the interaction between river flow and Man's desire to harness the river during its passage through Toulouse. The Garonne offers a well documented example of the historical changes in a large river in an urban area (Fortuné, 1988a).

In Toulouse, the Garonne (a 7th order river) has just been joined by its last Pyrenean tributary, the Ariege. Its hydrological characteristics are those of a mountain river with a very irregular flow regime: the mean annual flow is 200 $m^3 s^{-1}$, with a mean minimum flow of 51 $m^3 s^{-1}$ for the 50 year flood (3.5 m depth in Toulouse).

Old documents indicate that destruction of bridges by floods has been frequent in Toulouse. In fact, no post Roman bridge resisted floods in Toulouse before the construction of the Pont Neuf, begun in 1543 and opened in 1660. Its construction required elaborate techniques such as raised piers as well as holes in the brickwork to facilitate the flow during floods. Even, this bridge showed undermining as early as 1764. Later, two suspension bridges were erected, in 1842 and in 1852, but both were destroyed by the 1875 flood. These works formed the first obstacles to the flow of the river in Toulouse.

The construction of mill dams was a second obstacle. Since the Middle Ages, the two mills of Chateau Narbonnais and Bazacle have had the greatest long-term effects on the river-bed. These mills provided the Toulouse region with flour. In return, the mill owners were given rights to manipulate the river flow and to use the banks. A stone dam, completed towards the end of the 12th century was associated with each mill. The river responded to these obstacles by regular sedimentation and erosion, and by catastrophic destruction during flooding.

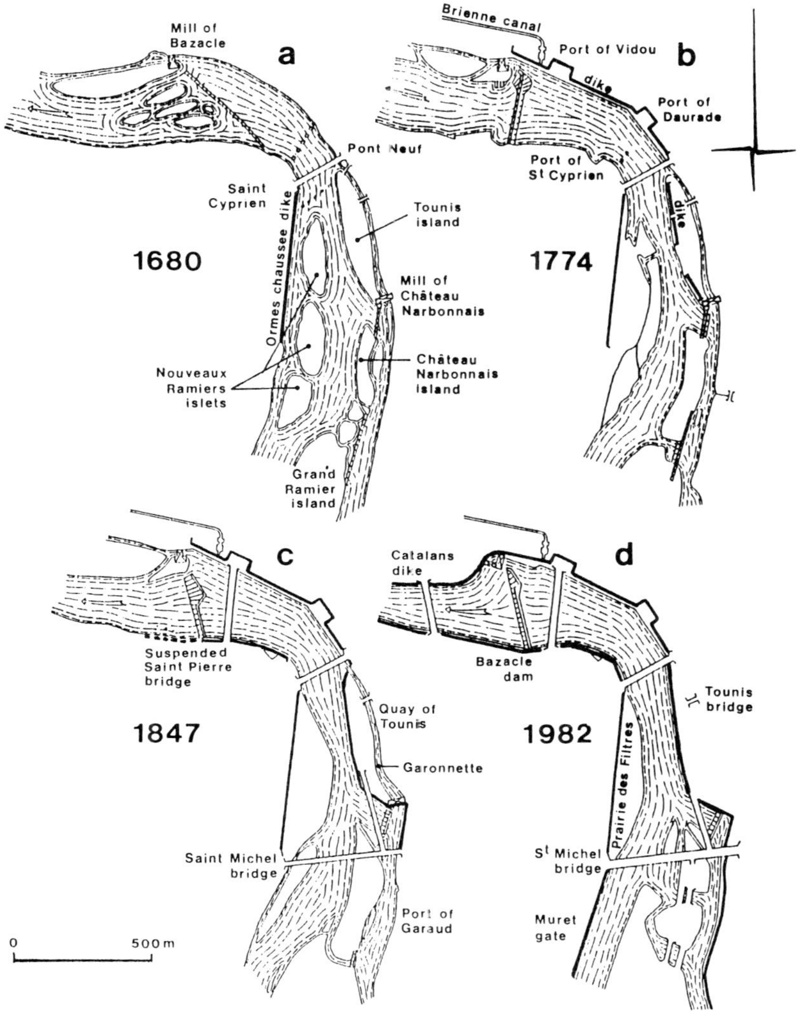

FIGURE 5. Historical changes to the Garonne during its passage through the city of Toulouse, based of the following plans: Jouvin de Rochefort (1680), Chalmandrier (1774), Bellot (1847), I.G.N. (1982) (from Fortuné, 1988a: reproduced by permission from John Wiley & Sons Ltd)

The construction of the first Bazacle dam and its passage for navigation caused erosion of the left bank and deposition of alluvium on the right bank, with the resulting formation of small islands (Figure 5). After its destruction during the flood of 1709, the dam was reconstructed in 1719. Built in stone, at a right angle to the axis of the river, and made higher, the passage for navigation was reduced. Large alluvial deposits formed upstream of the dam, and consequently, the level of the river bed was raised. These modifications exposed the left bank even more to floods, and were associated with rapid and catastrophic rises in water levels (Figure 5). The stone dam of 1719 was thus responsible for the left bank flooding in 1750, 1770, 1772. Water also regularly flooded the island of Tounis and

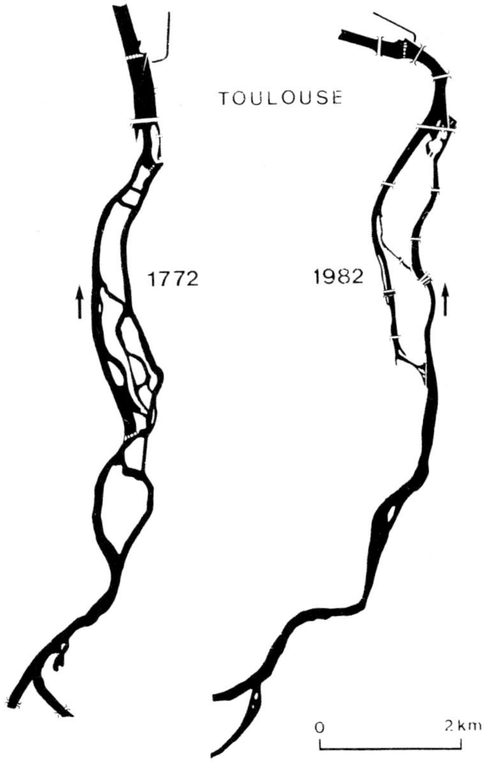

FIGURE 6. Evolution of the Garonne immediately upstream of Toulouse, based on maps edited in 1772 and 1982

via the suburb of Saint-Cyprien finally re-entered the Garonne downstream of the Bazacle dam. Disastrous flooding became even more common in the 19th century and the records show further disasters in 1827, 1835 and 1875.

With the building of bridges to span the river, and dams to harness the current, it soon became evident that embankments were necessary to prevent erosion and flood damage. For example, protection for the island of Tounis was instigated in 1667 and at the same time the alluvial islets were removed so as to slow down the current (Figure 5). This measure partially failed, because alluvial deposits then formed further upstream. The three islands aggraded to berth a single island, which after 1750 became attached to the left bank. A reinforcement of the dyke protecting the left bank was necessary between 1770–1780 and during the same century dykes were further raised to control floods and to protect the right bank. This work began on the island of Tounis in 1751 and included the construction of two ports on the right bank and one on the left. These protective measures were completed in 1850 with the Tounis dyke. Other protective dykes were constructed downstream during the 20th century. In 1951, the secondary arm separating the island of Tounis was filled in and transformed into a street! (Figure 5).

Channel changes by Man in the city of Toulouse can be identified from as early as 1650. Increased flood levels resulted from the construction works and the consequent channel changes. This required further works for flood and erosion control. However, major changes occurred in two periods between 1680–1750 and 1840–1880. During the first period, the number of islets decreased from 17 to nine, and the length of the shoreline from 11.4 to 9.5 km. At the present time, the number of islets is only four and the shoreline 7.2 km.

Two points are important when considering the evolution of the Garonne River immediately upstream from Toulouse during the same period (Figure 6). Firstly, a reduction of the number of channels took place, increasing the force of flow. Secondly, the right bank was eroded, and alluvium deposited elsewhere. These exaggerated the effects of the large and destructive floods that followed dam construction in the city of Toulouse, increasing erosion and sedimentation, and increasing flood levels.

In Toulouse, river navigation was important despite the floods and several active ports developed. These ports received goods from the Pyrenees and from the Atlantic Ocean. They were also centres of regional commerce and the river became the hub of the city, and greatly contributed to economic development towards the end of the 18th century. Thus, the attempted management of the Garonne through Toulouse has led both to prosperity and also to unforeseen and costly consequences. From an ecological point of view, the reduction of the length of river bank through the city, and along the river immediately upstream from it, has reduced the number of aquatic habitats for fish and other fauna in the urban area.

THE RIVER AND ECONOMIC DEVELOPMENT

Navigation

Historically, the Garonne was an important route for trade and transport between Toulouse and Bordeaux. However, navigation was more important downstream where it allowed commercial activities along the tributaries from the Massif Central, especially the Lot and the Tarn. Upstream of Toulouse, navigation was more difficult. Two main difficulties had to be overcome: the channel had to be maintained and the banks of the river protected. Works on these two problems were continuous and attempts to coordinate them began at the end of the 17th century, under Colbert, who inspected the river and suggested measures to maintain the bed for navigation. Under the natural conditions, only 120–150 days per year were suitable for navigation; rocks and sand bars deposited by floods were major obstacles. Water mills were another obstacle to navigation. They were thought to have been responsible for 90 per cent of the shipwrecks in the Garonne and were prohibited in 1830. Fish weirs and traps installed during upstream migrations were yet another obstacle. In spite of these difficulties navigation in the Garonne remained important as it was essential for economic exchange to and from the region.

Boats on the Garonne were generally made of oak; they were long and narrow and with flat bottoms. Equipment included a mast to fix towing lines. The loading gauge of these boats changed upstream: 'sapines' with 20 to 40 men on board 30 tons gauge were used to Auvillar, and then 'macalets' with 15 men sailed up to the confluence with the Tarn. Others such as 'ramoneurs' and 'miolles' reached Toulouse, with 5 to 10 men on board, and gauged 15 to 20 tons. Upstream of Toulouse 'gabarots', 10 m long and gauged

10 tons, sailed up the Ariege and up the Garonne to the Salat. The small freight of these boats were compensated for by the level of activity during the periods of navigation. The 1837 statistics record the following boats sailing up river between Agen and Auvillar: 301 sapines, 17 macalets, 1070 miolles, 303 ramoneurs, and 5598 gabarots. In 1841, at the height of navigation, 147 000 tons were transported—97 000 t downstream and 50 000 t upstream.

Men working in navigation on the Garonne were organized into a corporation: the mariners. They lived with their families, along the River. They did not usually own their boats, but were wage-earners, and were fed on their travels. As well as their main employment they did other jobs linked with the river: protection of the banks, digging out the channel during the low water, and fishing.

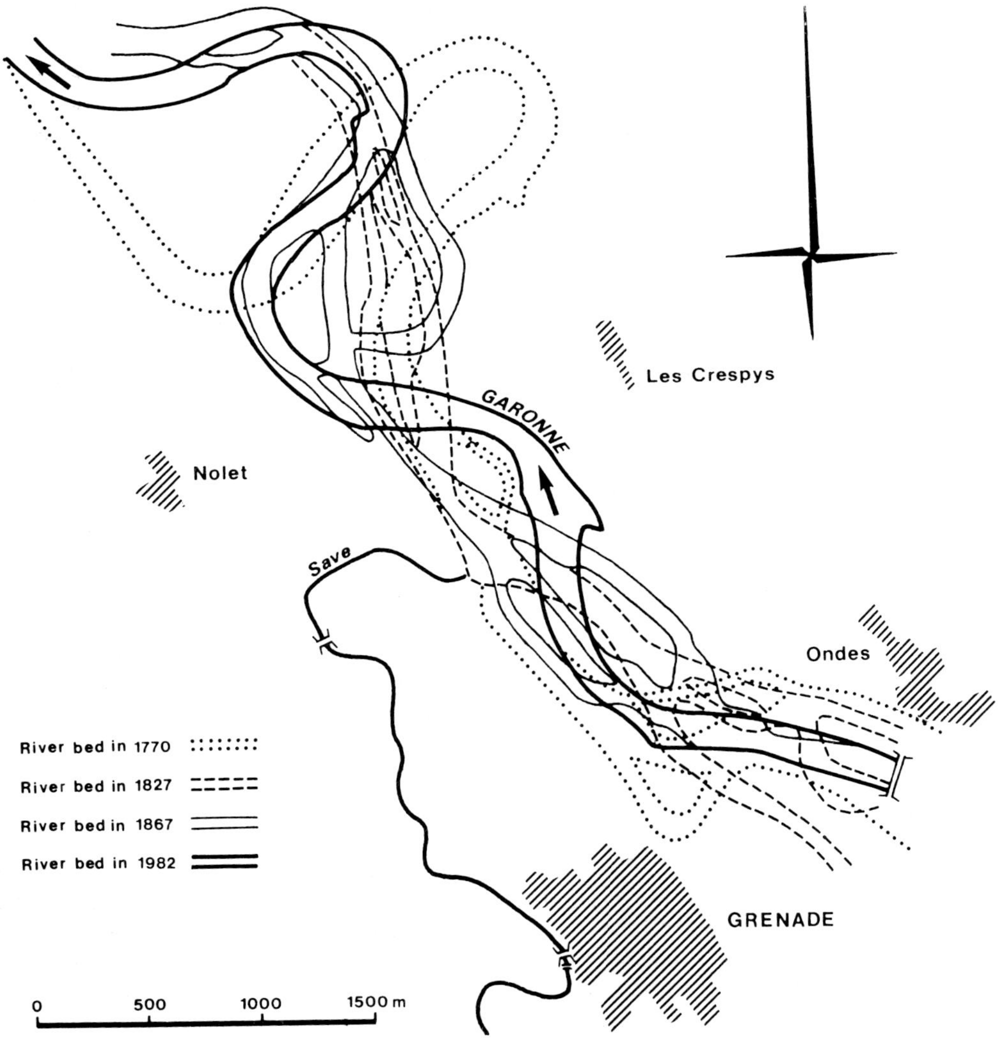

FIGURE 7. Lateral changes to the River Garonne in part of section 4

The only means of navigation upstream in the river was towing by men. For this, a tow path was constructed and the banks were consolidated. This led to changes in the riparian zone. For example elm trees had been planted as a means of protection against spates on the Garonne during the Middle Ages. In the 18th century with the increased interest in navigation the elms were replaced by more flexible cultivated willows which were cut every two years. Upstream of the confluence with the Tarn, the river was very active and often changed its course (Figure 7): navigation required continuous channel maintenance. Eroding banks were stabilized with timber stockades and willow plantations were used to stabilize the alluvium left after floods. Two towing paths had to be made because of the huge differences in flow during the year—one on the elevated parts of the banks for periods of high water levels and one nearer the river bed for times of low water. Such changes were all completed by the middle of the 19th century before the decline in use of the river for navigation began.

On the whole, suitable conditions for navigation were possible from autumn to spring, only stopped by short periods of exceptional floods. During summer, low water rendered transport difficult or impossible. However, whenever suitable conditions for navigation existed it was an important activity. Upstream transport from Bordeaux to Toulouse took about 10–15 days, downstream transport took 2–3 days. As a result, two rotations a month were possible.

The 18th century witnessed the peak of navigation in the Garonne and benefited economic development during the 19th century. The use of steamers allowed regular transport of passengers from Bordeaux to Toulouse: 250 000 people in 1840 and 450 000 in 1850. In fact, navigation dominated the economic life of the Garonne Valley from the beginning of the 18th to the middle of the 19th century. An important part of the activity of the region extended along the banks of the River. First, trades linked with water developed including boat construction, manufacture of ropes and sails, towing, etc. Secondly, ports grew up as centres for commercial exchange along the river.

Two particular examples may be cited of the influence of Man's activities. First, ropes and sails were necessary for navigation. Thus, hemp cultivation was extended into the alluvial lands of the valley. This hemp was retted in the Garonne in autumn, blocking the bed and the fermenting hemp led to elevated fish mortalities in some areas. Secondly, small market towns, which had been located on the levels above the floodplain, moved nearer to the river banks thus increasing activity along the river. However, during the 19th century, when navigation declined, people from these market towns dispersed over the alluvial plain and greatly changed the landscape with developments in agriculture.

Agriculture

The fertile alluvial valleys of the Garonne and its tributaries constitute a network with high agricultural potential. The river has allowed outlets for commercial products and vineyards, plum orchards and tobacco cultivation developed along the river system. Thus at least in the early days of agricultural development there existed a sort of symbiosis between navigation and agriculture. However, there have always been conflicts between these two activities. In the 18th century the interests of navigation were always given priority at the expense of agriculture or indeed other activities connected with the River such as mills or fisheries. This situation continued as long as navigation was of major importance in the region.

Gradually, agricultural activities came to dominate the economy of the valley. At the end of the 17th century, corn and wheat were not grown on the floodplain because of the risks of flooding. Such crops were found on the terraces and plateaux. However, the introduction of maize allowed agriculture to develop on the floodplain. Maize does well on humid soils, and also is sown in May and is thus not threatened by spring floods. Agricultural activities were constantly adapted to meet demand: corn for export, wheat, rye and maize for local needs. Cultivation spread into the riparian zone to meet the market demands. Vineyards have been found in the Garonne valley since Roman times, linked to river transport and exportation abroad via Bordeaux. Similarly, the international trade of wood, developed in the 17th century with the help of river transport. Other agriculture products—hemp, flax, plums and other fruit, and market gardening—were all linked to the presence of the river. All these crops were grown nearer to the banks of the river and for the last twenty years irrigation has become the key to agricultural development in the Garonne valley.

Poplar plantations are also important in the floodplain. Introduced Italian poplar trees had been planted from as early as the middle of the 18th century; next Californian poplar trees were planted and later 'F214' and 'Robusta' varieties; these last two are now the most common forms. The plantations replaced the natural communities of fine silt and clay patches in the riparian zone. They help fix alluvial soils and stabilize wet border zones. In 1979 poplar plantations covered 1755 hectares of the 3570 hectares of riparian wood in the section between Toulouse and its confluence with the Tarn.

Industrialization

Several factors have aided the region's industrialization such as the density of human population, openings of foreign markets, gravel exploitation, and hydroelectric power. The valley of the Garonne has been a densely populated area, in comparison to its surroundings, since at least the 16th century. In 1806, the population of the two main cities, Bordeaux and Toulouse, were 95 327 and 51 689 respectively. The third city was Montauban (23 895), followed by Agen (10 548), Moissac (10 331) and Castelsarrazin (10 144), all riverside towns (Poussou, 1982). This demographic predominance of the valley was characteristic of the area between the 16th century and the beginning of the 20th century when the river was an essential urban, industrial and commercial axis. During the 17th and 18th centuries the increasing economic importance of the floodplain drew people who had previously been living on the terraces. This was possible only by a continued effort to prevent flooding by extensive management of the banks. At the beginning of the 19th century, about 30 river ports were active between the two main centres of Toulouse and Bordeaux. Agen was the third most important port with river traffic of more than 4000 boats and 300 rafts (about 147 000 tons) in 1841. Such river traffic itself employed between 10 000 and 12 000 people in addition to the numerous ships carpenters, innkeepers, hauliers, etc. Very diverse industry existed along the river including textiles (falx, hemp, and wool), metallurgical industries, sawmills, and flour mills. However, this industry remained small, and did not expand in the 19th century.

During the last 300 years, the Valley has changed under the influence of the general European socioeconomic development. At the beginning of the 18th century, the valley became open to foreign markets, with Bordeaux becoming an important port on the Atlantic coast. Regional trade between the two main cities Toulouse and Bordeaux also increased with developing agriculture, along the valley. The role of the Garonne as a navigation axis increased during the first part of the 18th century but then slowed and finally ceased. This was partly because of the construction in 1856 of a canal, parallel with the river between Toulouse and Bordeaux, and also because of the increase from 1857 in the use of rail travel for transport from the Atlantic to the Mediterranean. Also, the Garonne did not become as important an industrial axis as other European rivers such as the Rhine and Rhône at the beginning of the 20th century. Nevertheless, economic development has been important and as a consequence has profoundly influenced the ecology of the river corridor of the Garonne since the end of the 17th century.

Hydroelectric management of the valley offered another aid to the economic development in the Garonne region. In Toulouse, the Bazacle power station supplied electricity to the city as early as 1889. Hydroelectric power stations were built on the upper parts of the Garonne and its tributaries from the end of the 19th century, and this continued from 1890 to 1970. Now, about ten power stations use the Garonne itself in sections 2 and 3, where the gradient is still steep. The capacity of these stations exceeds regional demand and power is exported to neighbouring areas. Chemical and electrochemical industries have developed along the Garonne using the hydroelectric power. A nuclear power station is under construction at the confluence with the Tarn using water for cooling. Here, a dam was constructed in 1971, and a system of canals supplies the power station.

Gravel has been extracted from the Garonne during the early 19th century. Until the beginning of the 20th century extraction from the river-bed and its banks was active but apparently did not cause major damage. However, in order to maintain the navigation channel after floods certain parts of the river became important extraction areas, but it is only recently, due to an increase in gravel extraction, that severe consequences have occurred to the environment. A gravel production of about 3.5 million tons in 1965 has doubled in ten years, and is now about 10 million tons in the region of the Garonne and its Pyrenean tributaries. This increase corresponds to a higher demand to meet the needs of urbanization and roads. Gravel pits have grown up along the River in a very disorderly manner. At first they were found near urban centres, but then they spread into the floodplain as mechanization allowed exploitation below the level of the water table. Since 1981, new works are subject to strict planning control in order to reduce the consequences to the environment.

ECOLOGICAL CONSEQUENCES

The effects of anthropogenic changes may be classified in three groups (Petts, 1987): isolation of the main river from its alluvial plain, introduction of barriers, and changes of instream habitats.

In the Garonne, the isolation of the main river from its alluvial plain is not complete. There are embankments in some places in section 4, and it is only downstream of the confluence with the Tarn (section 5) that the river-bed was regulated and stabilized in 1825–35. Large embankments such as those of the Rhine or Rhône are not found along the Garonne, and any that do are generally small (Deffontaines, 1932). Only a few isolated embankments built in the 18th century to protect new agricultural land in the fertile floodplain area (Vital, 1984). Thus, complete isolation of the river from its alluvial plain did not take place along the Garonne, at least until section 5. However, navigation and agricultural activities have affected the riparian woods along the river, causing a fragmentation of the forest cover and a substitution of woodland species by agricultural species and poplar plantations (Décamps *et al.*, 1987).

In the 18th century, with increased interest in navigation, elm trees previously planted as protection against floods were replaced by more flexible cultivated willows which were cut every two years. Eroding banks were stabilized with timber stockades, and willow plantations helped to stabilize the alluvium left after floods. In 1979, poplar plantations covered about 35 per cent of the riparian zone in section 4, replacing the various natural forest communities. Three of these communities are found on lowest areas of the floodplain: willow, alder and ash-elm communities. On the higher areas of the floodplain an oak–elm community once dominated. Agriculture flourished all along the Garonne valley during the 19th century. Cultivated areas previously restricted to the terraces spread across the floodplain towards the river restricting the area covered by riparian woods and accelerating the fragmentation of the forest cover in the valley.

Urban growth during the 19th and 20th centuries accelerated gravel working in the river-bed and from the floodplain. In many places the channel bed was lowered, molassic rock substrate revealed, and consequently the water-table subsided. Observations suggest a subsidence of about 1 m over the last 20 years (Beaudelin, 1987). As a consequence, some older trees in the riparian woods died and poplar trees which are shallowly rooted seem to have been particularly affected. Another result is that several dead arms of the river have dried up prematurely during the last two decades and this has reduced the spawning sites for fish. These impacts have all been intensified during the 20th century, particularly during summer low water levels when flow is regulated by hydro-electric dams on the river and its tributaries and also by an increase in the abstraction of water for irrigation.

In the riparian woodland along the Garonne valley, the tendency has not been to replace a reversible pattern of succession in the forest cover by an irreversible pattern (Pautou and Décamps, 1985). However, historical changes have resulted in (1) a fragmentation of the forest cover forming a mosaic of wooded patches in agricultural land, with dynamic boundaries, and (2) a substitution of species within the continuous corridor of riparian woods. Because of the non-continuous isolation of the river from its floodplain, floods are still capable of transforming the fluvial landscape as suggested in Figure 2

One of the main effects of Man on the river is probably an exaggeration of low water conditions in dry summers, as a consequence of lowering the water-table (Décamps, 1984). This is an important factor when considering the future evolution of fluvial landscapes in the climatic conditions of Southern Europe because the dry season constitutes a limiting factor in the development of willows, alders and elms in riparian wood communities.

Pinay (1986) has clearly shown the efficiency of riparian woods as natural filters regulating nutrient transfers to the river. During the 20th century, river developments have led to the removal of riparian systems by filling, dredging, or draining (Pinay and Décamps, 1988). These activities have resulted in a decrease, or even a disappearance, of the buffering capabilities of the riparian zone *vis-à-vis* nitrogen transfers between the drainage basin and the river.

The introduction of barriers on the river and the transformation of instream habitats, have had spectacular consequences on fish. Since the Middle Ages, many accounts give evidence of the importance of fish in the nutrition of people living along the river Garonne.

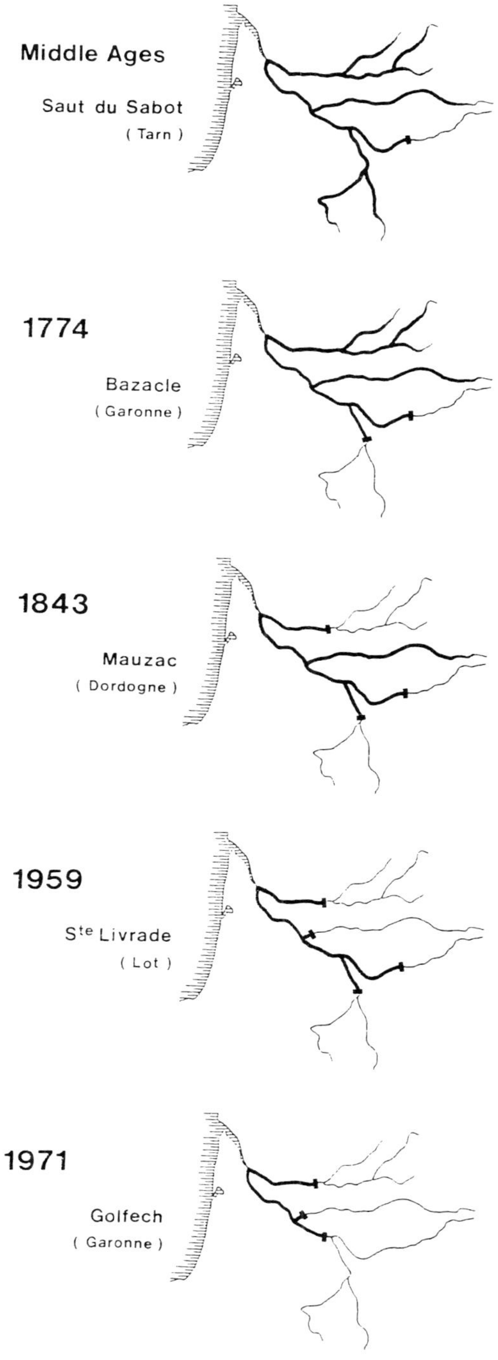

FIGURE 8. Restriction of spawning grounds for migrating fish because of dam construction in the Garonne and Dordogne basins since the Middle Ages

Fish markets existed in towns which bordered the river. Fishermen were organized as a profession, with rights and duties, and with their own locations for fishing. Various written documents recount the importance of fishing and fish handling for food and trade. However, there are no accounts of the size of the catches. The richness of the Garonne and its tributaries depended on the upstream migration of fish. Every year, in spring and the beginning of summer for most species, and in autumn–winter for salmon, fish migrated up the Garonne to their breeding places. This was section 5 for sturgeon, section 3 for lamprey and shad, sections 2 and 1 for salmon. Shad were the main catch for the last of the professional fishermen in Toulouse in the 1930s, and 50 years ago a small firm still dealt in caviar near the sturgeon's breeding place.

Constructions which have restricted spawning grounds for migrating fish are shown in Figure 8. In the Middle Ages a natural barrier existed on the Tarn (Figure 8a). The construction of the Bazacle dam in 1774 prevented migrating fish from entering the Pyrenean river network above Toulouse (Figure 8b). In 1843 and 1959, dams constructed on the Dordogne and Lot respectively reduced again the potential areas for spawning (Figures 8c, d). Finally, in 1971, the Golfech dam restricted potential spawning sites to downstream sections of the Garonne and Dordogne. However, the construction of a fish ladder in 1987 at the Golfech dam has increased the opportunities for salmon and shad to migrate upstream again at least as far as the Bazacle dam in Toulouse.

Human activities affecting instream habitats have made the river-bed uniform. Management of the river-banks—including embankments, regulation of the river, and removal of islands—has progressively reduced the aquatic ecotones where most of the fish species find their habitats. More recently, gravel extraction has resulted in a bare bed with sandstone substrate. Thus, sand and gravel spawning sites have decreased with dramatic consequences particularly on sturgeon populations. As a consequence of gravel extraction, the level of the water table has decreased and most dead arms of the river in sections 4 and 5 are no longer flooded, contributing to a decrease in habitat diversity and of the spawning grounds for fish. Another consequence of human activities, particularly since 1950, has been a general deterioration in water quality for life in the river. This is particularly noticeable downstream of urban and industrialized areas.

CONCLUSION

The river itself and the fluvial landscape may be distinguished when considering the history of Man's influence on the River Garonne. In the river itself, one main effect of anthropogenic changes is the phenomenom of channelization. This has taken place because of the development of navigation which was necessary for the region's economy (Fortuné, 1988b). Ecological consequences were a diminution of the land water interface, i.e. the length of ecotones between the terrestrial and the aquatic environment along the river, and at the same time, an increasing uniformity of the instream habitats. A second consequence is an exaggeration of the contrasts between hydrological conditions, i.e. between floods and low water. In more modern times pollution has affected water quality, and the construction of dams has introduced barriers to the migrating fish.

The history of the Garonne in Toulouse since 1680 provides an example of how a river reacts to Man's modifications and how in turn further modifications were necessary to protect against floods. This example also shows how important a role the river played in

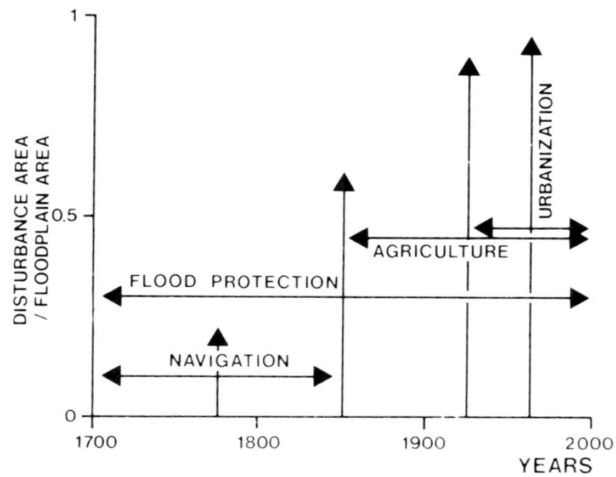

FIGURE 9. Increase in the proportion of disturbed areas within the floodplain as a consequence of Man's use of the River Garonne valley since 1700

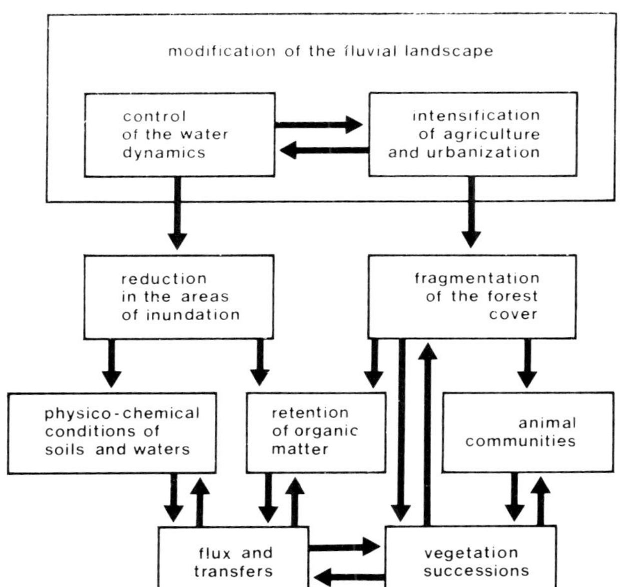

FIGURE 10. Consequences of the evolution of the fluvial landscape along the Garonne

the economic life of the area around the city. The interactions between economic development and ecology are also evident in the fluvial landscape (Décamps and Naiman, 1988). Man's activities for navigation, flood protection, agriculture, industry and urbanization have progressively increased the percentage of disturbed area within the floodplain (Figure 9). These activities have profoundly modified the evolution of the fluvial landscape along the River Garonne (Figure 10). On one hand, an intensification of agriculture and urbanization have caused a fragmentation of the forest cover, with consequences for retention of organic matter, vegetation succession, and animal communities (Décamps *et al.*, 1987). On the other hand, control of the water dynamics has caused a reduction in the areas affected by flooding, and this influences the physicochemical conditions of the soil and water, retention of organic matter, flux and transfers.

Two distinct periods of use of the river and its banks may be distinguished in the history of Man's effect on the fluvial landscape: navigation during the 18th century and agriculture during the 19th century (Deffontaines 1932). These have been followed in the 20th century by a third period with intensification of agriculture, industrialization and urbanization. The isolation of the Garonne from its floodplain has not been complete because of the limited number of embankments. As a result the conditions necessary for reversible trends in forest succession have not been lost on the floodplain but rather wooded areas have become fragmented, cultivated communities have replaced natural ones and there is an accentuation of water shortage during dry summers. This water shortage is specific to large rivers in Southern Europe.

Thus, a general model predicting the influence of Man on the riparian dynamics of the fluvial landscape must pay more attention to the gradient of climatic conditions from north to south (Décamps *et al.*, 1988). Central European models (such as those constructed for the Rhine–Rhône conditions) cannot be extrapolated for Western and Southern European fluvial landscapes like the Garonne. Moreover, as shown by Décamps *et al.* (1988) where economic activity is dominated by agriculture and not industry, the historical use of the floodplain results in areas where reversible trends in forest succession are possible, existing side by side or intermingled with areas where they are not. This is an important difference which should not be overlooked when comparisons are made between large rivers from industrialized and agricultural areas.

REFERENCES

Anonymous 1744 (?). *Copie du Plan envoyé au Roi par la ville de Toulouse, où l'on voit son ancien et son nouvean lit, les ouvrages qui ont occasionné les mutations et les ravages qui en ont été les suites. Confluence de l'Ariège—Chaussée du Bazacle.*

Anonymous 1772. *Plan de la rivière Garonne—depuis le faubourg Saint-Michel de la ville de Toulouse jusques au-dessous de la chaussée de Braqueville—sur lequel ont été tracés par des lignes rouges les bords du canal du Moulin du Château, tels qu'ils étaient lors de la levée du plan côté no 5, fait en l'année 1772-183, 30 octobre*, National Library, Paris.

Anonymous 1868 (?). *Carte au 10 000e de la Garonne sur laquelle les limites de l'inondation de 1855 sont reportées en 4 feuilles: no 27 Toulouse, 26 Blagnac, 25 Grisolles, 24 Verdun sur Garonne.*

Beaudelin, P. 1987. Les méfaits des extractions des galets de la Garonne. *Adour-Garonne*, **34**, 10–13.

Bellot, P. J. 1847. *Plan géométrique de la ville de Toulouse dressé d'après les plans du cadastre*, Raynaud, Toulouse.

Cassini 1753–1815. *Carte géométrique de la France à l'échele d'l ligne pour 100 toises (environ 1/86 400), réalisée de 1750 à 1815.*

Chalmandrier 1774. *Toulouse, la ville au XVIIIème siècle, dedicated to Monseigneur the bishop Loménie de Brienne, 1/13 000*, National Library, Paris.

Décamps, H. 1984. Biology of regulated rivers in France. In Lillehammer, A. and Saltveit, S. J. (Ed.), *Regulated Rivers*, Stockholm.

Décamps, H., Fortuné, M., Gazelle, F. and Pautou, G. 1988. Historical influence of man on the riparian dynamics of a fluvial landscape. *Landscape Ecology* **3**, 163–173.

Décamps, H., Joachim, J. and Lauga, J. 1987. The importance for birds of the riparian woodlands within the alluvial corridor of the river Garonne, S.W. France. *Regulated Rivers*, **1**, 301–316.
Décamps, H. and Naiman, R. J. 1989. L'écologie des fleuves. *La Recherche* (in press).
Deffontaines, P. 1932. *Les hommes et leurs travaux dans les pays de la moyenne Garonne (Agenais, Bas-Quercy)*, Lille.
Etat-Major 1859. *Carte de la France au 1/80 000 dite Carte d'Etat-Major-feuilles de la Haute-Garonne-levées de 1818 à 1881.*
Fortuné, M. 1988a. Historical changes of a large river in an urban area: the Garonne River, Toulouse, France. *Regulated Rivers*, **2**, 179–186.
Fortuné, M. 1988b. *Usages passés et Ecologie de la Garonne*, Thèse de Doctorat, Toulouse.
Hubschman, J. 1975. Le terrasses récentes de la Garonne et leur évolution, *Bull. Ass. franç. Etude Quaternaire*, **3/4**, 137–147.
I.G.N. 1982. *Carte au 1/25 000*, National Geographic Institute, Paris.
Jouvin de Rochefort 1680. *Plan 1/50 000*, National Library, Paris.
Lambert, R. 1982. Les crues de la Garonne. In Garrisson-Estèbe and Ferro, M. (Eds), *Une histoire de la Garonne*, Ramsay, Paris, 41–48.
Pautou, G. and Décamps, H. 1985. Ecological interactions between the alluvial forests and hydrology of the upper Rhône. *Archiv f. Hydrobiol.*, **104**, 13–37.
Petts, G. 1987. Ecological management of regulated rivers; a European perspective. *Regulated Rivers*, **1**, 363–369.
Pinay, G. 1986. *Relations sol-nappe dans les bois riverains de la Garonne. Etude de la dénitrification*, Thèse de Doctorat, Lyon..
Pinay, G. and Décamps, H. 1988. The role of riparian woods in regulating nitrogen fluxes between the alluvial aquifer and surface water: a conceptual model. *Regulated Rivers*, **2**, 507–516.
Poussou, J. P. 1982. Une vallée riche et peuplée. in Garrisson-Estèbe and Ferro, M. (Eds), *Une historie de la Garonne*, Ramsay, Paris, 361–390.
Tavernier 1631. *Plan Tavernier*, National Library, Paris.
Vital, P. 1984. *Requiem pour une Garonne défunte*, Wallada, Bordeaux.

Historical Change of Large Alluvial Rivers: Western Europe
Edited by G. E. Petts
©1989 John Wiley & Sons Ltd

16

Recent Historical Changes on the Belgian Meuse

J.-C. MICHA and M.-C. BORLEE

Unité d'Ecologie des Eaux Douces (UNECED), Facultés Notre-Dame de la Paix (F.N.D.P.), Rue de Bruxelles 61, B-5000 Namur, Belgium

INTRODUCTION

The Meuse flows through three main countries: France (492 km), Belgium (194 km) and the Netherlands (239 km). Forty per cent of its catchment area is in Belgium. Since the creation of the unit of freshwater ecology in 1974 at Namur University, we have been interested both in the ecology of the small salmonid rivers (Molignée, Samson, etc.) and also in the only large river in the country, the Meuse. Although the study of a large river is much more difficult than that of a small one, it was, in our view, necessary to attempt the task because for Belgium, and Wallonia in particular, the River Meuse is a waterway of major importance. In fact, it serves as a major routeway for different types of transport, provides water for industries and domestic supply, and receives large quantities of effluent water that it dilutes and assimilates to varying degrees.

Our study of the River Meuse began with the installation of a nuclear power station at Tihange. In this context, we studied the impact of thermal, radioactive and chemical waste on the water of the Meuse and on its biocenoses (Genin and Micha, 1977; Meurisse-Genin and Micha, 1978; Detollenaere and Micha, 1978; Detollenaere and Micha, 1980; Mattheeuws *et al.*, 1981; Gillet *et al.*, 1984; Detollenaere and Micha, 1986; Gillet and Micha, 1987). We very soon realized that this study, which was limited to the part upstream and downstream of this power station, needed a broader context. Thus, investigations were expanded to include the whole of the Belgian part of the Meuse from the French to the Dutch border. In the first instance, our task was confined to drawing on the records of various administrations since 1800 and to consulting various bodies which have been regularly concerned with the Meuse since that period. After describing the catchment area and the modes of occupation of the Meuse valley, we shall examine three successive periods of development of the channel bed and the floodplain of the Belgian part of the Meuse. Their hydrological, physicochemical and ecological consequences are then examined. Finally, we discuss cause and effect relations, with a view to making forecasts and recommendations for the future.

THE CATCHMENT AREA OF THE MEUSE AND ITS OCCUPATION BY MAN

Hydrographic network and general characteristics of the Meuse

The River Meuse rises in the Plateau of Langres in France at an altitude of 402 metres and flows for 925 km before entering the North Sea south of Rotterdam (Netherlands), at which point it merges into the Rhine–Meuse–Escaut complex of channels. Its catchment area (Figure 1) is spread over five countries (France, Belgium, the Netherlands, Germany, Luxemburg) and covers 36 011 km^2 of which 13011 km^2 are in Belgium. Its main tributaries are the Chiers, the Semois, the Lesse, the Sambre, the Ourthe, the Ur, the Rur and the

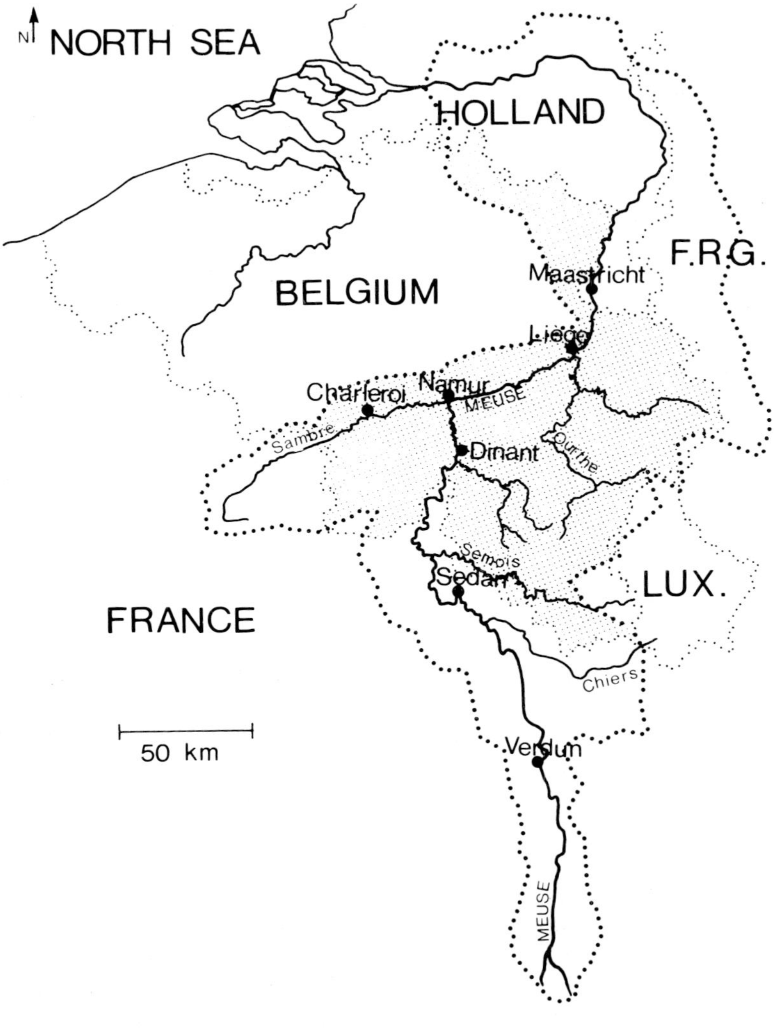

FIGURE 1. The catchment area of the Meuse and its main tributaries, [legend symbol]. Belgian area of the basin

Niers. Its upper course has hardly any tributaries because of successive losses (Meurthe, Moselle, etc.) occasioned notably by regressive erosion of the Rhine. Possessing a surprising shape in the form of a corridor 150 km long and from 15 to 40 km wide as far as Verdun, the basin of the Meuse then spreads out, reaching a maximum width of about 180 km near Namur. Its basin narrows again a last time on a line with Maastricht where it reaches 70 km. The Meuse then flows into the sea through a series of weirs built to control flooding.

Before 1940, the mean slope of the Meuse was 0.45‰ over the whole of its course and 0.23‰ for the 194 km of the Belgian reach. The natural slope of the river has, however, been modified for navigational purposes. The Meuse is a river with a temperate zone rainfall regime: its flow is at its highest in winter and at its lowest in summer. The annual average flow is 280 $m^3 s^{-1}$ upstream from Liège but the minimum and maximum daily figures can differ considerably from this average: 10 to 20 $m^3 s^{-1}$ from April to September 1976, more than 2000 $m^3 s^{-1}$ in July 1980. The composition in major mineral elements shows a predominance of calcium ions (100 $mg\,l^{-1}$) and of bicarbonates (160 $mg\,l^{-1}$) which gives to the water an alkaline pH (7–8). The average conductivity is of the order of 500 $\mu S\,cm^{-1}$.

Occupation of the region

Data relative to the occupation of the region is limited to the French and Belgian areas. In France in the last century the catchment area of the Meuse was given over to grassland and sheep rearing. A number of small industries existed on the Chiers. At Sedan, the textile industry prospered. Between Charleville and Givet, metallurgical industry, nail factories, iron foundries, hardware businesses, etc. stood alongside each other (Piérot and Mahaut, 1981). In Belgium, the south of the Sambre and Meuse line was covered in large part by forests which underwent intense deforestation to meet the needs of new industry and those of breeding and agricultural development. Thus, the pasture that covered 20.7 per cent of the Walloon area in 1846 represented 30.3 per cent in 1910 (Génicot *et al.*, 1973). Industrial plants had spread throughout the whole of the basis: blast furnaces, rolling mills, collieries, stone quarries, glassworks, paper mills, tanneries and textile industries, appeared along the Meuse and its tributaries, drew the workers towards the towns: Liège, Charleroi, etc (André *et al.*, 1930).

At the present time, in France, one quarter of the catchment area is taken up by forests, while cultivation and pasture account for the remainder (George, 1972). Industry was mainly concentrated in the Chiers basin (Figure 2) but there is currently a marked decline. Finally, the scant population is relatively scattered, except for several sizeable towns: Longwy (pop. *c.* 25 000), Sedan (*c.* 25 000), Charleville (*c.* 65 000), Givet (*c.* 10 000). In Belgium, the catchment area is divided between forests of deciduous trees and spruce (Ardennes), intensive farming (cereals and beetroot), pasture and orchards. Industrial activity is concentrated along the Sambre, downstream from Charleroi and along the Meuse upstream and downstream from Liège (Figure 2). However, industrial activity in these areas, as in France, has been in marked decline for more than a decade. As for housing, this is clearly spread out between the French border and Namur, the only sizeable town being Dinant (*c.* 15 000). From Namur, the population is concentrated along the river in towns of greater or lesser importance (Andenne: *c.* 25 000; Huy:*c.* 25 000; Liège: *c.* 200 000).

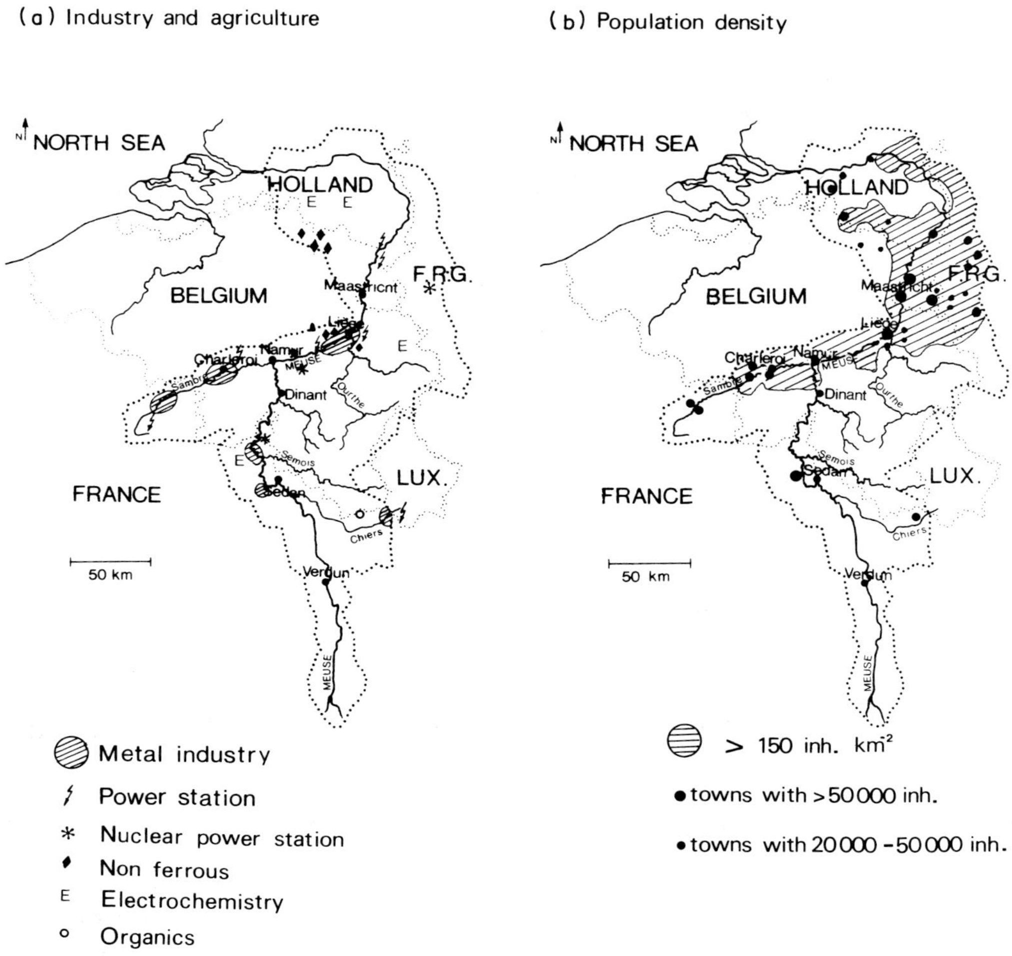

FIGURE 2. Map of the Meuse basin with zones of industrial (a) and urban (b) concentrations (adapted from Van Craenenbroeck, 1986)

Changes

Changes to the Meuse have been made over two successive periods. The 'historical' period begins after the last ice age and ends in 1800, the time when the Meuse was brought into the European network of river communications (Questienne and Dubois, 1978). Suttor (1986) notes that during this period the essential characteristics of the river did not change. This chapter is concerned with the second period. This modern period extends from 1800 to the present day and falls into three main phases:

—first canalization from 1800 to 1880;
—adaptation and stagnation from 1880 to 1918;
—modernization from 1918 to the present day.

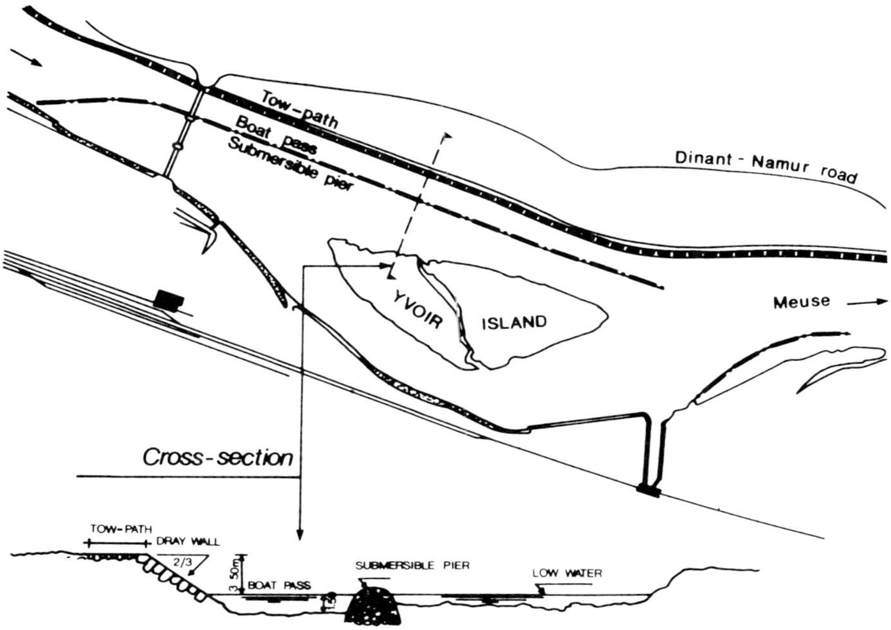

FIGURE 3. Improvements to the navigation on the Meuse by Guillery (1942, in Works involving alterations to the Namur part of the Meuse since 1837, unpublished, 20 pp)

First canalization (1800–1880)

At the end of the Napoleonic Empire and after the treaties of 1815 binding our region to the Netherlands, the Treaty of Vienna, in its Annex XXXI, charged riverside dwellers with the work of improving the regime and 'facilitating' navigation. In 1819, these charges fell to the provinces. It was only on December 31st, 1838, that the State handed the management of the Meuse over to the Public Works Department and a ministerial decree of August 11, 1839 commissioned the engineer Guillery to undertake work for the improvement of navigation by transforming the deep stretches into navigable passes (Berck, 1978). These passes were made by constructing longitudinal underwater piers which maintained the water at a minimum height of 1.50 m at low water (Figure 3). Guillery improved the towpaths by raising them to a height of 3.50 m above low water mark in order to prevent them being flooded and by increasing their width to 2 metres. These were the first systematic changes carried out on the Meuse (Marote, 1897). However, with the development of boat transport, in particular following the building of canals (Liège–Maastricht with draught of 2 m), the pass system of Guillery was abandoned in favour of the canalization of the channel of the Meuse. In 1853, the building of mobile weirs with lateral locks began, and ended in 1880 when canalization reached the French border. The Belgian part of the Meuse then numbered 23 weir-locks from Hastière to Visé.

Adaptation and stagnation (1880–1918)

During this period, the government was less concerned with the river than with the development of the railways, so very little important work was carried out on the Meuse. Nevertheless, on December 21 and 22 1880, the largest rise in the water level of the century revealed the shortcomings of the work that had been done. The Minister of Public Works, therefore, created the unified service of the Meuse in order to improve the waters and the navigability of the river (Marote, 1897).

Encroachments on the channel Marote (1897) observes that the cross-sectional area of the flow at the weirs in the upper part of the Belgian Meuse vary from 235 to 315 m^2 and that the smallest sections are situated downstream (Tailfer and La Plante: 240 to 250 m^2)! Moreover, whereas four bridges existing in 1830 between the French border and the Netherlands, there were 22 by 1880: these further constricted the flow in response to growing river management problems, the Commission of the Ministry of Public Works (Marote, 1897) proposed that the curves be straightened and that the channel be normalized by regulating its slope and by fixing standards of minimum width:

—80 to 90 m from the French border to the confluence of the Lesse;
—90 to 100 m from the confluence of the Lesse to that of the Sambre;
—100 to 110 m from the confluence of the Sambre to that of the Hoyoux;
—110 m to 120 m from the confluence of the Hoyoux to the Saint Lambert Valley;
—130 m from the Saint Lambert Valley to the confluence of the Ourthe;
—150 m from the confluence of the Ourthe to the Dutch border.

By the end of the 19th century, the canalization of the Meuse was gathering momentum.

Encroachments of the floodplain In his study on the improvements of the Meuse, Marote (1897) further notes that the floodplain bed has been largely sacrificed and that the main encroachments are the result of urbanization, of road and rail expansion.

The presence of water has always been one of the main reasons for the growth of villages and towns. Gradually, imperceptibly, the scattered villages met to form a continuous urban ribbon. Lack of space in the valley forced the population to encroach on the river's channels. Classic examples of towns that developed in this way are Dinant and Liège.

Riverside dwellers have long taken advantage of the alluvial deposits for farming purposes. The rich soil of the alluvial plain of Namur (Monmart, 1985) and that of the right bank of Huy have fed several generations. Although today it has disappeared beneath urban expansion, it was nevertheless bound to become impoverished because canalization was gradually preventing the flooding that fertilized the neighbouring land.

Industrialization and population growth required improved means of transport. Roads have always existed but initially they were only paths, lanes, tracks. They were modernized and became waterproof, covered with materials which only increased the number of surfaces along which water could flow and which brought more water to the river than before. Apart from the towpaths, the valley roads were gradually built from the late 18th century (Ministère des Travaux Publics, 1880):

1785: Road from Huy to Namur, in the valley of the Meuse;
1812: Road from Huy to Liège in the valley of the Meuse;
1852: Tow-Road from Waulsort to Moniat.
They are unfortunately situated in the flood zone.

Another human invention which constantly competed with the river was the railway. The later encroached on to the floodplain, raised walls and embankments, straightened and narrowed the river-bed with ballast, and even joined islands to the floodplain. Numerous examples exist of such changes along the Meuse.

Modernization (from 1918 to the present day)

With the development in trade and industry, transport networks had to become more efficient and to remain competitive the river navigation system had to be modernized. The chronology of the process of modernization undertaken since 1918 leads to the division of the Belgian Meuse into the lower and upper Meuse, from the Dutch border to Namur and from Namur to the French border respectively. It was in fact the lower Meuse which was the first to be developed; the upper Meuse only recently changed its appearance from that created during the period of initial canalization.

The new programme to improve navigation on the Meuse began around Liège in January 1923. It had three goals:

1. To improve navigation on the Meuse by reducing the number of locks and to increase their size and moorage space on the river. Between Ben-Ahin (omitted) and the Liège Foundries (included), the seven weirs with manual locks were reduced to three with electromechanical locks. In the parts of the river between the locks, the mooring places were brought to a depth of 2.60 m but later attained 3 m and even more when dredged;
2. To improve the flood regime by regulating the channel bed, increasing depths and removing old weirs, with raised foundations and poor openings, which constituted obstacles to the free flow of the high waters;
3. To build hydroelectric power stations downstream to develop the river's energy potential utilizing the drop in water level between the new weirs (Lambermont, 1939; Ministère des Travaux Publics, 1880).

However, this program was only implemented in 1927 when in response to the disastrous flooding of 1926, the State set up a commission and special funds for major works.

Weirs of a new type whose channels are closed by lift-up sluice gates and swinging flush-boards, electromechanically operated, were built from 1930 (Lambermont, 1939). Simultaneously, a new campaign to remove the islands of the lower Meuse, which constituted an obstacle to the free flow of the water, was launched.

Thus it was that from 1900 until the present day the number of islands on the Belgian Meuse (Table 1) was reduced from 56 to 25 and the current tendency of the Administration of Water Ways is still to try to remove them when undertaking new alterations. This is currently the case at Waulsort where a new weir is being built.

Since the nineteen sixties, navigation has progressed: motor barges have been replaced by push boats with four barges, each 76.5 m by 11.4 m. The locks must, therefore, be at

TABLE 1. Comparison of the number of islands on the Belgian Meuse at the beginning of the 20th century and today

Situation	Around 1900	1987
Upper Meuse	24	16
Lower Meuse	32	9
Total	56	25

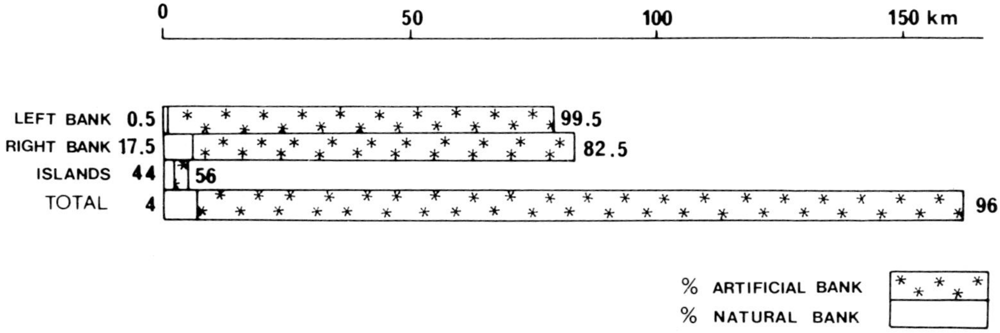

FIGURE 4. Results of the changes to the banks of the lower Belgian Meuse (adapted from Verniers, 1988)

least 185 m long (two barges one in front of the other) and 24 m wide (two barges next to each other). That is why, in order to be able to have four barges (4 × 2250 tons, i.e. 9000 t) on the Albert Canal and the Meuse upstream from Liège to Namur, not only had new locks to be built (200 m by 25 with a draught of 3.5 m) but also the river had to be graded in consequence. This plan to raise capacity to 9000 tons has by now been almost completed between Liège and Namur.

The results of this modernization in the Dutch border–Namur sector are that:

—the number of weir-locks has been reduced from 15 to 7 in 1988 and will drop to 6 in 1989 when the last old manual weir at Argenteau will be demolished (Micha, 1985);
—the difference in level between the new stretches of unbroken water has considerably increased: 4 to 6 m instead of 1 to 3 m (Micha, 1985);
—the Meuse has been or will be systematically deepened to 5 m to allow a continuous draught of 3.5 m;
—all the natural banks have been straightened (Figure 4) with the elimination of gently sloping zones and the building of vertical or slightly sloping concrete banks;
—the width of the Meuse meets the recommended size of 100 to 150 m according to the situation of the channels: this involves the straightening of certain sectors considered as too wide and consequently filled in (cf. Sclayn downstream from Namur).

It was only in 1983 that the modernization programme set up in 1960 passed Namur and continued upstream towards the French border. This modernization has two aims: (1) to

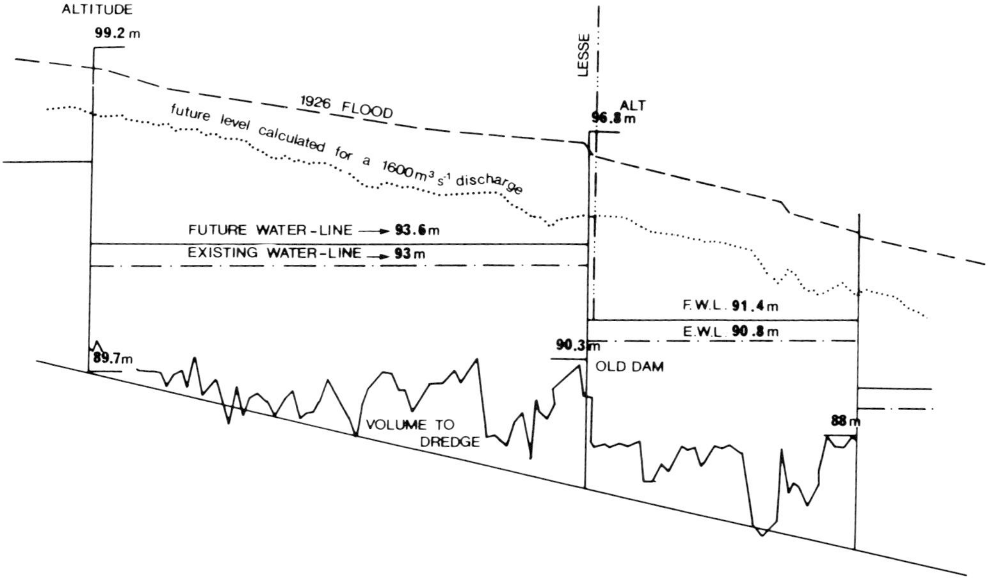

FIGURE 5. Modification of the profile of the upper Belgian Meuse: current and future situation between Waulsort and Dinant

make the Meuse suitable for navigation up to 1350 T; and (2) to effectively combat flooding. Fortunately, the modernization of the upper part of the Belgian Meuse, due to the current reduction in river traffic, does not involve adapting it to navigation of 9000 T, but only to 1350 T. This means that the nine old manual weirs are to be retained and simply replaced by new weirs with electromechanical flush-boards. However, the slight rise in the water level of the unbroken stretches of water and the systematic dredging of the bed (Figure 5) involves strengthening and protecting the banks at least upstream and downstream from the weirs. Moreover, the increase in width from 80 to 100 m according to the position of the channels necessitates the river being widened or filled in the case of excessive width. The new appearance of the upper Belgian Meuse must allow, on the basis of calculations of hydraulic axes, a diminution of 1.5 m to 2 m in the level of flood water from that reached in 1926 (Figure 5). On the whole, the changes resulting from this modernization should not drastically modify the aquatic environment of this sector of the Meuse where a certain diversity of habitats looks like continuing to exist in the future.

IMPACT OF THE CHANGES

The latest changes in the catchment area of the Meuse, notably the encroachments of the floodplain and the channel, inevitably result in modifications to the river's ecosystem. In order to have an idea of some of its effects, we have independently researched the recent historical evolution of some parameters characteristic of the ecosystem of the Meuse. Without wishing to be exhaustive, we have chosen to present the data

TABLE 2. Evolution by decades of the gross and specific annual mean flows worked out for the Meuse downstream from Liège between 1910 and 1980 (adapted from Vereestraeten, 1972 and completed by data from the Service of Hydrological Studies)

Years	Gross annual mean flow ($m^3 s^{-1}$)	Specific annual mean flow ($l s^{-1} km^{-2}$)
1911–1920	317	15.2
1921–1930	307	14.8
1931–1940	272	13.1
1941–1950	235	11.3
1951–1960	219	10.5
Average from 1911–1960	270	13.0
1961–1970	251	12.1
1971–1980	184	8.8

available relating to the hydrology, the physicochemistry of the waters and certain biocenoses.

Hydrological evolution of the Meuse

The comparison of the gross average annual flow of the lower Meuse downstream from Liège, made by Vereestraeten (1972) and completed by us, for the period from 1910 to 1980 (Table 2) shows a clear decline: from 317 $m^3 s^{-1}$ for the decade 1911–20 to 184 $m^3 s^{-1}$ for the decade 1971–80. A careful comparison of the rainfall and water flow made by Vereestraeten (1972) indicates that between 1900 and 1960, the flow of the Meuse downstream from Liège was reduced by 16.4 per cent while the rainfall in the catchment area dropped by only 2.7 per cent. This is probably due to human interference with the hydrological equilibrium of the Meuse. Abstractions account for 8.1 $m^3 s^{-1}$ (Smitz, 1988):

—via the canals: in France: 0.8 $m^3 s^{-1}$ towards the Rhine and the Seine
in Belgium: 0.5 $m^3 s^{-1}$ towards the Escaut
1.5 $m^3 s^{-1}$ towards the Albert Canal
—via industry: in France: 1.5 $m^3 s^{-1}$ for the nuclear power station at Chooz
in Belgium: 2 $m^3 s^{-1}$ for the nuclear power station at Tihange
1 $m^3 s^{-1}$ for the Liège industry
—via the towns for the distribution of drinking water:
in Belgium: 0.8 $m^3 s^{-1}$ at Tailfer

To these withdrawals must be added abstractions from the alluvial aquifer and the withdrawals from tributaries and underlying ground water.

Physical and chemical evolution of the water

Little information exists from former times regarding the physical and chemical characteristics of the Meuse, the only exception being the data relating to suspended matter. Thus, we shall limit outselves to this problem, completing it nevertheless with data relative

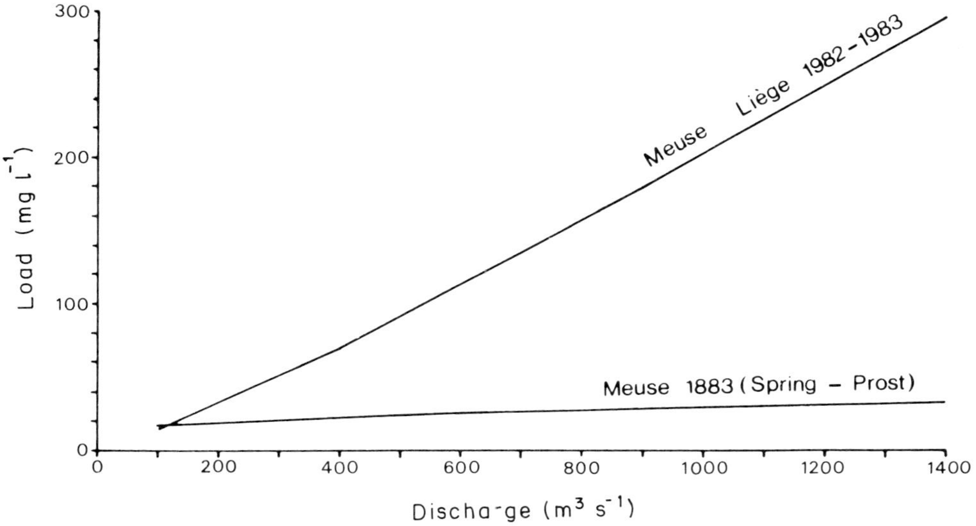

FIGURE 6. A century's evolution in the relationship between the suspended matter in the Meuse and the river's flow at Liège (adapted from Lemin *et al.*, 1987)

to the river's eutrophication while focusing solely on the evolution of the water's nitrogen concentrations.

The variation of suspended matter in the Meuse over the past one hundred years has been examined by Lemin *et al.* (1987) who compared their recent data to that assembled by Spring and Prost in 1883. The total amount in suspension has clearly increased since 1883, especially for flows exceeding 100 $m^3 s^{-1}$ (Figure 6). Thus, for a flow of 1000 $m^3 s^{-1}$ the quantity of suspended matter being transported according to the lines of regression obtained was 30 $mg l^{-1}$ in 1883 and 204 $mg l^{-1}$ in 1983. On the basis of mean flows, the concentration of suspension matter in the Meuse at Liège has tripled over the last century. The total discharge of suspended matter into the Meuse between Namur and Liège from the quarries (*c.* 30 000 $m^3 yr^{-1}$), from metallurgical industry (*c.* 15 000 $m^3 yr^{-1}$) and from thermal power stations (*c.* 5000 $m^3 yr^{-1}$) is in the order of 50 000 $m^3 yr^{-1}$ but this is largely offset by current dredging in the order of 80 000 $m^3 yr^{-1}$. Consequently, the cause of the spectacular increase in concentrations of suspension matter probably results from the channelization of the river which can no longer overflow its banks and deposit sediment on its alluvial plain (Lemin *et al.*, 1987).

Evolution of nitrogen concentrations

For nitrates, the oldest data that we found only go back about twenty years and have been kindly supplied to us by Mr Van Craenenbroeck. Average concentrations (Figure 7) have tended to increase in the same way from 1962 (*c.* 1.5 $mg l^{-1}$) to 1974 (*c.* 3.3 $mg l^{-1}$) at the three stations where samples were taken. Subsequently, the concentrations in nitrates seem to stabilize and even slightly diminish from 1981 onwards. Nevertheless, it is probable that, the nitrate concentrations of the Belgian Meuse are far in excess of what they were. As

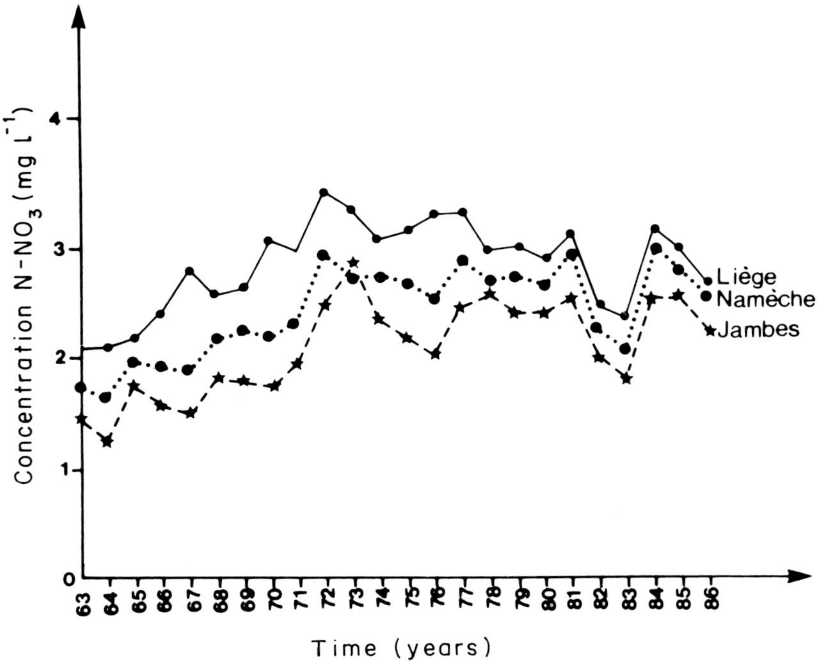

FIGURE 7. Evolution of the annual concentrations of nitrates in the three stations of the Belgian Meuse over the last twenty years (adapted from Van Craenenbroeck, personal communication)

for ammonia, the average concentrations over the last twenty years have remained low in the upper Belgian Meuse (Figure 8, Jambes station) but have clearly increased since 1967 in the lower Belgian Meuse, changing from some $\mu g\,l^{-1}$ to $0.8\,mg\,l^{-1}$ in 1976 downstream from Namur (Namèche) and Liège (Figure 8) to decrease thereafter to some $0.4\,mg\,l^{-1}$, which is still high, and indicating a clear alteration in the aquatic environment.

Evolution of floral and faunal communities

The temporal change of aquatic biocenoses in the Meuse is not easy to trace for there is very little regularly collected data available from the past. On the other hand, the contrasting evolution of the upper and lower Belgian Meuse is clearly revealed by various studies (Meurisse-Genin *et al.*, 1987; Descy and Servais, 1988; Gillet, 1988; Verniers, 1988). It may be assumed that the upper Meuse provides a model of the historical character of the lower Meuse which has been modified most dramatically during this century. By taking a few representative links from the trophic chain of the Meuse ecosystem, we shall present the temporal and/or spatial evolution of the phytoplankton, macrophytes, macroinvertebrates and fish.

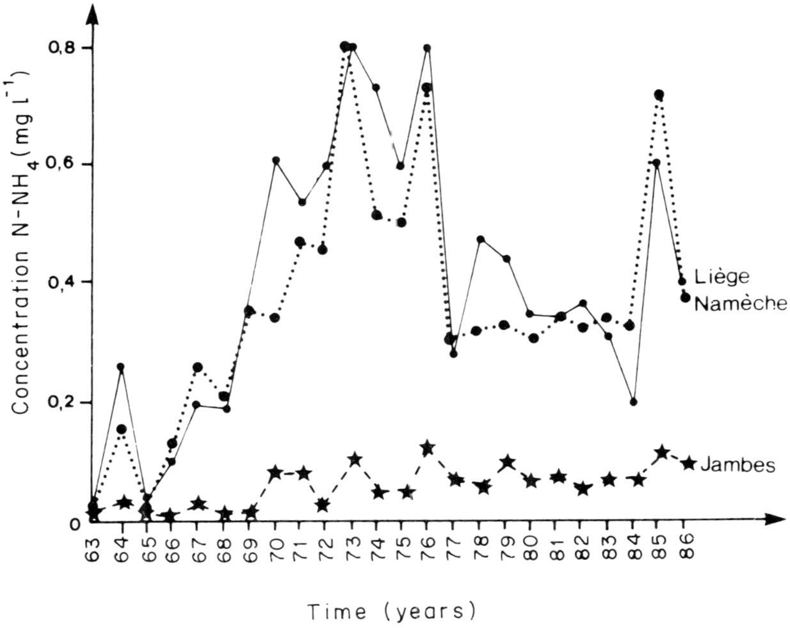

FIGURE 8. Evolution of the annual concentrations of ammonia in three stations of the Belgian Meuse over the last twenty years (Van Craenenbroeck, personal communication)

Phytoplankton

Phytoplankton furnish an account of the history of the body of water over a certain distance. Its current production represents the greatest part of the primary production of the Meuse. Two important studies were carried out on the phytoplankton of the Belgian Meuse in the summers of 1946 and 1953 (Symoens, 1957) and in the autumn of 1981 (Descy, 1983). Although it is very difficult to compare two sets of data (Table 3) assembled at different times, using different techniques, at different seasons (summer–autumn), and expressed in different ways, it is nevertheless possible to make certain conclusions by limiting attention to the sector upstream from Namur. In 1946, as in 1981, the Diatoma remain the dominant group of the Meuse phytoplankton, but the dominant species have changed: *Asterionella formosa*, well known in mesotrophic to eutrophic water, has clearly decreased and *Stephanodiscus hantzschii*, an eutrophic species, dominates. That would be a reflection of the eutrophication which is also marked by a significant development, though occasional, of the Cyanophyceae and Euglenophyceae, and the current absence of *Dinobryon divergens*, a species of oligotrophic tendency, which was fairly well represented in the time of Symoens (1957). Moreover, it appears that the Chlorophyceae or green Algae are more abundant at present, which, after the fashion of the other groups, explains the reproduction of river phytoplankton in the channel bed of the Meuse.

TABLE 3. Comparison of the readings for phytoplankton in the upper Belgian Meuse for the summers of 1946 and 1953 (adapted from Symoens, 1957) and for autumn 1981 (Descy, 1983)

	Summers 1946 and 1953 Index of abundance	Autumn 1981 Relative abundance (%)
CHLOROPHYTA		(30.9)
Chlorella vulgaris	0	9.2
Coelastrum microporum	—	5.7
Pediastrum boryanum	ac	+
Pediastrum duplex	ar	+
Scenedesmus acuminatus	0	3.6
Scenedesmus quadricauda	X	2.1
CHRYSOPHYTA		(63.0)
Chrysophyceae		
Dinobryon divergens	ac	—
Synura uvella	—	0
Bacillariophyceae (Diatomeae)		
Asterionella formosa	c	+
Cyclotella meneghiniana	—	0.9
Diatoma vulgare	r	+
Fragilaria crotonensis	c	0.3
Melosira italica	ac	1.5
Melosira varians	ac	+
Nitzschia acicularis	X	+
Nitzschia actinastroides	X	0
Stephanodiscus hantzschii	—	50.1
Synedra acus (and var. *radians*)	ac	0
CYANOPHYTA		(6.0)
Microcystis cf. *aeruginosa*	—	6.0
EUGLENOPHYTA	rr	0.2
PYRROPHYTA		(0.3)
Cryptomonas sp.	0	0.2

— = not listed
c = common
ac = fairly common
ar = quite rare
r = rare
rr = very rare
x = present
+ = single case
0 = not indicated

Macrophytes

Among the fixed vegetation, macrophytes (filamentous algae, aquatic mosses and vascular plants) develop essentially in a shallow zone along the channel margins (max. 1.5 m). For the macroscopic algae, the qualitative and quantitative studies of Descy (1976) have shown their sensitivity to water pollution which is revealed by marked changes in the composition of the stocks. Upstream from Namur, the most abundant species at present is the green Algae *Cladophora glomerata*. It is accompanied in calm waters by Chlorophyceae *Ulothrix* spp. and a Xanthophyceae: *Vaucheria* sp. The lotic environments, mainly the foundations of weirs, are also colonized by Rhodophyceae: *Lemanea fluviatilis* and *Bangia atropurpurea*. More limnophile red Algae: *Thorea ramosissima* and *Hildenbrandia rivularis* are also found

FIGURE 9. Evolution of the macrophytic stocks of the Belgian Meuse from a postcard of 1910 and a photo of 1988 at Dave (upstream from Namur). (See also pp. 284 & 285)

FIGURE 9

FIGURE 9

TABLE 4. Upstream-downstream evolution of the aquatic vascular macrophytes of the Belgian Meuse (adapted from Descy, 1987)

	French Meuse (Ham-Givet)	Upper Belgian Meuse (Heer-Agimont)	Central Meuse (Namur-Liège)	Meuse at Liège	Downstream from Visé
Helophytes	abundance	by sectors, decreasing	very marked decreased	disappearance	local restauration
'Reeds'					
Phragmites australis	X	X	X	—	—
Phalaris arundinacea	X	X	X	—	X
Others:					
Iris pseudacorus	X	X	X	—	X
Sparganium erectum	X	X	—	—	—
Typha latifolia	X	X	—	—	—
Acorus calamus	X	X	—	—	—
Butomus umbellatus	X	X	—	—	—
Carex div. sp.	X	X	X	—	—
Scirpus lacustris	X	X	—	—	—
Hydrophytes	abundance	spradicity	absence	absence	presence
Myriophyllum spicatum	X	X	—	—	—
Ceratophyllum demersum	X	X	—	—	—
Polygonum amphibium	X	X	—	—	—
Potamogeton pectinatus	X	?	—	—	X
Nuphar lutea	X	X	—	—	—
Ranunculus fluitans	X	—	—	—	—

X = presence — = not indicated

there. Downstream from Namur, it should be pointed out that only *Cladophora glomerata* develop well, and this they do the whole length of the lower Belgian Meuse.

For the bryophytes or aquatic mosses, Empain (1977) has shown that they are tied to the shallow zones of fast-moving current situated downstream from the weirs. The weirs and their foundations on the upper Meuse are colonized by numerous species: *Cinclidotus danubicus*, *C. nigricans*, *Octodiceras fontanum*, *Platyhypnidium riparioides*, *Fontinalis antipyretica*, *Fissidens crassipes*. Upstream from Namur, the *Cinclidotus* dominate, but after the Sambre confluence, they progressively give way to *Platyhypnidium riparioides*. In general, the reduction in delicate species in favour of species tolerant of pollution is a sign of the longitudinal degradation of the quality of the river water.

The communities of vascular aquatic (hydrophytes) and subaquatic (helophytes) plants in the Belgian Meuse was well developed before the 'modernization' for navigation. However, we have found only a few systematic studies; postcards and photographs (Figure 9) best testify to the temporal evolution. Spatial evolution (Table 4) presents a striking contrast existing between the French and the Belgian part of the Meuse. Thus it is that upstream from the Franco-Belgian border, one encounters diverse formations of helophytes (small island of *Scirpus lacustris* downstream from Givet) and hydrophytes (small island of *Myriophyllum spicatum* downstream from Givet). For Descy and Servais (1988) the main and obvious reason for the floristic diversity in the French Meuse (Table 4) is the maintenance of the diversity of habitats which have been very little affected by the alterations to the channel bed of the Meuse and its banks. The scarcity of vascular plants downstream from Namur is mainly the result of the disappearance of these habitats. In the Meuse around Liège, the total disappearance of these plants is also linked to water pollution. The status of this macrophytic vegetation in the Belgian Meuse is, therefore, at the present time catastrophic.

Benthic macroinvertebrates

The sources available relating to the benthic macroinvertebrates which allow a temporal comparison are limited to the upper Belgian Meuse (Waulsort) and come from the samplings of Damas (1939) during the 1930s and of Meurisse-Genin *et al.* (1987), Mol (1987) and d'Udekem (1987), for the present period. In the nineteenth thirties, Damas (1939) reported 52 taxa of benthic macroinvertebrates in the Meuse at Waulsort. Today's authors (d'Udekem, 1987; Meurisse-Genin *et al.*, 1987; Mol, 1987) report 90. By lining up the taxonomic level with that of the less precise data, in general those of Damas (Table 5), the number of taxa reported by Damas is 50 and that of the recent authors 88; 27 taxa are common to both lists. The greater number of taxa currently reported is in large part attributable to a more thorough sampling and to the progress made in taxonomy; this means that one must take great care when comparing and interpreting the two sets of data. This is the case with the Planarians *Dendrocoelum lacteum* and *Dugesia gonocephala* (Table 5) which are absent from the current Waulsort samplings but which are still found in the neighbouring stations in more or less sporadic fashion (Meurisse-Genin *et al.*, 1987). Their absence in current samplings at this station is either due to chance or to a certain reduction in strength between the two periods. Another Planarian *Dugesia tigrina* is, on the other hand, solely represented in the current samplings. It is, moreover, well represented over the whole of the Belgian Meuse (Meurisse-Genin *et al.*, 1987). American in origin, this species probably colonized the Meuse quite a long time after its arrival in Europe in 1925 and in the Paris region in 1945 (Pattee and Gourbault, 1981). This Planarian which is perfectly thermophilic, is favoured by the increase in temperature resulting from the thermal discharges of the nuclear power station at Tihange; this may be interpreted from its abundance at Amay and Ivoz-Ramez, stations where the heating of the waters is at a maximum on account of the accumulation of thermal effluents from Tihange and Awirs (Meurisse-Genin *et al.*, 1987).

The apparent absence at Waulsort of the Oligochaeta *Branchiura sowerbyi* and of the Hirudinea *Piscicola geometra* can, like that of the Planarians *Dendrocoelum lacteum* and *Dugesia gonocephala*, be attributed to chance or to a relative reduction in strength. In fact these species are occasionally found in other stations on the Meuse (Meurisse-Genin *et al.*, 1987).

With the exception of *Unio crassus*, present at both periods, the different species of Unionidae Molluscs are reported either only in 1939 or only in 1987 (Table 5). A modification between the two periods is plausible (Borlée, 1988). Among the Gasteropod

TABLE 5. List of the macroinvertebrates at Waulsort before 1939 and from 1980 to 1987 (current list) (adapted from Damas, 1939; d'Udekem, 1987; Meurisse-Genin *et al.*, 1987; Mol, 1987)

	< 1939	80–87		< 1939	80–87
TRICLADIDA			INSECTS		
Dendrocoelum lacteum	+	–	EPHEMEROPTERA		
Dugesia gonocephala	+	–	*Ephemera*	+	+
Dugesia lugubris	+	+	*Ephemerella*	+	+ (2spp.)
Dugesia polychroa	+	+	*Heptagenia*	+	+
Dugesia tigrina	–	+	*Ecdyonurus*	+	–
Polycelis nigra	+	–	*Caenis*	+	+
Polycelis cf. *tenuis*	–	+	*Baetis*	+	+ (4spp.)
			Raptobaetopus tenellus	–	+
OLIGOCHAETA					
Lumbricidae	–	+	PLECOPTERA		
Lumbriculidae	–	+	*Leuctra* cf. *fusca*	–	+
Naïdidae	–	+			
Branchiura sowerbyi	+	–	ODONATA		
other Tubificidae	–	+	*Calopteryx*	+	–
			Platycnemis	+	+
HIRUDINA			*Coenagrion*	+	–
Erpobdella octoculata	+	+	*Aeschna*	+	–
Glossiphonia complanata	+	+	*Somatochlora metallica*	–	+
Glossiphonia cf. *heteroclita*	–	+	*Gomphus vulgatissimus*	–	+
Piscicola geometra	+	–			
			HETEROPTERA		
MOLLUSCS LAMELLIBRANCHA			*Aphelocheirus aestivalis*	+	+
Unio crassus	+	+			
Unio pictorum	–	+	COLEOPTERA		
Unio tumidus	+	–	*Riolus*	+	+
Anodonta anatina	–	+	*Stenelmis*	+	–
Anodonta cygnaea	+	–	*Elmis*	–	+
Pseudanodonta complanata	–	+	*Esolus*	–	+
			Oulimnius	–	+
Sphaerium	+ (3spp.)	+	*Normandia*	–	+
Pisidium	–	+ (3spp.)	Hydrophilidae	–	+
Dreissena polymorpha	+	+	Helodidae	–	+
MOLLUSCS GASTEROPODA			MEGALOPTERA		
Physa acuta	–	+	*Sialis lutaria*	–	+
Lymnaea	+ (2spp.)	+	*Sialis fuliginosa*	–	+
Ancylus fluviatilis	+	+	*Sialis nigripes*	–	+
Bithynia tentaculata	+	+			
Bithynia leachi	–	+	TRICHOPTERA		
Potamopyrgus jenkinsi	–	+	*Rhyacophila*	+	+

TABLE 5. (*continued*)

	<1939	80–87		<1939	80–87
Viviparus viviparus	+	+	Glossosomatidae	–	+
Theodoxus fluviatilis	+	+	*Hydroptila*	–	+
Lithoglyphus naticoïdes	–	+	*Agraylea*	–	+
Anisus vorlex	+	–	*Hydropsyche*	–	+ (4spp.)
Anisus carinatus	+	–	*Cheumatopsyche lepida*	–	+
Valvata piscinalis	+	–	*Polycentropus*	+	+
Other Planorbidae	+	–	*Cyrnus trimaculatus*	–	+
			Neureclipsis bimaculata	–	+
CRUSTACEANS ISOPODA			*Psychomyia pusilla*	–	+
			Tinodes waeneri	–	+
Asellus	+	+ (2spp.)	*Ecnomus tenellus*	+	+
CRUSTACEANS AMPHIPODA			*Athripsodes + Ceraclea (syn. Leptocerus)*	+	+ (3spp.)
Gammarus fossarum	–	+	*Oecetis notata*	–	+
Gammarus pulex	+	–	*Leptocerus + Setodes*	+	–
Gammarus roeseli	–	+	*Anabolia*	+	–
Echinogammarus berilloni	+	–			
Orchestia cavimana	–	+	DIPTERA		
Corophium curvispinum	–	+	Chironomidae	+	+
			Simuliidae	+	+
CRUSTACEANS DECAPODA			*Atherix*	–	+
			Ceratopogonidae	–	+
Orconectes limosus	–	+	Empididae	–	+
			Anthomyidae	–	+
ARACHNIDA			Tipulidae	–	+
Hydracarina n. det.			Limoniidae	–	+
			Total taxa	50	88

– = absence + = presence

Molluscs, the Hydrobiidae *Potamopyrgus jenkinsi*, probably of Australian origin, was introduced via the British Isles. Spreading from salt waters to fresh waters, *P. jenkinsi* arrived in the Meuse through brackish waters or waters in contact with them, like the Campine canals linking the Meuse to the Escaut, where it was reported in 1927. Of fairly recent introduction, it is possible that by 1939 it had not yet colonized the Meuse as far as Waulsort. The *Anisus* have been exclusively collected by Damas (1939). These organisms live in a lentic environment and more particularly in riverine vegetation, their absence from current samplings could be linked to the considerable reduction in this type of vegetation. *Valvata piscinalis* was only found in the Liège region (Monsin) in 1987 whereas Damas (1939) found it at Waulsort.

Among the Ephemeroptera, it appears symptomatic that only *Ecdyonurus*, the most rheophile, is no longer found at Waulsort today and seems to be becoming rare in the Meuse (Meurisse-Genin *et al.*, 1987).

If the great diversity of Coleoptera listed at the present time can be attributed to their small size and the extent of the sampling, it appears surprising, on the other hand, that Damas (1939) did not collect any Megaloptera, insects which are much larger in size and, what is more, easy to collect. They were in fact already abundant in the Meuse as is testified to by the large harvests of just one of the three species present in the Meuse that were made at Dinant at the end of the last century (Stroot, 1986).

For the Diptera, Damas only reports the two families most represented at the present time, which again may be attributed to too limited sampling. The same tendency can be noted with the Trichoptera. We should point out, however, that Damas (1939) cites two taxa which are not represented today at Waulsort. These are *Setodes*, now divided into two genera *Leptocerus* and *Setodes*, and *Anabolia*. The last-mentioned genus is no longer reported in the Belgian Meuse (Meurisse-Genin *et al.*, 1987). However, both species of the old genus *Setodes*, which could be caught in a large river, *Setodes punctatus* and mainly *Leptocerus interruptus*, have not been caught in the Belgian Meuse since 1942 at Jambes and 1943 at Hastière (Stroot, 1985). It looks as if both these species have totally disappeared from the Belgian Meuse.

Apart from the few cases mentioned when comparing Damas (1939) and the current studies, there seem to be other species which have disappeared or which have become rare along the Belgian Meuse. Among these figures three Decapodal Crustaceans. First the Chinese crab, *Eriocheir sinensis*, developed in the Belgian Meuse around the 1940s (Leloup, 1938; Huet, 1941) and has now disappeared. Secondly, the European crayfish, *Astacus astacus* probably lived in the Meuse in the last century and its numbers were drastically reduced by the crayfish plague (Gérard, 1986) thus wiping out this species in the Meuse. Thirdly, a freshwater prawn, *Atyaephyra desmaresti*, a Mediterranean species, made its way up the canals to the north of Europe and became abundant in the Belgian Meuse between 1977 and 1980 (Meurisse-Genin *et al.*, 1987) but recently disappeared (Borlée, 1988).

Finally, the relatively recent spread of two Crustaceans must be noted: *Orconectes limosus* which recently became established in the Belgian Meuse (Gérard, 1986) to attain its full development between 1959 and 1961, and *Corophium curvispinum* which is reported for the first time at Huy in 1981 and currently proliferates in the upper Belgian Meuse (d'Udekem, 1987).

Fish

The hydrographic characteristics of a river do not remain constant from its source to its estuary, a fact which allows one to distinguish along its course to the sea several successive zones generally characterized by the predominance of one species (fish-breeding zones of Huet, 1949). Thus, the Belgian Meuse corresponds to the barbel zone; if one takes into account the various human activities taking place in the catchment area and in the channel of the river (flow regulation, canalization, etc.) downstream from Namur the Belgian Meuse is situated in a zone falling somewhere between a barbel zone and a bream zone.

An inventory of the fauna shows that 47 species of fish have lived in the Belgian part of the Meuse, of which 39 are indigenous, seven have been introduced and become acclimatized, and one is of indeterminate status (Philippart *et al.*, 1988). Since the end of the last century, seven species of anadromous fish: the marine lamprey (*Petromyzon marinus*), the river lamprey (*Lampetra fluviatilis*), the sturgeon (*Acipenser sturio*), the allis shad (*Alosa alosa alosa*), the twaïte shad (*Alosa alosa fallax*), the houting (*Coregonus oxyrhynchus*) and the atlantic salmon (*Salmo salar*) have disappeared following over-exploitation, pollution and the building of weirs. A certain number of species are rapidly decreasing and are considered endangered. These are mainly rheophile species, the barbel (*Barbus barbus*), the nase (*Chondrostoma nasus*), the dace (*Leuciscus leuciscus*), which require gravelly, shallow and well-oxygenated habitats to spawn or limnophile species, such as the pike (*Esox lucius*), which require vegetal support to lay their eggs.

In contrast, man has introduced into the Meuse different species since the end of the last century (the catfish: *Ameiurus nebulosus* and *Ictalurus melas*, the pumpkinseed: *Lepomis gibbosus*) which are acclimatized but which remain few in number and have little impact on the other indigenous species. At the beginning of the century, the pike perch (*Stizostedion lucioperca*) was introduced and it found its way via the canals into the Belgian Meuse where it became acclimatized. The restocking, carried out since 1979 around Dinant, and its spreading via the canals account for its recent development which since 1983 has been accompanied by serious problems of larvae bucephalosis in cyprinids (Galvez and Micha, 1987).

One the whole, during alterations to the river, the fish-breeding stocks of the Belgian Meuse have undergone considerable modifications with not only the disappearance of indigenous species and the appearance of immigrant species but also a marked reduction in rheophile species (barbel: *Barbus barbus*, nase: *Chondrostoma nasus*, etc.) to the benefit of limnophile species (roach: *Rutilus rutilus*, bream: *Abramis brama*, etc.) where fish biomass currently predominate. For additional information relating to fish fauna in the Belgian Meuse, the recent publication by Philippart *et al.* (1988) may be consulted.

DISCUSSION

Since the last century, the ecosystem of the Meuse has undergone, due to the increasing activity of man, modifications of increasing importance: marked reduction of the water flow, a drastic increase in the suspended material being transported, a degree of eutrophication of the water, and the disturbance of the original floral and faunal communities. The causes of this evolution of the Meuse are multiple but they can be itemized as different types of human interference in descending order of importance:

1. Occupation of the catchment area: since 18th century, the area covered by deciduous forests has been reduced to the advantage of industrial and agricultural development (Génicot *et al.*, 1973). Industrialization has been accompanied by the concentration of population in a few large riverside towns (Charleville in France, Liège in Belgium, Maastricht in the Netherlands, etc.) This had led notably to water withdrawal and the diversion of underground and surface water to supply the population and industry, with the consequence that there has been a marked fall in river flow (16.4 per cent between 1900 and 1960 according to Vereestraeten, 1972). Moreover, the absence of water-purifying stations for the large riverside towns has led to an alteration in the quality of the water (increase in concentration of ammonia downstream from Liège) and a marked eutrophication (Van Craenenbroeck, 1986).

2. Encroachment on the floodplain: gradually the rich soil of the alluvial plain, at one time intensively cultivated, has disappeared following urban expansion (Monmart, 1985), and the construction of roads and railways. The results of this encroachment are almost immediate: increasing flooding and requiring renewed channel engineering work to reduce the possibility of disaster.
3. Encroachment on the channel bed: in order to reduce flooding and to improve navigation, an undertaking which was at its height at the beginning of the 20th century, the Ministry of Public Works removed islands, the number of which was reduced from 56 to 25 between 1900 and 1988, straightened the banks by canalizing the river to a width of between 80 and 150 m depending on the location, and constructed a series of remarkably imposing weir-locks in the lower Belgian Meuse (downstream from Namur). Consequently, the amount of suspended material being transported by the Meuse for an average flow has tripled between 1883 and 1983 (Lemin *et al.*, 1987), and the floral and faunal communities are disrupted. Native species have disappeared (including seven species of anadromous fish) and immigrant species appeared via canals, etc.
4. Destruction of habitats: the encroachment on the channel bed has, in particular, led to the destruction of a whole series of habitats which are indispensable for the flora and fauna. Thus aquatic and semi-aquatic plants, which only develop in shallow water (less than 1.5 m) along the gently sloping banks of the river, have been virtually eliminated by the construction of vertical, often concrete, banks. The marginal zones also served as spawning grounds for species of limnophile fish; these species have been severely impacted by channel engineering works. Moreover, the systematic deepening of the Meuse to 3 or 5 m, according to the inter-weir stretches and immediately downstream from the weirs, leads to the destruction of the gravelly rheophile zones, thus no longer allowing the species of invertebrates and those of the fish linked to them either to develop or to reproduce (cf. the barbel *Barbus barbus*, an endangered species).
5. Water pollution: domestic and industrial effluents have, for several decades, led to a high degree of pollution of the waters, though this has been mainly localized downstream from Liège. But since the 1960s, a general eutrophication of the river has been taking place and is particularly apparent in:
   —Phytoplanktonic changes with the disappearance of an oligotrophic Chrysophycea (*Dinobryon divergens*) and a decrease in species of mesotrophic Diatoma (*Asterionella formosa*) to the advantage of eutrophic species (*Stephanodiscus hantzschii*).
   —A reduction in the diversity of macroscopic algae with a marked predominance of *Cladophora glomerara* downstream from Namur.
   —A reduction in the diversity of bryophytes, the *Cinclidotus* species giving way downstream to a more resistant species: *Platyhypnidium riparioides*.
6. Overexploitation of fish-breeding stocks: overintensive fishing led to the extermination of sturgeon at the end of the last century and contributed to that of the atlantic salmon.
7. Introduction of foreign species: certain species which have been introduced since the last century (the Mollusc *Potamopyrgus jenkinsi*, the pumkinseed *Lepomis gibbosus*) remain few in number and offer no major problem for the indigenous biocenoses. Others, however, pose a problem. A case in point is the American crayfish (*Orconectes limosus*) which has become very abundant in the Meuse and has contributed to the almost total extermination of the indigenous European crayfish, *Astacus astacus*. The pike perch

Stizostedion lucioperca which was introduced recently into the upper Belgian Meuse, has caused serious problems of larva bucephalosis among the cyprinids in this sector.

All this early and recent interference on the part of Man has, therefore, resulted in a general and rapid deterioration in the ecosystem of the Meuse. It is more than time that those responsible are made aware of this fact and respect the few zones that still remain along the upper Belgian Meuse, between the French border and Namur. As for the lower Meuse, alterations to make way for navigation of 9000 tons have destroyed all the important habitats. Thought should be given to restoring these sectors in the future by recreating shallow riverside zones suitable for aquatic macrophytes, for the macroinvertebrates which are linked to them, and for the reproduction of many species of fish. In conclusion, the example of human interference on the Meuse reveals the necessity of taking into account environmental problems; this would involve a global policy of economic-ecological management.

ACKNOWLEDGEMENTS

For having provided us with historical documents on the changes to the Meuse, we wish to express our thanks to the Namur and Liège Services of the Meuse and in particular MM. Dermience, Schiepers and Lefevre. We also extend our thanks to Mr. Van Craenenbroeck of the A.W.W. at Antwerp for his original data relating to the chemical quality of the water of the Meuse and to Mr. Schoofs for allowing us access to old postcards.

Finally, all our gratitude goes to J.-P. Descy, A. Gillet, G. Verniers and Ph. Stroot for their information, questions and ideas relating to the evolution of the aquatic biocenoses of the Meuse.

REFERENCES

André, E., Baix, F., Beapain, P., Brouwers, D., Courtois, F., Deloz, A., de Mévius, Everard, G., Fivet, E., Gaussin, A., Haire, A., Hustin, F., Legrand, J., Masonin, M., Pechon, L., Pieltain, F., Schmitz, J., Servais, M., Tellier, J., Thibaut de Maiziere, A., Visart de Bocarme, R. and Walravens, L. 1930. *La Province de Namur 1830–1930*, Wesmael-Charlier, Namur, vol. 1, 367 pp., vol. 2, 362 pp.

Berck, F. 1978. *La navigation à vapeur sur la Meuse. Notes et documents*, Musée du fer et du charbon, Section du musée de la vie wallonne, Liège, 24 p.

Borlée, M.-C. 1988. *Evolution historique récente de la Meuse belge: interventions humaines et modifications de l'écosystème*, Mémoire pour l'obtention du grade de licencié en sciences zoologiques (inédit), F.N.D.P., Namur, 112 p.

Damas, H. 1939. *La faune de la Meuse belge*, Ass. Fran, Avanc. Sci., 63ème Congrès, Liège, 150–159.

Descy, J.-P. 1976. *Etude quantitative du peuplement algal benthique en vue de l'établissement d'une méthodologie d'estimation biologique de la qualité des eaux courantes*, Application au cours belge de la Meuse et de la Sambre, Recherche et Technique au service de l'Environnement, CEBEDOC, Liège, 159–206.

Descy, J.-P. 1983. Les communautés algales de la basse Meuse belge et la qualité des ses eaux: un bilan à l'automne 1981. *Bull. Soc. Roy. Bot. Belg.*, **116**(1), 31–50.

Descy, J.-P. and Servais, P. 1988. Conséquences sur la flore. In Micha, J.-C. and Pilette, S. (Eds), *L'impact de l'homme sur l'écosystème Meuse*, A.E.E. Presses Univesitaires de Namur, Namur, Belgium, 140 pp.

Detollenaere, A. and Micha, J.-C. 1978. *Effects of thermal pollution of a nuclear power on aquatic organisms of the river Meuse*, C.S.E. Eur. 6766, Harwood Ac. Pb. PR, 207–210.

Detollenaere, A. and Micha, J.-C. 1980. Effets des rejets thermiques d'une centrale nucléaire sur les poissons de la Meuse. *Revue des Questions Scientifiques*, **151**, 1, 3–12.

Detollenaere, A. and Micha, J.-C. 1986. Impact des rejets thermiques de la centrale nucléaire de Tihange sur les poissons de la Meuse. *Trib. Cebedeau*, **516**, 39, 9–26.

d'Udekem d'Acoz, C. 1987. *Etude semi-quantitative et quantitative des macroinvertébrés benthiques des radiers de barrage de la haute Meuse*, Mémoire pour l'obtention du grade de licencié en science zoologiques (inédit), F.N.D.P., Namur, 101 pp.

Empain, A. 1977. *Ecologie des populations bryophytiques aquatiques de la Meuse, de la Sambre et de la Somme*, Relation avec la qualité des eaux, écophysiologie comparée et étude de la contamination par métaux lourds, inédit, Liège, 179 pp.

Galvez, M. and Micha, J.-C. 1987. Introduction, extension et répartition du sandre (*Stizostedion lucioperca* L.), en Belgique. *Trib. Cebedeau*, **521**, 33–42.

Génicot, L., Manil, G., Bonenfant, P., Amand, M., Joris, A., Descy, G., Ruwet, J., Moureaux, P., Stiennon, S., Demoulin, R. and Boland, A., 1973. *Histoire de Wallonie.*, Edit. universitaire PRIVAT, Toulouse, 502 pp.

Genin, M. and Micha, J.-C. 1977. Contamination radioactive de la faune aquatique dulcicole. *Ann. Assoc. Belge Radioprot.*, **3**, 3, 217–234.

George, P. 1972. *La France.*, Presses universitaires de France, Paris, 269 pp.

Gérard P. 1986. Les différentes espèces d'écrevisses en Belgique et leur répartition géographique. Ministère de l'agriculture. Administration de la recherche agronomique. Station de recherche forestières et hydrologiques. *Groenendael. Hoeilaart. Travaux série D.*, **54**, 255 pp.

Gillet, A., Reydams, A., Meurisse, M. and Micha, J.-C. 1984. Incidence de repeuplements dans la population de gardons *Rutilus rutilus* (L.) en Meuse. *Cybium*, **8**(3), 51–61.

Gillet, A. and Micha, J.-C. 1987. Biologie et radiocontamination de 3 espèces animales (*Dreissena polymorpha* P., *Rutilus rutilus* (L.) et *Perca fluviatilis* (L.)) représentatives de différents maillons trophiques de l'écosystème Meuse. *Ann. Assoc. Belge Radioprot.*, **12**, 2–3, 139–155.

Gillet, A. 1988. Conséquence des rejets polluants sur la faune de la Meuse. In Micha, J.-C. and Pilette, S. (Eds), *L'impact de l'homme sur l'écosystème Meuse.*, A.E.E. Presses universitaires de Namur, Namur, Belgium, 148 pp.

Huet, M. 1941. *L'extension du crabe chinois* (Eriocheir sinensis *M. Edw.*) *en Belgique*, Administration des Eaux et Forêts. Station de recherche de Groenendael, Communications, série D, 2–5.

Huet, M. 1949. Aperçu des relations entre la pente et les populations piscicoles des eaux courantes. *Stat. de Rech. de Groenendael fasc.*, **3–4**, 332–351.

Lambermont, J.-F. 1939. *Le régime des eaux, les crues, la signalisation de celles-ci et les travaux de protection contre les inondations*, Ass. Franç. Avanc. Sci. 63ème Congrès, Liège, 93–111.

Leloup, E. 1938. Contribution à l'étude de la faune belge. La propagation du crabe chinois en Belgique pendant l'année 1937. *Bulletin du musée royal d'histoire naturelle de Belgique*, **XIV**, 14, 1–25.

Lemin, G., Koch, G., Hurtgen, C and Pissart, A. 1987. Les transports en suspension de la Meuse, l'Ourthe et la Hoëgne. *Bull. Soc. Géographique de Liège*, **22–23**, 39–61.

Marote, M.-E. 1897. *Les travaux d'amélioration de la Meuse*, extrait du 4e fascicule des Annales des Travaux Publics en Belgique, Goemaere, Bruxelles, 34 pp.

Mattheeuws, A., Genin, M., Detollenaere, A. and Micha, J.-C., 1981. Etude de la reproduction du gardon *Rutilus rutilus* (L.) et des effets d'une élévation provoquée de la température en Meuse sur cette reproduction. *Hydrobiologia*, **85**, 271–282.

Meurisse-Genin, M. and Micha, J.-C. 1978. Impact des centrales nucléaires sur les écosystèmes dulcicoles. *Les Naturalistes Belges*, **59**, 149–158.

Meurisse-Genin, M., Reydams-Detollenaere, A., Stroot, Ph. and Micha, J.-C. 1987. Les macroinvertébrés benthiques de la Meuse belge: bilan de 5 années de recherche (1980–1984). *Arch Hydrobiol.*, **109**(1), 67–88.

Ministère des Travaux Publics 1880. *Voies navigables de la Belgique. Recueil de renseignements*, Ed. Weissenbruch, Bruxelles, Tome premier, 248 pp.

Ministère des Travaux Publics 1987. *Annales des Travaux Publics de Belgique, 1837–1987.* Fascicule 2/3 avril–juin, 326 pp.

Micha, J.-C. 1985 Obstacles physiques à la remontée du saumon atlantique dans le bassin mosan en Belgique. In *Réintroduction du Saumon atlantique dans le bassin de la Meuse*, Ministère Région Wallonne, Bruxelles, 115 pp.

Mol, A. W. M. 1987 *Caenis beskidensis* Sowa new to Belgium, with remarks on the Ephemeroptera of the river Meuse. *Ent. Ber, Amst.*, **47**(4), 60–64.

Monmart, A. 1985. *Les grandes inondations à Jambes de 1740 à 1984*, Institut St Berthuin, inédit, 50 pp.

Pattee, E. and Gourbault, N. 1981. Turbellariés Triclades Paludicoles (Planaires d'eau douce). *Bull. Soc. Linn. Lyon*, **50**, 9, 279–304.

Philippart, J.-C., Gillet, A. and Micha, J.-C. 1988. *Fish and their environment in large European river ecosystems. The river Meuse*, Water management in the alluvial valley of large river, research thème n.5., Workshop held in Liège (Belgium), Sciences de l'eau, **7**(1) 115–154.

Piérot, F. and Mahaut, C. 1981. *La Meuse fleuve d'Europe*. Wahle, E. avec collaboration Brisco, P., Liège, 334 pp.

Questienne, P. and Dubois, N. 1978. *La navigation en Wallonie*, Musée de la vie wallonne, Ministère de la Culture Française et Foire Internationale de Wallonie, Liège, 84 pp.

Smitz J. 1988. Les prélèvements. In Micha, J.-C. and Pilette, S. (Eds), *L'impact de l'homme sur l'écosystème Meuse*, A.E.E. Presses universitaires de Namur, Namur, Belgium, 140 pp.

Stroot, Ph. 1985. *Actualisation du catalogue des Trichoptères de Belgique*, Soc. r. belge Ent. (Suppl.) Bruxelles, 61 pp.

Stroot, Ph. 1986. Révision des Mégaloptères de la collection belge de l'I.R.S.N.B. *Sialis nigripes*. Ed. Pictet, 1865, Belg. n. sp. (Megaloptera, Sialidae). *Bull. Annls Soc. r. belge Ent.*, **122**, 195–201.

Stroot, Ph. 1987. An attempt to evaluate the state of the caddis fly fauna of Belgium. *Series Entomologica*, **39**, 79–83.

Suttor, M., 1986. *La navigation sur la Meuse moyenne des origines à 1650*, Centre belge d'histoire rurale publication n°86, Liège–Louvain, 162 pp.

Symoens, J.-J. 1957. Les eaux douces de l'Ardenne et des régions voisines: les milieux et leur végétation algale. *Bull. Soc. Roy. Belg.*, **89**, 111–314.

Van Craenenbroeck, W. 1986. Characteristics of the water quality of the Meuse, *The radiological exposure of the population in the Meuse basin Commission of the European Communities*, Report Eur 10670 EN, Brussels, 166 p.

Vereerstraeten, J. 1972. Le bassin de la Meuse. Etude de géographie hydrologique. *Rev. bel. Géogr.*, **94**, 1970, 1.2.3., Soc. r. bel. géogr. 339 pp.

Verniers, G. 1988. Impact des aménagements et de la navigation sur l'écosystème Meuse. In Micha, J.-C. and Pilette, S. (Eds), *Impact de l'homme sur l'écosystème Meuse*, A.E.E. Presses universitaires de Namur, Namur, Belgium, 140 pp..

Historical Change of Large Alluvial Rivers: Western Europe
Edited by G. E. Petts
©1989 John Wiley & Sons Ltd

17

Historical Changes of the River Weser

D. BUSCH

*Wasserwirtschaftsamt Bremen, Theodor-Heusse-Allee 21,
D-2800 Bremen 1, West Germany*

and

M. SCHIRMER, B. SCHUCHARDT and P. ULLRICH

*Arbeitsgruppe Limnische Ökologie, Fachbereich 2 (Biologie), Universität Bremen,
Leobener Straße, D-2800 Bremen 33, West Germany*

INTRODUCTION

The River Weser (northern Germany) originated, as did most European rivers, during the final stages of the last glacial period and is formed by the confluence of the Werra and Fulda rivers. Its annual discharge reflects the north German climate. Although the vegetation communities of the river-bed, riparian zones and floodplains have been changed by human activities throughout the last 4000 years, a dramatic change in the river's ecology was caused by the increases in agriculture, flood control, industrialization, and channelization during the 19th and 20th centuries. Canalization drastically modified channel morphology but water quality changes, culminating in major pollution during this century probably have had the greatest impact. High salinity due to waste water effluents from potash industries influenced the life of the formerly typical fauna and flora. The decrease in, and extinction of, its fish populations reflected the consequences of the new use of the river for shipping and waste water disposal. Pollutant residues in sediments and aquatic organisms became relatively high. Eutrophication and damming still lead to dramatic oxygen deficits in summer, although improvements in sewage treatment have been made. This paper reviews the historical changes of the Weser hydrosystems, assesses the effects of these on the present river, and briefly considers the future for management.

The River Weser has a total length of 432 km and drains an area of 46 136 km^2. Geographically the river may be divided into three sections: the Ober- (Upper), Mittel- (Middle) and Unterweser (Lower; the inner estuary, subject to tides: see Figure 1). The water quality is generally considered as alpha–beta–mesosaprobe, but partly alpha–mesosaprobe (Gewässer-Gütekarte, 1985).

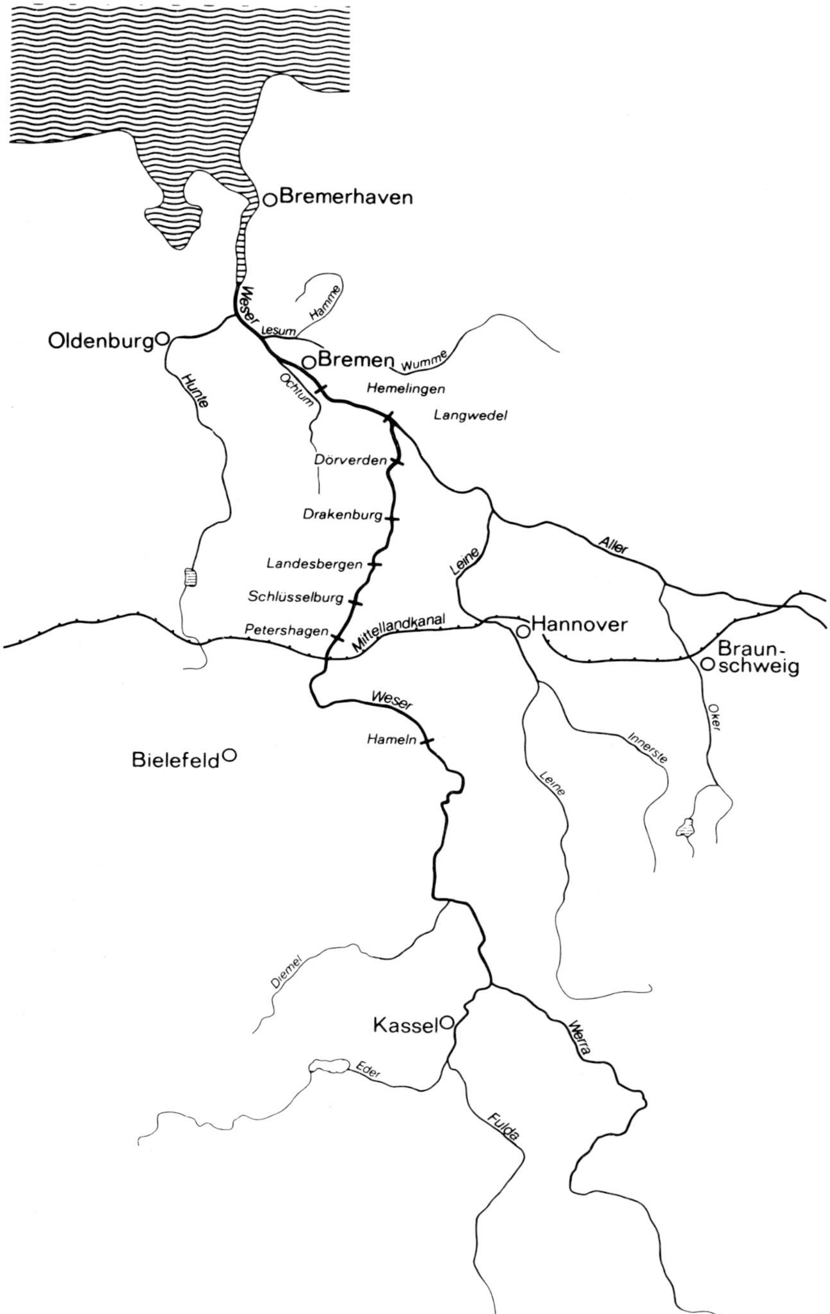

FIGURE 1. (*caption opposite*)

SOME HISTORICAL ASPECTS OF RIVER REGULATION

Mittel- and Oberweser

The historical changes in the post-glacial hydrosystem of the River Weser have endogenic and anthropogenic causes. Several phases and processes are distinguishable. For about 8000 years the Weser hydrosystem changed very slowly in response to moderate climatic variations, which caused the immigration and emigration of plant and animal species. This process was modulated by a continuous but local variation of biotopes induced by channel migration, sedimentation and erosion (Albrecht and Kirchhoff, 1987). Between the New Stone Age and the Bronze Age, the settling of humans triggered clay sedimentation onto the sandy and gravel floodplains. Deforestation and cultivation spread across the loess areas on the valley slopes, accelerating erosion and raising sediment yields. The result was a drift in the vegetation and animal life which preferred more fertile soil conditions (Strautz, 1963; Fittkau and Reiss, 1983; Behre, 1985; Natermann, 1941).

The spread of settlements between 500 and 1300 AD and especially after the 30 year war at the end of the 17th century, changed not only the mountain area of Weserbergland but also the floodplains, more drastically than the endogenic forces (Table 1). Various authors describe this radical change in vegetation communities and animal variety (Tüxen and Lohmeyer, 1950; Krause and Schröder, 1979; Hesmer and Schroeder, 1963).

Fishermen and millers had used the potential energy of the river by building dams since the 5th century (Loebe, 1968). Besides these local influences, the hydrosystem remained untouched. Later the river and its main tributaries were used as shipways and this required channel maintenance work on the shoreline and riverbed which brought conflicts with the farming communities next to the riparian zone. Regular maintenance of tow paths since the 17th century destroyed large areas of the riparian vegetation which had guaranteed relatively effective protection against river bank erosion.

Organized regulation started around 1800 and resulted in a new structure for the Weser (Figure 2). A semistable 'fairway' structure with groynes, longitudinal dykes and controls on the bed of the river destroyed the delicate balance between land and water; shallows and sidearms were filled in with dredged material or were silted up. The river channel became narrower and deeper within a few decades (Protokolle, 1838, 1865, 1870).

Around 1900 the river channel seems to have been stable, and maintenance work was minimal. The guaranteed draught was 0.7–1.0 m in the Oberweser and around 1.25 m in the Mittelweser (Keller, 1901). However, thirty years later the progress in industrialization forced new steps in channelization to meet the greater loading capacity of barges (Weser–Ems-Wasserstrassenbeirat, 1927). It has taken only 250 years to change the lentic Mittelweser irreversibly into a semilotic channelized river with seven dams and locks between Minden and Bremen (Loebe, 1960). The natural river-land ecotone has been destroyed: there is

FIGURE 1. Weser with tributaries and dams
Explanations:
– – / – – Location of dams in the Weser River
Oberweser (Upper-) Part of Weser from confluence of Werra and Fulda to Mittellandkanal Crossing
Mittelweser (Middle-) Part of Weser from Mittellandkanal to dam at Bremen–Hemelingen
Unterweser (Lower-) Part of Weser from Bremen–Hemelingen to Bremerhaven

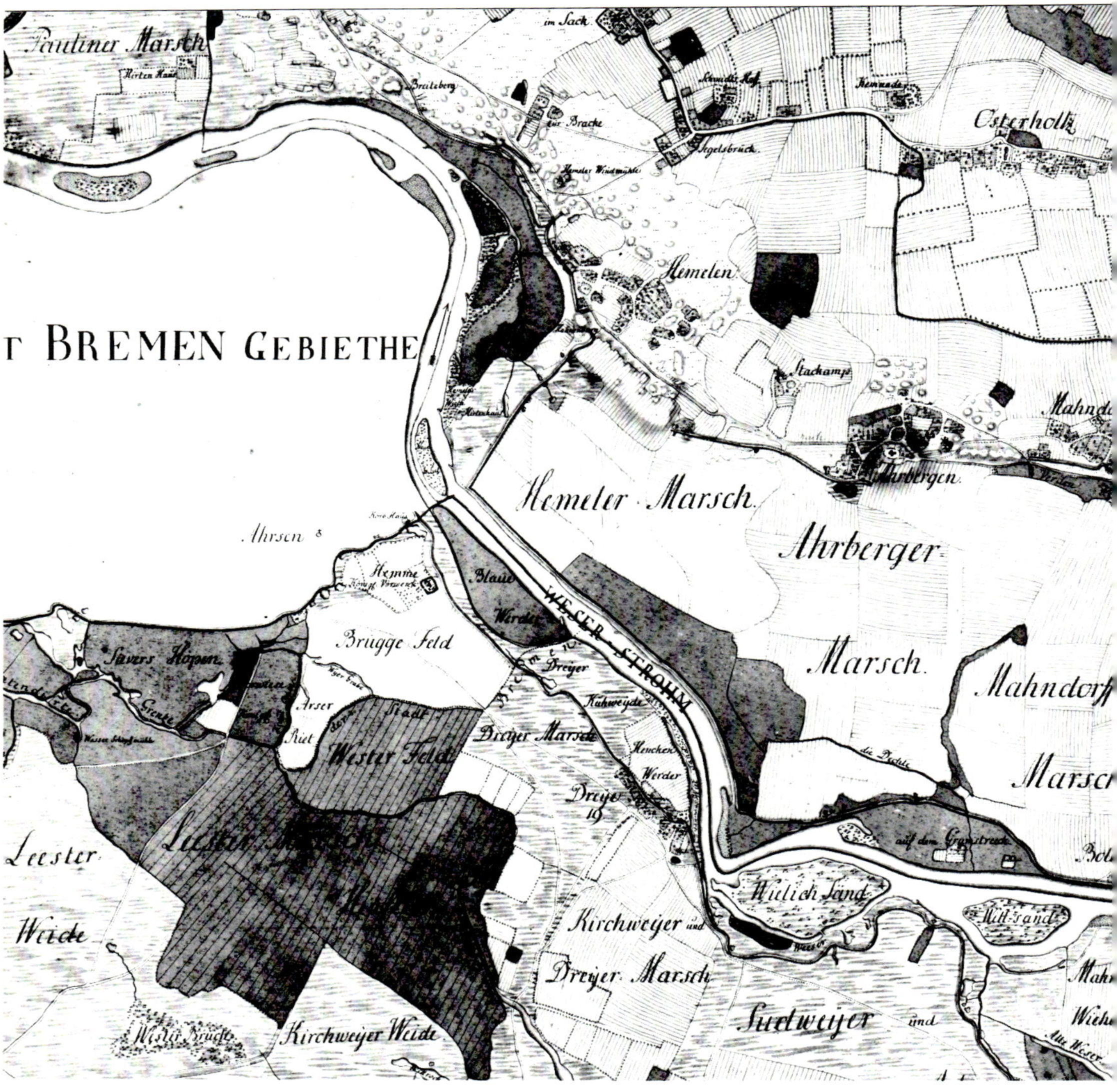

FIGURE 2. The Mittelweser upstream from Bremen around 1760 and today (km 351–360) (Above) Kurhannoversche Landesaufnahme 1773, Reprint 1960. (From Niedersächsisches Landesverwaltungsamt–Landesvermessung Hannover). (Opposite) Topographical Map, 1984, 2919 Bremen-Ost

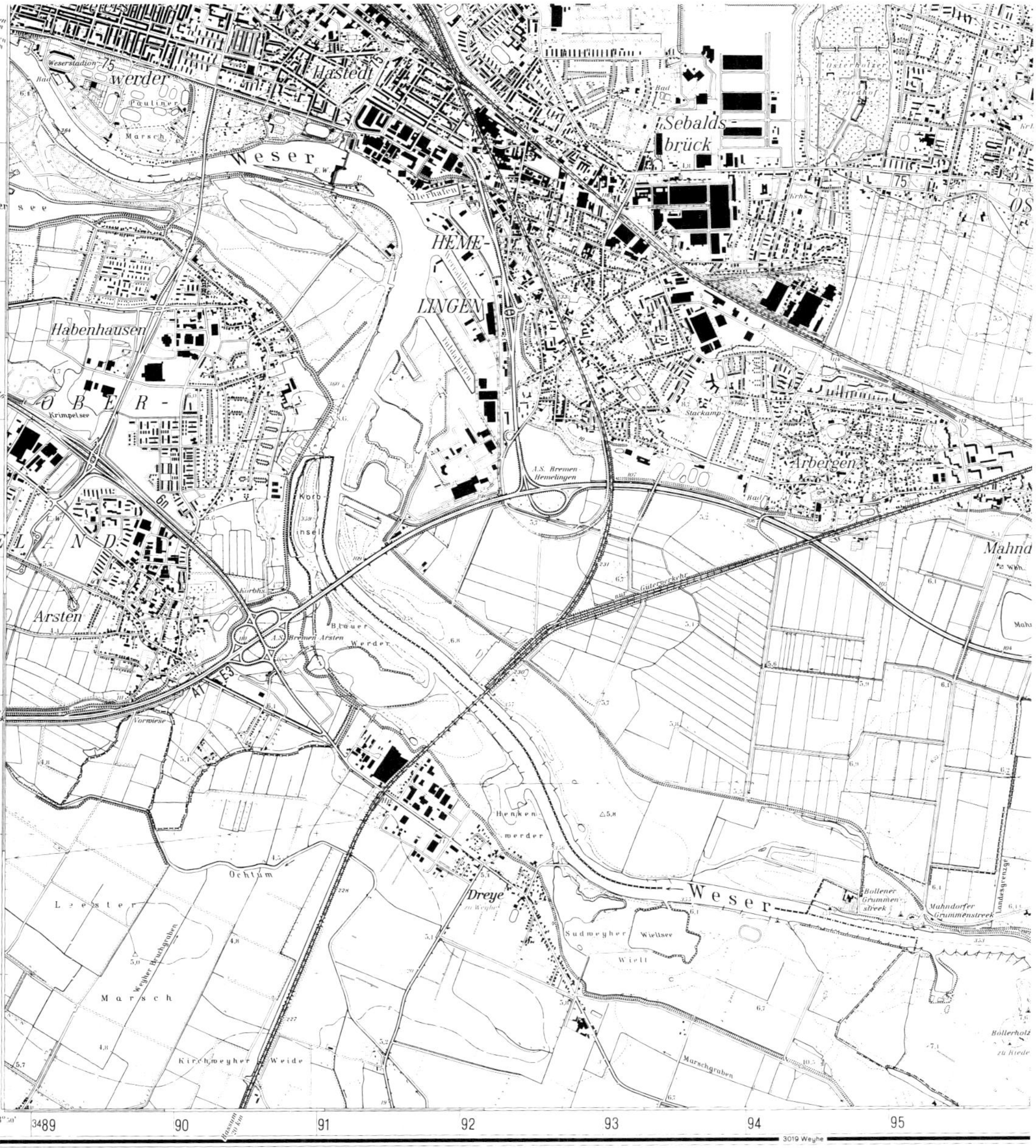
Weserstadion
werder
Paulinen
Marsch
Hastedt
Sebalds-
brück
Weser
HEME-
LINGEN
Habenhausen
O B E R -
Krimpelsee
L A N D
Arsten
Arbergen
Stackamp
A.S. Bremen-
Hemelingen
Korb-
insel
Güterverkehr
Blauer
Werder
A.S. Bremen Arsten
Henken-
werder
Ochtum
Dreye
Leester
Marsch
Weyher Neuergraben
Kirchweyher Weide
Sudweyher
Wiellsee
Wiell
Marschgraben
Weser
Bollener
Grummen-
streek
Mahndorfer
Grummenstreek
Landesgrenze
Mahnd
Bollerholz
3489
90
91
92
93
94
95
3019 Weyhe

FIGURE 2

TABLE 1. Phases of human interference in Weser river system (various authors, see text)

Period Area	Post glacial period	2000 BC–500 AD	500–1300 AD	1300–1700	1700–1800	1800–1900	1900–1987
River–aquatic zone	Vegetation and animal life slowly modified by climatic oscillation	Beginning of organized fishing in the Weser	Ecological condition locally modified by fish and mill dams	Beginning of organized commercial shipping	First connection of islands with shore, loss of gravel and sandbanks by regulation, first dams with locks	Dredging and regulation destroy former erosion and sedimentation process	Channelization, damming, and pollution destroy lentic structuring and diminish biocoenosis
Riparian zone	Alterion of local biotope by river bifurcation	*	First settlements on river banks ('Schlagten')	Towing paths, willow planting and cutting regularly	→	Shore works change former character of littoral (diminishing reed fields)	Artificial and petrified shoreline decrease typical vegetation
Floodplain	Formation of new meanders due to erosion and sedimentation	Slow, progressive increase in the sedimentation of clay influences plant communities	Relative fall of ground water level due to rise in clay sedimentation and change of flood-plain forests to meadow communities	Winter dyking in order to farm, smaller floodable areas	→	Summer dyking closes floodplain to regular flooding	Intensive farming, draining, flood control and traffic lines obliterate remnants of former coherence
Valley slopes				Spreading of farmland and settlements (garden plants)	→	→	
Mountain area	Vegetation and animal life slowly modified by climatic oscillations	First destruction of forested areas by the tilling of loess grounds (increase in erosion)	→	Wood cutting and ore mining destroy coherent forest areas	Total forest area shrinks by 3/4	Forest area reaches its minimum	Intensive farming, foresting and settling diminish last elements of former slope and mountain vegetation
	Endogenous river and landscapeand landscape dynamics	→	increase of anthropogenic influence	→			River system dynamics fully exploited and tamed

→ duration and amplification of processes
* no direct interference by human activities

virtually no typical riparian vegetation, but rather meadows border the waterline (Table 1); and Phragmite reed no longer exists except at rare locations between Minden and Bremen (Garz *et al.*, 1987).

The width of the Mittelweser and the size of 'Europaships' (1350 tons, 2.5 m draught) do not allow passage of two barges at all parts of the sinuous and narrow 'fairway'. Therefore the regulation of the Mittelweser to a draught of 2.5 m, even during the lowest summer water level, was started in 1987. Although a discussion about the requirements for the reestablishment of at least parts of the former hydrosystem qualities along large German rivers had just started (Bundestagsdrucksache, 1986) the water and shipping authorities only needed to compensate for the actual dredging activities by small and local compensation measures (BFG, 1986).

Channelization of the Unterweser

Today the Unterweser is one of most regulated rivers used as an ocean-going shipway in Europe (Weserabflussfibel, 1983). The first attempts in regulation go back to the 17th century (Keller, 1901). The depth of the Unterweser decreased heavily after the middle ages, due to increasing erosion upstream and the initiation of dyking in the estuarine area: in 1750 its depth was only 0.8 m downstream of Bremen. To ensure economic competitiveness of the harbours of Bremen, some attempts had been made in the 18th and 19th centuries to deepen the river-bed. However, they were of limited success (Rohde, 1970). The first deepening showing substantial results was done from 1887 to 1895. Franzius (1895), the leading engineer used three main principles which are still common:

1. Shaping the Unterweser into a funnel to increase the tidal exchange;
2. Diminishing river channel bifurcation and;
3. Concentrating the tidal currents to the navigable channel by constructing embankments, groynes, etc.

Tidal exchanges and tidal currents increased, deepening the river-bed, because the tides intruded much further up the river. In addition, intensive dredging activities (Rohde, 1970) supported this process. Subsequently, seagoing vessels drawing 5 m, could reach the harbours at Bremen.

The same principles were used during four further phases of channel deepening and dredging: 1913–1916 to 7 m; 1925–1928 to 8 m; 1953–1959 to 8.7 m; and 1973–1977 to 10.5 m (called 9m deepening due to the changed reference level: sea-chart zero).

This sequence of engineering works drastically changed the morphological, hydrological and ecological conditions within the Unterweser:

1. The tidal range at Bremen increased from about 0.3 m around 1880 to 4 m today and altered littoral habitats (Grabemann *et al.*, 1983; Busch *et al.*, 1984).
2. The river surface as well as the riparian area were reduced to about one third (Figure 3). Most of the ecological highly valuable backwaters and reed- and mud-flats were lost by filling with dredged material (Heinrich and Mühlner, 1980). At the same time dykes intruded forward into the navigable channel. Today, nearly no adjoining marshes and wetlands are to be found.

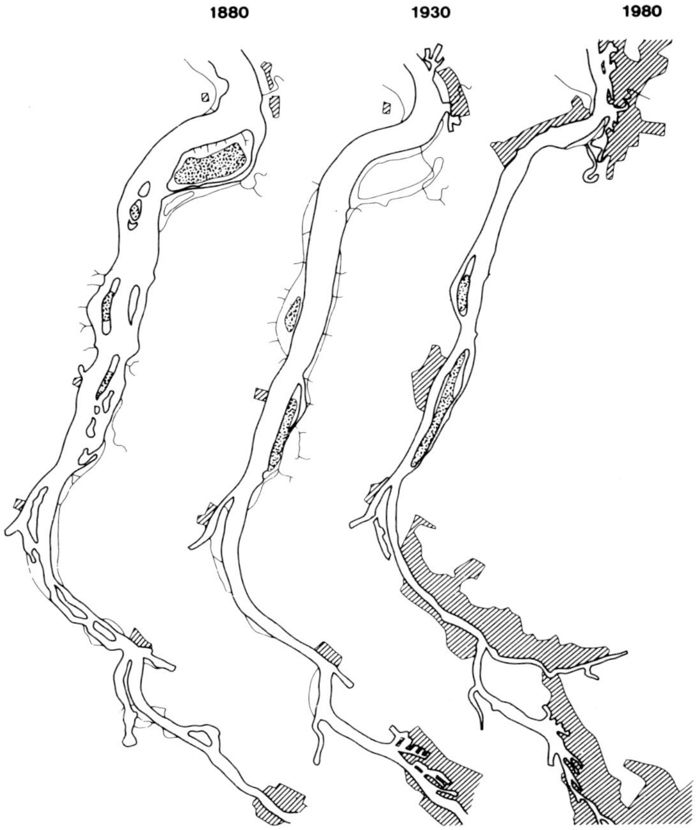

FIGURE 3. Reduction of the Unterweser area after channelization around 1880, 1930, and 1980 (From Busch *et al.*, 1984)

3. About 60 per cent of the eulittoral embankments of the navigable channel are covered by various packing materials, less than 40 per cent are left uncovered (Schuchardt *et al.*, 1984). Lotic zones and submerged plants are virtually missing.
4. The river profile was changed into a channel-like one and, thus, became very monotonous, due to the extreme deepening operations (Figure 4).
5. In 1912 a tidal barrage at Bremen-Hemelingen was built to stop erosion by the falling low water levels. The barrier interrupted spawning migrations of some fish species and caused a decrease of population densities (Nolte, 1976).
6. Due to strong siltation, especially in the tidal harbours of Bremen, (siltation rate 1981–86: 23.2 cm yr^{-1} or about 355 000 m^3 yr^{-1} (Hafenbauamt, 1987), much maintenance dredging continues. The dredged material (mainly silt) had to be disposed of, because of its high pollution load. Disposal sites were located in adjoining marshes behind the dykes and, thus, changed the ecological conditions there drastically.

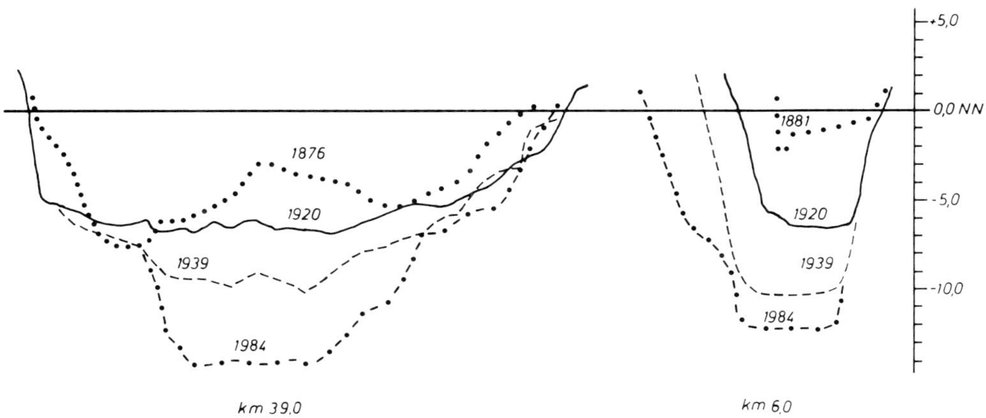

FIGURE 4. Profiles of the Unterweser at Brake (km 405.7) and Bremen Harbour (km 372.7) during different stages of channelization (Redrawn after Rhode, 1970 and data from Wasser- und Schiffahrtsamt Bremen). (km 405.7 = Unterweser km 39; km 372.7 = Unterweser km 6)

7. Both estuarine fauna and flora may be harmed by contaminants released into the water and the local O_2-depletion caused by maintenance dredging operations (Johnston, 1981; Engel *et al.*, 1984).

POLLUTION IN THE RIVER WESER

Effects of Salt Pollution

The hydrosystem of the River Weser, and especially its headwater tributary the Werra, have been badly polluted by salt wastewaters from the potash industry at Thüringen (GDR) and Hessen (FRG), since the beginning of this century. Salt pollution began about 1860 (Table 2) and has reached a load of about 35 000 td^{-1} in 1986, i.e. an average load of 410 kgs^{-1} (Buhse, 1987). Compared with the salt pollution of the Rhine, the Werra carries the same salt load, but has only 1/40 of the Rhine's discharge and, consequently, markedly higher salt concentrations. After confluence of the Werra and Fulda tributaries the salt concentration is diluted by one half. At Bremen-Hemelingen (km 361, tidal barrage) the concentrations of chloride lie between 0.5 and 1.4 $kg^{-1} m^{-3}$, with a yearly average of about 0.9 $kg\,m^{-3}$ (Arge–Weser Zahlentafel, 1985).

The severe salt pollution of the Weser and its daily and even hourly changes in composition and concentration is admittedly a heavy stress factor to the biota, especially after the strong increase in the 1960s. The pollution not only concerns the fishery, but impacts upon the whole aquatic life of the river. First, drastic changes in its phytoplankton composition were recorded in the early 1920s (Budde, 1927). In the 1950s there was a further increase of halophile species and a steep decrease of limnetic forms (Ziemann, 1967). At the same time Albrecht (1954) still found about 64 invertebrate species in the Werra. With increasing salt concentrations the populations of the most limnetic invertebrates collapsed and there have been some attempts to introduce euryhaline species as a new feeding base for fish (Meijering, 1980). The most successful species was the crustacean *Gammarus tigrinus*, which

TABLE 2. Historical development of salt pollution in the Weser River

1856	Discovery of potassium in the Thüringen and Hessen area, largest potassium deposits in central Europe
1861	Start of potassium mining and inauguration of the first potash factory
1911	For the first time the salt load of the Weser reaches critical concentrations: the drinking-water production of Bremen becomes endangered for the first time
1913	As a result of negotiations, a concentration limit for the salt-load of the Weser at Bremen (250 mg $Cl^- l^{-1}$) is fixed
1914	Agreement on new concentration limits: in the Weser at Bremen to be 350 mg $Cl^- l^{-1}$; Werra 2500 mg $Cl^- l^{-1}$
around 1925	Increase in salt-production and start of disposing of salt-polluted waste-waters in caverns
1947 to 1951	New treaties on the concentration limits and salt pollution loads: 62 per cent from production in Thüringen (GDR) and 38 per cent from Hessen (FRG)
1949 to 1950	The heavy fish mortalities in the Werra due to salt pollution are documented for the first time
since 1952	Increasing violation of concentration limits by factories in the GDR
1968	Thüringen factories stop disposing of salt wastewaters in caverns because of damaging drinking water sources: further increase in the salt concentration of the Werra
1976	Maximum salt pollution reached in the Werra: 40 000 mg $Cl^- l^{-1}$, yearly average rate 17 000 mg $Cl^- l^{-1}$
since 1980	New negotiations between the GDR and FRG about the salt pollution of the Weser
since 1982	Despite the availability of a technical solution for the salt pollution, negotiations between the GDR and FRG show no results
1987	Implementation of new technology in GDR factories fails after disagreement about licencing authorities

was introduced in the Werra in 1957 (Schmitz, 1960). By 1963 this species was found in the whole Ober- and Mittelweser (Tesch and Fries, 1963) and it has also established a stable estuarine population (Schuchardt *et al.*, 1987). Presently about 95 per cent of the invertebrate biomass in the Werra and Oberweser is due to this species (BFG, 1987), and it provides the dominant food component for various fish species (Fries and Tesch, 1965).

Even more brackish-water species can be found in the Mittelweser, i.e. barnacles in a section of 130 km downstream from Minden (Busch *et al.*, 1987 unpublished) and the estuarine crustacean *Corophium lacustre* (Herbst, 1982).

Pollution by degradable organic compounds, nutrients and cooling water

Due to the mainly 20th century increase in the human population, the increasing use of detergents and artificial fertilizers, industrial development, centralization of the water

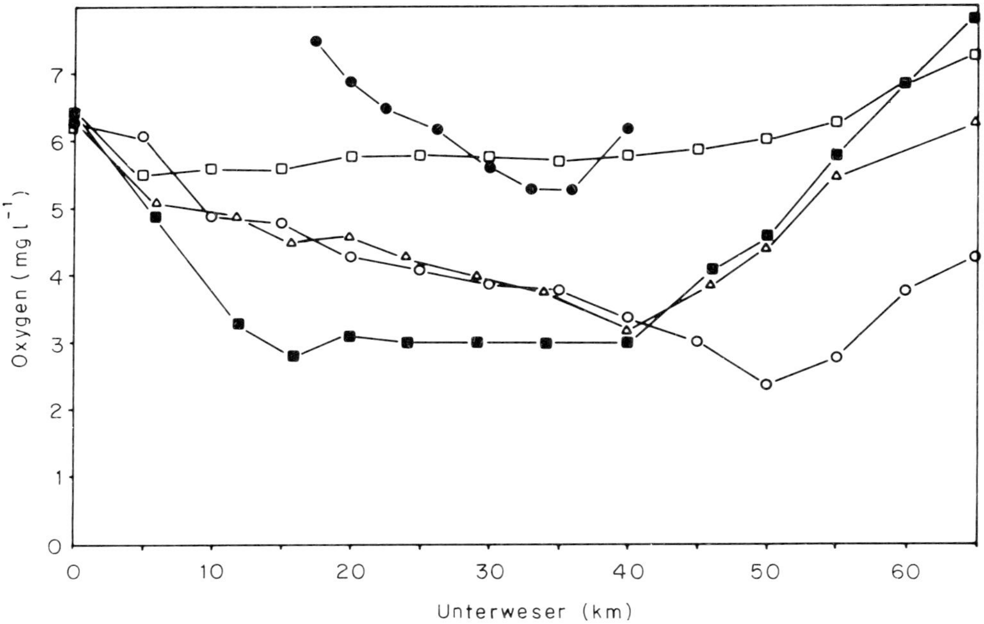

FIGURE 5. Summer oxygen levels in the Unterweser

	Date	Temperature	Discharge	Source
● dark circles	8/1929	18–19°C	**	Schräder (1941)
■ dark squares	6/1964	**	127 $m^3 s^{-1}$	Zietz (1975)
△ open triangles	6/1974	**	180 $m^3 s^{-1}$	Zietz (1975)
○ open circles	7/1979	19–20°C	200 $m^3 s^{-1}$	Schirmer *et al.* (1983)
□ open squares	8/1986	21°C	133 $m^3 s^{-1}$	Wasserwirtschaftsamt Bremen

** no data available

TABLE 3. Heavy metal load of the Weser at Bremen Hemelingen in 1985 (River km 362)

Metal	Annual load in tons (1) 1985	Average concentration in $\mu g\, l^{-1}$ (2) 1985
Cadmium	6.3	0.6
Lead	59.9	6.4
Chromium	50	5.9
Nickel	72.5	8.0
Mercury	<2.2	0.14
Copper	56.8	6.1
Zinc	457	53

1. Calculated by Jahte upon data from Arge Weser Zahlentafel (1985)
2. Arge Weser Zahlentafel (1985)

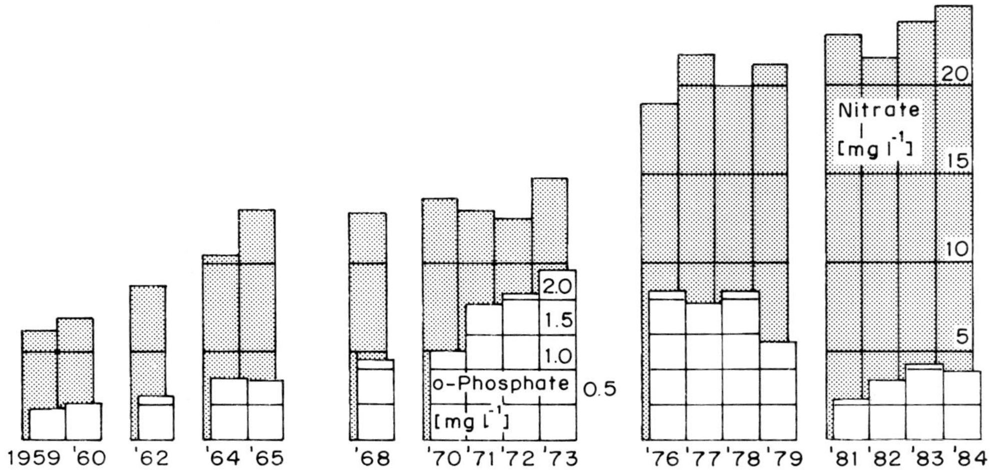

FIGURE 6. Concentration of NO_3–N and PO_4–P in the Weser at Bremen: annual average during 1959–1984 (Source: Albrecht and Kirchhoff, 1987)

supply and wastewater discharges there has been an increase in water pollution within the Weser (Schuchardt *et al.*, 1985; Albrecht and Kirchhoff, 1987; Weserlastplan, 1982). Delays in the construction of sewage treatment plants are significant and have many consequences for the hydrosystem of the Weser. They are easy to demonstrate, e.g. by changes in summer oxygen levels in the Unterweser downstream of Bremen (Figure 5). The oxygen deficits result mainly from two processes (Schirmer *et al.*, 1983):

1. The mineralization of organic matter from sewage effluents or autochthonously produced organic matter (mostly measured as BOD-5).
2. The nitrification of NH_4^+ from sewage effluents.

Within the Mittelweser, the intensified primary production (eutrophication) during summer daylight leads to very high oxygen levels due to increased nutrient levels and the altered hydrographic situation after the construction of seven dams (Figure 1). This coincides with very low oxygen values at night: i.e. down to 0.9 mg l^{-1} (Albrecht and Kirchhoff, 1987).

In addition, the Weser currently receives up to 12 600 MJ s^{-1} of thermal energy originating from the cooling water of three nuclear and seven non-nuclear powerplants and one steelwork (Weserlastplan, 1982). During the summer of 1986 downstream from a powerplant at Landesbergen (Mittelweser) water temperatures were up to 30°C across the whole profile of the Weser (Busch *et al.*, 1986, unpublished). This indicates a transgression of the officially fixed limit of 28°C set by the water authorities in the Wärmelastplan (1974) and the Weserlastplan (1982).

Within the last few years, the BOD-5 burden of sewage effluents has been reduced due to the construction of sewage treatment plants with biological treatment. For example, the BOD-5 burden of the Unterweser has been reduced from 87 000 kg d^{-1} in 1972 (Wöbken, 1973) and 90 500 kg d^{-1} in 1980 (Wöbken and Kunz, 1980) to about 12 300 kg d^{-1}

TABLE 4. Heavy metal concentration in tissues of aquatic organisms of the Weser at Bremen in ppm ($mg\ kg^{-1}$ dry weight)

Metal	1 Bream (liver) (1985)	2 White bream (liver) (1985)	3 Eel (liver) (1985)	4 *Gammarus tigrinus* whole animal (1986)	5 *Enteromorpha* whole algae (1984)
Cadmium	0.764 (±0.588)	0.763 (±0.364)	1.185 (±1.879)	0.669 (±0.102)	4.23 (±1.54)
Lead	0.329 (±0.126)	0.342 (±0.354)	2.330 (±0.957)	3.039 (±0.664)	48.64 (±15.0)
Chromium	0.295 (±0.209)	0.181 (±0.089)	0.081 (±0.069)	3.386 (±0.913)	—
Nickel	0.838 (±0.685)	0.808 (±0.712)	0.276 (±0.172)	4.415 (±0.988)	19.61 (±6.85)
Copper	42.4 (±25.0)	36.5 (±21.5)	26.1 (±8.1)	—	—
Zinc	116.5 (±25.00)	108.0 (±21.5)	126.8 (±16.7).	—	387.2 (±107)
Mercury*	(0.893 (±0.540)	0.972 (±0.414)	0.680 (±0.256)	—	—

* Mercury concentrations in muscle tissue
—no data available (±) Standard deviation
Sources: 1 + 2 Busch *et al.*, in press; 3 Busch, Cetinkaya, Ullrich unpublished; 4 Busch unpublished; 5 Jahte, 1985

in 1986 (Personal communication from the Water authority of Bremen), resulting in increasing oxygen levels within the Unterweser since 1985 (Figure 5). In the Ober- and Mittelweser, a reduction of the BOD-5 burden had been accomplished somewhat earlier, mainly in the 1970s (Neumann and Schöneborn, 1982). However, this did not result in a decrease in the NO_3 concentration (Figure 6). The decrease in the PO_4 concentration, due to an alteration of the detergent composition in the 1960s, is too small to affect primary production. Eutrophication, especially in the Mittelweser, is still a great problem (Albrecht and Kirchhoff, 1987), e.g. as shown by the diurnal changes in the oxygen concentrations during the summer between 1st and 3rd July 1986 oxygen levels varied daily from $7\ mg\ l^{-1}$ to $15\ mg\ l^{-1}$ (Wasserwirtschaftsamt Hildesheim, personal communication).

Heavy Metals and Chlorinated Hydrocarbons

The further pollution of the Weser is due to heavy metals and chlorinated hydrocarbons. This results in elevated concentrations of both groups of pollutants in sediments and in the tissues of aquatic organisms. Three main sources of contaminants can be distinguished:

1. Municipal waste waters,
2. Industrial waste waters,
3. Leachate from top soil layers of land-fill sites associated with ore mines and smelter plants in the Harz area (FRG), which was one of most important metal producing centers in the Middle Ages.

The third factor especially is the reason for the very high burden of heavy metals in the basin of the River Aller, which is the most important tributary of the Weser. The high

TABLE 5. Tissue concentrations of polychlorinated hydrocarbons in fish caught at Bremen in 1985 in ppb ($\mu g\ kg^{-1}$ fresh weight)

	White bream* (*Blicca bjoerkna*)	Bream* (*Abramis brama*)	Eel† (*Anguilla anguilla*)
DDT (total)	46.7 (±74.7)	106.1 (±108.0)	133 (±148)
HCH (total)	18.04 (±9.9)	21.2 (±10.7)	26.1 (±10.2)
Lindan	6.7 (±5.5)	12.9 (±6.9)	21 (±8.7)
Chlordan	1.0 (±3.1)	2.9 (±6.3)	<1
Endosulfan (total)	1.2 (±4.7)	1.1 (±2.9)	<1
Dieldrin	5.9 (±6.7)	18.0 (±17.3)	15.9 (±19.4)
HCB (total)	1.7 (±1.7)	4.2 (±2.2)	12 (±5.7)
PCB (total)	156.0 (±140)	321.0 (±22.1)	424 (±212)

* Busch *et al.* (in press)
† Busch, Cetinkaya, Ullrich (unpublished)

Cadmium burden is remarkable (Table 6) even in comparison with other heavily polluted rivers such as the Rhine and Elbe. With the exception of various PCBs (polychlorinated biphenyls) the level of chlorinated hydrocarbon pollution is similar in the Weser, Rhine and Elbe. The high PCB burdens possibly have their sources in disused mining sites in the Harz area as well as in the recent potash mining in the Thüringen area (Albrecht and Kirchhoff, 1987). Poisoning with PCBs seems a serious problem for the Weser ecosystem (Tables 5–7) (Busch *et al.*, 1988).

With regard to the various pollutants, it is important to recognize that even without posing an acute risk to human health all pollutants have a prejudicial effect on the animal community. The populations of fish and invertebrates are endangered by substances like cadmium and PCBs, which are highly concentrated in the tissues of their target organs. These concentrations may lead to disfunctions of liver and kidney (cadmium) as well as to raised mortality rates of larvae and to genetic mutations.

HISTORICAL CHANGES IN THE WESER FISHERY

Unterweser

The existence of fishweirs along the whole Weser and its tributaries from the 5th century on are confirmed by historical documents (Keller, 1901). Traditionally the Weser has been known among European rivers for its abundance of fish. As early as the Middle Ages, Bremen conferred Rights of Fishing through legal privileges to a cooperative guildlike union of fishermen called 'Bremer Amtsfischer'. These fishermen practiced their trade as a Right of the City of Bremen in its territory (which was the whole Unterweser and parts of the Mittelweser at times) until the 19th century. Large-scale yarn fishing was generally used. Atlantic salmon, sturgeon, shads and lampern were mainly caught. Several phases in the development of the fishery can be distinguished (Schuchardt *et al.*, 1985) (Figure 7). Overfishing, building of river barrages (at Hameln and Hannoversch Münden), and regulation activities in the Mittel- and Oberweser lead to a decline of the traditional fishery in the 19th century. In 1880 the Union of the Bremer Amtsfischer had already lost most of its privileges and shrunk to only 30 members, among whom only four worked

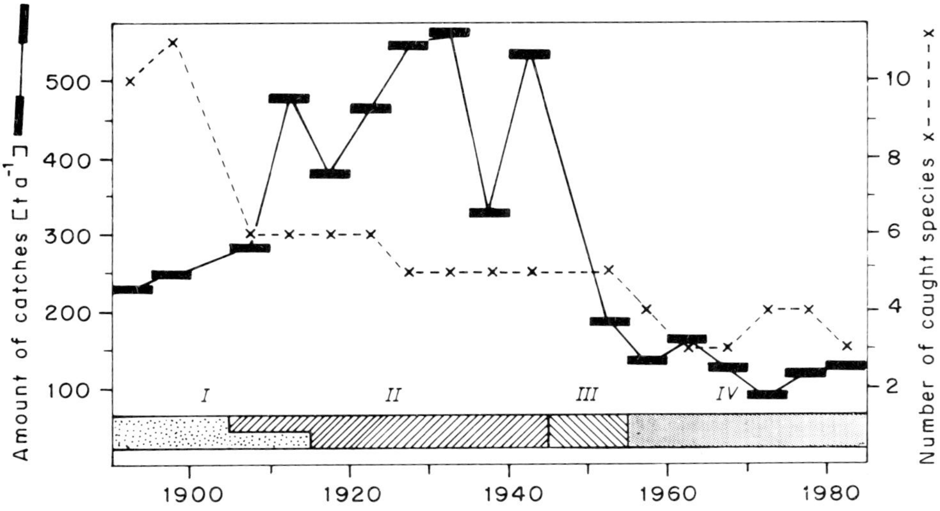

FIGURE 7. Four phases of fishery in the Unterweser (Source: Schuchardt *et al.*, 1985)

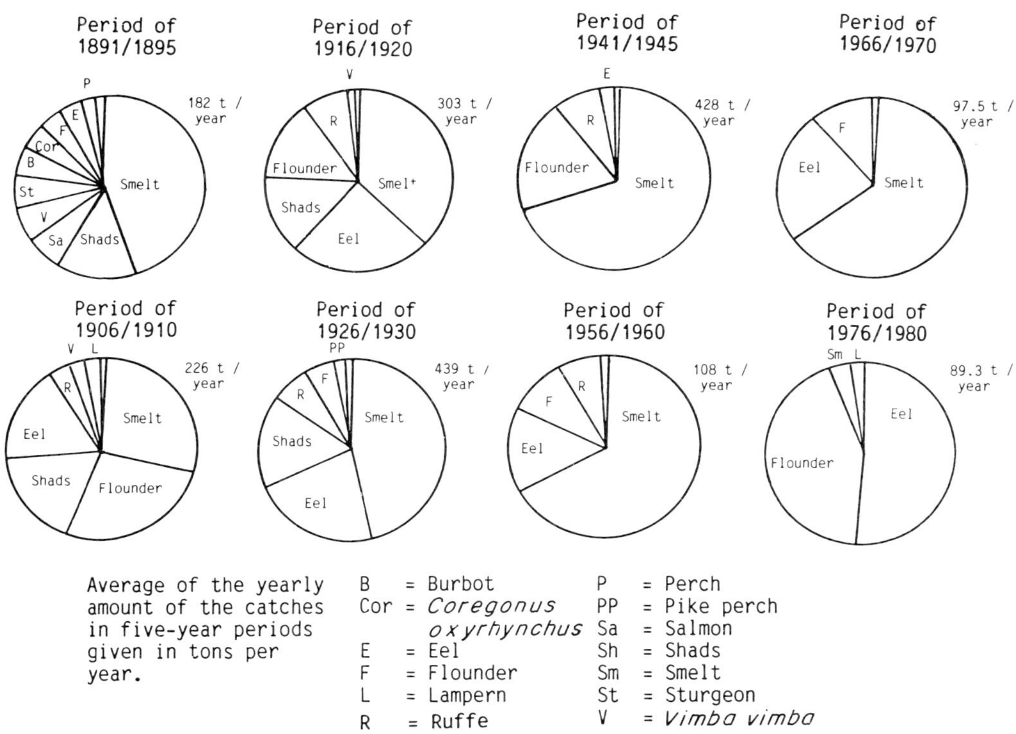

FIGURE 8. Changes in the fish catch in the Unterweser 1900–1980 (Source: Busch *et al.*, 1988)

TABLE 6. Decrease of the commercial base of the river fisher in the area of Brake during the 8.7 m deepening of the riverbed

	1931–1938		1953–1960	
	Catches (kg)	Proceeds (DM)	Catches (kg)	Proceeds (DM)
Smelt	93 000	27 915	20 000	5 884 (21%)
Eel	59 375	181 687	31 200	85 919 (47%)
Summ		209 602		91 806 (44%)

Price-niveau calculated for 1960; other species caught in less amounts have not been taken into consideration. After Nolte, 1974

fulltime in the Weser fishery. The first—historical—phase of the Weser fishery ended at that time.

The several phases of the river fishery are reflected in the changes in the amounts, species and composition of the catches, represented in Figure 8, which is based upon the catch statistics of the National Fishery Office in Bremerhaven (Busch *et al.*, 1988). At the beginning of the 20th century the changes continued, but with transition to other, more profitable species. During this second phase (1905–45) fishing methods changed to big hamen-nets and eelbaskets and, therefore, until 1926 the landing of fish increased. But effects of river channelization and industrialization in the Unterweser region prevented a new rise in the Weser fishery.

During this phase, fishermen caught mostly eel, smelt, ruff, shad, flounder, and lampern. Obviously, sewage pollution from rising numbers of inhabitants and industrialization had no significant effect on these fish populations; but the channelization of the Unterweser, started in 1887, did. Spawning grounds and feeding areas were heavily destroyed by the regulation works.

A third phase took place between 1945 and 1955. Whereas the spectrum of species caught was similar to that of the phase before, the amount of fish landed was less by half. Many fishermen had to move to more lucrative jobs in industry, because of worsening conditions in the river. Some consequences of the channelization for the Weser fishery were shown by a report of the National Fishery Office in Bremerhaven on the occasion of the 8.7 m deepening in 1953–59 (Nolte, 1974). Catches of eel and smelt in this area were compared for the periods of 1931/38 and 1953/60 and a significant decrease could be demonstrated (Table 8). Fishermen from the town of Brake (Unterweser) were for the first time successful in obtaining compensation for their commercial losses, but the trial lasted for more than 20 years. Because of decreasing catches and the decreasing commercial basis, nearly all the fishermen in this area left fishing (Busch *et al.*, 1982): the number of fishermen in the Brake region fell from 35 in 1953 to 13 in 1965, and to 3 in 1981.

The fourth and hitherto final phase of the development continues up to the present. It is characterized by two more deepenings of the river, a very strong increase in shipping and industrialization, as well as an abrupt rise in salt and sewage pollution of the river system. During this period, the populations of ruff, twaite shad and lampern collapsed (Tables 9 and 10).

TABLE 7. Recent and extinct species of fish in the Werra, Ober-, Mittel- and Unterweser

	Ober- and Mittelweser	Unterweser
Petromyzontidae		
Lampetra planeri (stream lampern)	*2	
Lampetra fluviatilis (lampern)		*2
Petromyzon marinus (sea lamprey)		*1
Acipenseridae		
Acipenser sturio (sturgeon)	*1	*1
Clupeidae		
Alosa alosa (allis shad)		*1
Alosa fallax (twaite shad)		*
Clupea harengus (herring)		*
Sprattus sprattus (sprat)		*
Salmonidae		
Salmo salar (atlantic salmon)		*4
Salmo trutta forma fario (lake trout)	*	*4
Salmo gairdneri (rainbow trout)	*	*4
Salvelinus spec. (char)	*5	
Coregonidae		
Coregonus oxyrhynchus (houting)		*1
Thymallidae		
Thymallus thymallus (grayling)	*	
Osmeridae		
Osmerus eperlanus (smelt)		*
Esocidae		
Esox lucius	*4	*
(pike)	*4	*
Cyprinidae		
Rutilus rutilus (roach)	*	*
Leuciscus leuciscus (dace)	*	*
Leuciscus cephalus (chub)	*	*

(*continued*)

TABLE 7. (*continued*)

	Ober- and Mittelweser	Unterweser
Leuciscus idus (ide)	*	*
Scardinius erythrophthalmus (rudd)	*	*
Tinca tinca (tench)	*	*
Gobio gobio (gudgeon)	*	*
Barbus barbus (barbe)	*	
Alburnus alburnus (bleak)	*	*
Blicca bjoerkna (white bream)	*	*
Abramis brama (bream)	*	*
Abramis ballerus		*1
Vimba vimba	*	*1*
Carassius carassius (crucian carp)	*	*
Cyprinus carpio (carp)	*	*
Siluridae		
Silurus glanis (catfish)	*5	*5
Ictaluridae		
Ictalurus nebulosus (brown bullhead)		*5
Anguillidae		
Anguilla anguilla (eel)	*4	*
Gadidae		
Lota lota	*3	*3
Gasterosteidae		
Gasterosteus aculeatus (threespined stickleback)		*
Pungitius pungitius (tenspined stickleback)		*
Percidae		
Perca fluviatilis (perch)	*	*
Stizostedion lucioperca (pike perch)	*4	*
Gymnocephalus cernua (ruff)	*3	*3

TABLE 7. (*continued*)

	Ober- and Mittelweser	Unterweser
Gobiidae		
Pomatoschistus microps		*
(common goby)		
Cottidae		
Cottus gobio	*	
(millers thumb)		
Pleuronectidae		
Platichthys flesus		*
(flounder)		

1 = extinct
1* = anadromous population
2 = threatened by extinction
3 = endangered
4 = restocking
5 = introduced
(from Buhse, 1987; Busch *et al.*, 1987)

Mittel- and Oberweser

Until the 20th century, the decline of river fishery in the Ober- and Mittelweser resulted from the same causes as that in Unterweser. Even overfishing of the Atlantic salmon is documented (Nolte, 1976). This resulted in the management of their recruitment. As early as 1857 a rearing station was founded at the barrage in Hameln. Demoll *et al.* (1926), with regard to the salmon catch in Hameln in 1893–96 and 1906, reported that more than 90 per cent of the migrating salmon were caught directly downstream of the barrage; only 10 per cent successfully passed through the fishladder. After the building of the tidal barrage at Bremen–Hemelingen, the salmon fishery in the Ober- and Mittelweser collapsed completely. Spawning migrations of other species like smelt were also stopped by this dam (Nolte, 1976).

Compared to the deepening of the river-bed and increasing sewage pollution, heavy salt pollution has had the worst influence on fish populations in the Mittel- and Oberweser during the 20th century. The biological consequences and economical effects of the salt pollution have been documented by Buhse (1987). For the unpolluted river system, Buhse estimated a 'natural' fish yield of about 200 kg ha^{-1} yr^{-1} for the Werra river and 150 kg ha^{-1} yr^{-1} for the Ober- and Mittelweser as a basis for estimating biological and economical losses. This 'natural' yield is a reconstruction of a stock base expected after the collapse of the salmon fishery in the 19th century. In the upper 80 km part of the Werra, Buhse (1987) found the fish yields decreased to 5 per cent of the natural yield. In the lower 80 km part of the Werra it dropped to 25 per cent between 1970 and 1978, and to 15 per cent after 1980. The upper part of the Oberweser (km 0–45) experienced a reduction in 1960–1975 of 33 per cent, and since 1977, 20 per cent. Between 1960–1965 the stock in the river from km 85 to km 172, was reduced to 40 per cent; between 1965 to 1970 it improved to 70 per cent; and since 1980 it was further reduced to 40 per cent.

TABLE 8. Reasons for the decline in fish stock of economic interest (Busch *et al.*, in press)

Species	Decline of catches since	Counter measures	Present status	Channelization of Weser		Construction of barrages	Spawning grounds became		Over-fishing	Pollution effects
				Mittel-	Unter-		destroyed	blocked off		
Sturgeon (*Acipenser sturio*)	1850	none	extinct	*	*	*	*	*	*	
Atlantic salmon (*Salmo salar*)	the 19th century	breeding since 1857, presently restocking by game fishermen in tributaries	rare	*		*		*	*	
Allis shad (*Alosa alosa*)	the 19th century	none	extinct	*	*	*	*	*	*	
Twaite shad (*Alosa fallax*)	1980	none	frequent		*		*			*
Eel (*Anguilla anguilla*)	1970	restocking by professional and game fishermen	frequent			*			*	*
Flounder (*Platichthys flesus*)	1915	none	frequent		*			*		*
Smelt (*Osmerus eperlanus*)	1950	none	frequent		*	*	*			*

* explanation of present status

The surviving fish show high rates of disease: epidermal ulcerations, discolouration of the liver, bloated consistency of the kidneys, reduced coelom and pericard fluid, enlargement of the gallbladder, reddened serum due to erythrocytolysis (Halsband, 1976) and a high rate of parasite attack (Garz *et al.*, unpublished). In 1973 Buhse (1987) found the following rates of illness:

Species	Rate of illness (per cent)
Bream (*Abramis brama*)	60.5
Pike (*Esox lucius*)	63
White bream (*Blicca bjoerkna*)	40
Roach (*Rutilus rutilus*)	49.5
Barbe (*Barbus barbus*)	30
Dace (*Leuciscus leuciscus*)	25
Eel (*Anguilla anguilla*)	5
Perch (*Perca fluviatilis*)	1.6

Several kinds of fish death result from the high salt concentrations. In 1971, for example, there was about 6000 kg of dead fish per hectare in the Mittelweser (Buhse, 1987). Buhse defines five types of fish mortality:

1. Poisoning by potassium ions, caused by the poisoning influence of ionic potassium, which lead to deregulation of nerve metabolism, starting at K^+ concentrations of about 200 mg l^{-1}.
2. Osmotic fish mortality, caused by strong variations in the salt concentrations, for which the stressed fish are not able to compensate.
3. Assimilation fish mortality. The planktonic algal community consists of only a few salt-tolerant species during the summer. In cases of algae blooms, the oxygen concentration in the water can reach up to 200 per cent saturation, and the pH increases to 9.5. This situation leads to colliquation necrosis of the fish branchia.
4. Suffocation fish mortality. At phases of algal mass development, very low oxygen concentrations are observed during the night in the water, therefore fish die from suffocation.
5. Supercooling fish mortality. Due to high salt-concentrations, the freezing point is lowered. In the Werra, water temperatures around −5.2°C have been found in the winter. These low temperatures lead to fish mortality due to the build-up of ice crystals within the blood of the fish.

These examples of (a) fish mortality, (b) extinct species and (c) high rates of diseases within the surviving fish, demonstrate how the formerly multifaceted and rich lotic ecosystem has been destroyed. Although the most important factor in the worsening of the ecosystem is salt pollution, the other factors, such a sewage load, thermal pollution and channelization, have contributed to system degradation.

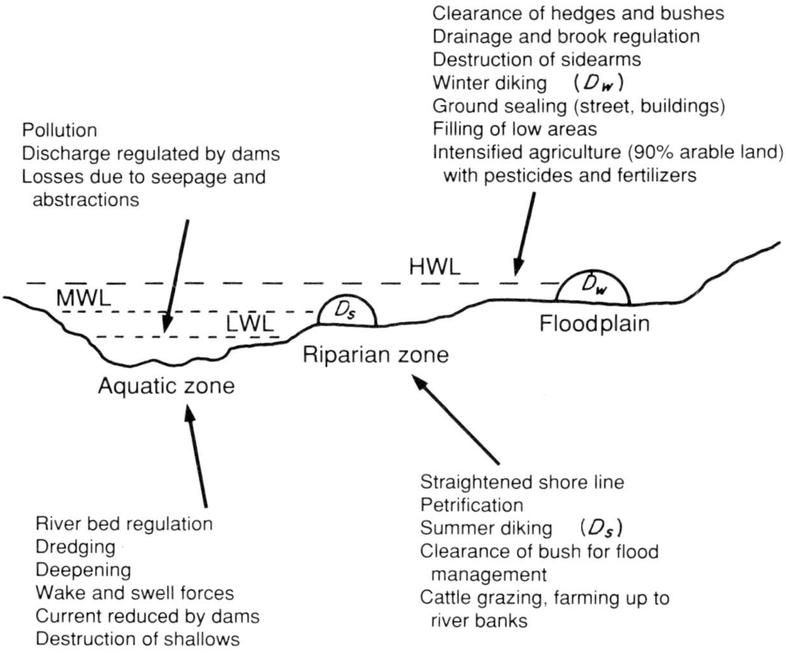

FIGURE 9. Typical factors inhibiting the self renewal of the Weser ecosystem

Is there a future for the Weser Ecosystem?

Dyking, regulation, channelization, and pollution have caused a dangerous and critical situation for the Weser hydrosystem (Albrecht and Kirchhoff, 1987). The invertebrates have suffered a dramatic reduction. The fish stocks can only be maintained by extensive restocking efforts (Buhse, 1987). The former river and valley landscapes, typical and richly structured, have been destroyed except in some small and localized areas (Table 1). The numbers of factors inhibiting the potential renewal and spread of the fauna and flora are immense (Figure 9). The discussion of engineering works along the Mittelweser to the 'Europe' ship size class (1.350 to 2.50 m draught) drastically showed, that the FRG has no legal obligation to compensate for damage to natural resources, which result from existing channelization and regulation before new changes are allowed. The environmental laws of the 1980s concerning protection of the remaining natural resources from further destruction by human activities, only prescribe compensation for newly planned activities (Bundesnaturschutzgesetz, Wasserhaushaltsgesetz — WHG).

The ongoing discussion on restoring regulated streams and brooks (BBU, 1979; LWA, 1980, 1986; DVWK, 1984) induced the opposition SPD party to make a statement in parliament about restoring channelized and overregulated rivers (Bundestagsdrucksache, 1984). But this effort failed in the debate about the German Water Pollution Protection Act (WHG) in 1984–1986 (Bundestagsdrucksache, 1986). Nevertheless some steps to compensate for the channelized and regulated river have already occurred at gravel-lakes all along the Weser. These artificial lakes made by excavation for building material give some relief to stressed plants and animal life still existing in Weser valley, especially if they are connected permanently to the river itself. However their site and shape are more or less totally prescribed by economic reasons (Berndt and Neumann, 1985; Garz *et al.*, 1987).

A real change requires the implementation of a coordinated programme to restoring local river and valley regions, combined with drastic effluent reductions in salt, heat, and other pollutants and a switch to different ship types with less wake and swell, or even a ban on big ship traffic along the Ober- and Mittelweser. Only by these measures might some of the remaining ecological qualities of the Weser be saved.

ACKNOWLEDGEMENTS

Acknowledgements for data and help to Wasserwirtschaftsamt Bremen, Wasserwirtschaftsamt Hildesheim, Wasser- und Schiffahrtsamt Bremen, Landeserwaltungsamt-Landesvermessung Hannover, and Dr M. Vicker.

REFERENCES

Albrecht, M. L. 1954. Die Wirkung der Kaliabwässer auf die Flora und Fauna der Gewässer unter besonderer Berücksichtigung der Werra und Wipper. *Fortschritte der Wasserchemie*, **7**, 50–80.

Albrecht, J. and Kirchhoff, N. 1987. Ökologie der Weser — Der Fluß als Lebensraum im Wandel der Zeit. In Bachmann and Hartmann (Ed.), *Schiffahrt, Handel, Häfen — Beiträge zur Geschichte der Schiffahrt auf Weser und Mittellandkanal*, Minden, 295–325.

Arge Weser Zahlentafel 1985. *ARGE Weser Zahlentafel*, ARGE Weser.

Aurich, M., Brönnle, K., Lucker, T., Madeheim, E., Ney, A., Shwarze, H. and Ullrich, P. 1986a. *Gewaesser der Mittelweseraue unter besonderer Berücksichtigung der mit der Weser verbundenen Baggerseen*, Exkursionsbericht FB 2, Universität Bremen.

Aurich, M., Brönnle, K., Gumprecht, R., Heidenreich, G., Jahn, N., Lucker, T., Madeheim, E., Moeller, P., Ney, A., Schwarze, H. and Ullrich, P. 1986b. *Belastung bremischer Gewässer: Hegemann- und Atlassee: Zwei mit der Weser verbundene Baggerseen im Vergleich*, Projektbericht FB 2, Universität Bremen.

BBU 1979. *Biotop-Zerstörung durch wasserbauliche Maßnahmen, Methoden des naturnahen Gewässerausbaus*, Arbeitsgruppe Wasser im Bundesverband der Bürgerinitiativen Umweltschutz e.V.. Worms.

Behre, K.-E. 1985. Die ursprüngliche Vegetation in den deutschen Marschgebieten und deren Veränderung durch prähistorische Besiedelung und Meeresspiegelbewegungen. *Verh. Ges. ökol.*, **13**, 85–96.

Berndt, J. and Neumann, D. 1985. Baggerseen mit Verbindung zu Fließgewässern — Ergebnisse einer ökologischen Studie am Niederrhein. *Natur und Landschaft*, **60**(1), 3–9.

BFG 1986. *Anpassungsmaßnahmen an der Mittelweser zur Befahrung mit 2.5 m abgeladenen Europaschiffen — Stauhaltung Petershagen, ökologische Untersuchungen*, Gutachten der Bundesanstalt für Gewässerkunde. Koblenz.

BFG 1987. *Anpassungsmaßnahmen an der Mittelweser zur Befahrung mit 2.50 m abgeladenen Europaschiffen — Stauhaltung Petershagen, Bestandsaufnahme der Wirbellosen*, Gutachten der Bundesanstalt für Gewässerkunde. Koblenz.

Budde, E. 1927. Eine Wasserblüte der Weser. *Arch. Hydrobiol*, **18**, 316–318.

Buhse, G. 1987. *Fischereibiologische Auswirkungen durch die Salzstörungen in Werra und Oberweser*, Göttingen.

Bundesnaturschutzgesetz 1976. Gesetz über Naturschutz und Landschaftspflege (BNatSchG) v.20.12.1976 (BGBI.I.S.3574, ber.1977 S.650).

Bundestagsdrucksache 1984. *Sofortprogramm zum Schutz des Wassers der SPD-Fraktion im Deutschen Bundestag*, Bundestagsdrucksache 10/1823 v.02.08. 1984, Bonn.

Bundestagsdrucksache 1986. *Beschlußempfehlungsbericht des Innenausschusses des Deutschen Bundestages zur Vorlage der Bundesregierung zur Novelle des Wasserhaushaltsgesetzes (WHG)*, Bundestagsdrucksache 10/5727 v.24.06.1986, Bonn.

Busch, D., Schröder, K. and Schuchardt, B. 1982. *Der Niedergang der Flußfischerei auf der Unterweser*, Projektbericht FB 2, Universität Bremen.

Busch, D., Schirmer, M., Schuchardt, B. and Schröder, K. 1984. Der Ausbau der Unterweser zum Großschiffahrtsweg und seine Auswirkungen auf das Flußökosystem und die Flußfischerei. *Neues Archiv für Niedersachsen*, **33**(1), 60–80.

Busch, D., Haesloop, U., Scheffel, J. and Schirmer, M. 1988. Fish and their environment in large European river ecosystems: the river Wesser, FRG. *Science de l'eau*, **7**, 75–94.

Demoll, R., Maier, H. N. and Wundsch, H. 1926. *Handbuch der Binnenfischerei Mitteleuropas, Band* 5, Stuttgart.

DVWK 1984. Ökologische Aspekte bei Ausbau und Unterhaltung von Fließgewässern. *Deutscher Verein für Wasserwirtschaft und Kulturbau, Merkblatt*, **204**.

Engel, H., Hellman, H., Kitchesch, V., Knöpp, H., Mayer, H.J., Müller, D., Mundschenk, H., Scheichert, U., Teuber, W. and Tippner, M. 1984. Hydrologische und ökologische Voraussetzungen zur Umlagerung von Baggergut in staugeregelten Bundeswasserstraßen. *BFG—Jahresbericht*, **1984**, 1–41.

Fittkau, J. and Reiss, F. 1983. Versuch einer Rekonstruktion der Fauna europäischer Ströme und ihrer Auen. *Arch. Hydrobiol*, **97**(1), 1–6.

Franzius, L. 1895. *Die Korrektion der Unterweser und Projekt zur Korrektion der Außenweser*, Leipzig.

Fries, G. and Tesch, F. W. 1965. Der Einfluß des Massenvorkommens von Gammarus tigrinus auf Fische und Niedere Tierwelt in der Weser. *Arch. FischWiss*, **16**, 133–150

Garz, T., Lucker, T., Schirmer, M., Schwarze, H. and Ullrich, P. 1987. Welche Zukunft hat die Weserlandschaft. *Die Weser*, **61**(2), 28–45.

Gewässergütekarte 1985. *Gewässergütekarte der Bundesrepublik Deutschland*, Länderarbeitsgemeinschaft Wasser, Stuttgart.

Grabemann, I., Krause, G. and Siedler, G. 1983. Langzeitige Änderungen des Salzgehaltes der Unterweser. *Dt. Hydrogr. Z.*, **36**, 61–77.

Hafenbauamt 1987. *Sedimentationsverhältnisse in Bremer Hafenbecken*, Hafenbauamt Bremen, unpubl.

Halsband, E. 1976. Untersuchungen über den Krankheitszustand der in der Weser lebenden Fische. *Infn. Fischw.*, **23**(10), 548–551.

Heinrich, T. and Mühlner, G. 1980. *Beurteilung der Röhrichtbestände an der Unterweser im Bereich des Landkreises Wesermarsch im Hinblick auf ihre Schutzwürdigkeit*, Brake, unpubl.

Herbst, V. 1982. Amphipoden in salzbelasteten niedersächischen Oberflächengewässern. *Gewässer und Abwässer*, **68/69**, 35–40.

Hesmer, H. and Schroeder, F.-G. 1963. Waldzusammensetzung und Waldbehandlung im Niedersächsischen Tiefland westlich der Weser und in der Münsterschen Bucht bis zum Ende des 18.Jahrhunderts. *Decheniana—Beihefte*, **11**.

Jathe, B. 1985. *Das Biomonitoring von Schwermetallen in der Unterweser mit Hilfe der Grünalge Blidingia minima*, Diplomarbeit FB 2, Universität Bremen.

Johnston, S. A. 1981. Estuarine dredge and fill activities: a review of impacts. *Environmental Management*, **5**(5), 427–440.

Keller, H. 1901. *Weser und Ems. Ihre Stromgebiete und ihre wichtigsten Nebenflüsse*, Berlin.

Krause, A. and Schröder, L. 1979. Vegetationskarte der Bundesrepublik Deutschland, Potentielle natürliche Vegetation. *Schriftenreihe für Vegetationskunde*, **14**.

LWA 1980. *Fließgewässer, Richtlinien für naturnahen Ausbau und Unterhaltung*, Landesamt für Wasser und Abfall Nordrhein Westfalen, Düsseldorf.

LWA 1986. *Bäche und Flüsse naturnah—Verbesserung des ökologischen Zustandes von Fließgewässern*, Landesamt für Wasser und Abfall Nordrhein Westfalen, LWA Schriftenreihe 43, Düsseldorf.

Loebe, K. 1960. *Unternehmen Mittelweser—Ein Werk für Bremens Zukunft*, Bremen.

Loebe, K. 1968. *Das Weserbuch*, Hameln.

Meijering, M. P. D. 1980. Die Werra—Lebensraum für Meerestiere? *Die Weser*, **54**(1), 7–10.

Nattermann, E. 1941. Das Sinken der Wasserstände der Weser und ihr Zusammenhang mit der Auelehmbildung des Wesertales. *Archiv für Landes- und Volkskunde Niedersachsen*, **1941**(9), 288–309.
Neumann, H. and Schöneborn, C. 1982. Die Weser—Belastung, Güte, Überwachung. *Berichte der ATV*, **33**, 725–760.
Nolte, W. 1974. *Gutachten über den Minderertrag der Braker Berufsfischer durch den 8.7 m Ausbau der Unterweser*, Bremerhaven.
Nolte, W. 1976. Die Küstenfischerei in Niedersachsen. *Forschungen zur Niedersächsischen Landeskunde, Band*, **105**.
Protokolle 1838, 1865, 1870. *Protokolle der technischen Kommission zur Untersuchung des Weserstromes*, Staatsarchiv Bremen, unpubl.
Rohde, H. 1970. Die Entwicklung der Wasserstraßen im Bereich der deutschen Nordseeküste. *Die Küste*, **20**, 1–44.
Schirmer, M., Hackstein, E. and Liebsch, H. 1983. Kritische Belastung des Sauerstoffhaushaltes der Unterweser durch kommunale und industrielle Abwässer. *Verh. Ges. ökol.*, **10**, 337–344.
Schmitz, W. 1960. Die Einbürgerung von Gammarus tigrinus SEXTON auf dem europäischen Kontinent. *Arch. Hydrobiol*, **57**, 223–225.
Schrader, T. 1941. Fischereibiologische Untersuchungen im Wesergebiet. Teil 2: Hydrographie, Biologie und Fischerei der Unter—und Außenweser. *Zeitschrift für Fischerei und deren Hilfswissenschaften*, **39**, 528–692.
Schuchardt, B., Busch, D., Schirmer, M., Schröder, K. 1985. Die aus Fangstatistiken rekonstruierbare Bestandsentwicklung der Fischfauna der Unterweser seit 1891: Ein Indikator für Störungen des ökosystems. *Natur und Landschaft*, **60**(11), 441–444.
Schuchardt, B., Beckmann, M., Knust, R. and Schirmer, M. 1984. Eulitorale Uferstrukturen an der Unterweser. *Drosera*, **2.**, 83–90.
Schuchardt, B., Haesloop, U., Schirmer, M. 1987. Zur Verbreitung von Gammarus tigrinus SEXTON in der Unterweser. *Drosera*, **87**(2), 129–134.
Strautz, W. 1963. Auelehmbildung und—gliederung im Weser—und Leinetal mit verschiedenen Zeitbestimmungen aus dem Flußgebiet der Elbe. *Beiträge zur Landespflege*, **1**, 273–314.
Tesch, F. W. and Fries, G. 1963. Die Auswirkungen des eingebürgerten Flohkrebses Gammarus tigrinus auf Fischbestand und Fischerei in der Weser. *Der Fischwirt*, **13**, 319–326.
Tüxen, R. and Lohmeyer, W. 1950. Bemerkenswerte Arten aus der Flora des mittleren Wesertales. *Jahresbericht der naturhistorischen Gesellschaft zu Hanover*, **100**, 53–75.
Wasserhaushaltsgesetz 1976. Gesetz zur Ordnung des Wasserhaushaltes (WHG) i.d.F.der Bekanntmachung vom 16.10.1976 (BGBI.I.S.3017) geändert durch Gesetz vom 14.12.1976 (BGBI.I.S.3341).
Weserabflussfibel 1983. *Weserabflußfibel*, Robin Wood e.v. Bremen.
Weser-Ems-Wasserstrassenbeirat 1927. Denkschrift über die Kanalisierung der Mittelweser von Minden bis Bremen. *Weser-Ems-Wasserstrassenbeirat Heft*, **19**.
Wärmelastplan 1974. *Wärmelastplan ARGE Weser*, Bremen.
Weserlastplan 1982. *Weserlastplan ARGE Weser*, Bremen.
Wöbken, K. 1973. Beitrag zu Fragen der Wassergüte der Unterweser. *Wasser und Boden*, **9**, 281–285.
Wöbken, K. and Kunz, H. 1980. Beitrag zu Wassergütefragen der Unterweser. *Wasser und Boden*, **8**, 372–377.
Ziemann, H. 1967. Die Wirkung der Kaliabwässer auf die Flora und Fauna der Gewässer unter besonderer Berücksichtigung der Werra und Wiper. *Fortschritte der Wasserchemie*, **7**, 50–80.
Zietz, U. 1975. Probleme der Gewässergüte in der Unterweser. *Deutsche Gewässerkundliche Mitteilungen, Sonderheft*, 87–94.

Historical Change of Large Alluvial Rivers: Western Europe
Edited by G. E. Petts
©1989 John Wiley & Sons Ltd

18

Ecological Changes of the French Upper Rhône River Since 1750

A. L. ROUX

Laboratoire d' Ecologie des Eaux Douces, U.A. CNRS 367, Université Claude Bernard Lyon 1, F-69622 Villeurbanne Cedex, France

J. P. BRAVARD

Laboratoire de Geographie Rhodanienne, U.A. CNRS 260, Université Jean Moulin Lyon 3, F-69239 Lyon Cedex, France

C. AMOROS

Laboratoire d'Ecologie des Eaux Douces, U.A. CNRS 367, Université Claude Bernard Lyon 1, F-69622 Villeurbanne Cedex, France

and

G. PAUTOU

Laboratoire de Biologie Végétale, U.A. CNRS 243, Université de Grenoble I, F-38402 Saint Martin d'Hères, France

INTRODUCTION

The floristic and faunistic composition of its populations demonstrate that the Rhône River can be designated a typical medio-European river, similar to the Rhine and the Danube. Like the Rhine, the Rhône flows through overdeepened glacial basins, the Rhine through Lake Constance and the Rhône through Lake Leman. In each case, deposition in the lake deprives the downstream river of its gravel load. But in contrast to the Rhine, the Rhône River undergoes a kind of rejuvenation due to the confluence with a major alpine tributary, the Arve River, just downstream from the outlet of Lake Leman.

Between the Arve confluence and Lyon, a reach of some 200 km, the diversity of the geological structures crossed by the river, and the effects of repeated glaciation and deglaciation, have created a wide diversity of geomorphological patterns. These include

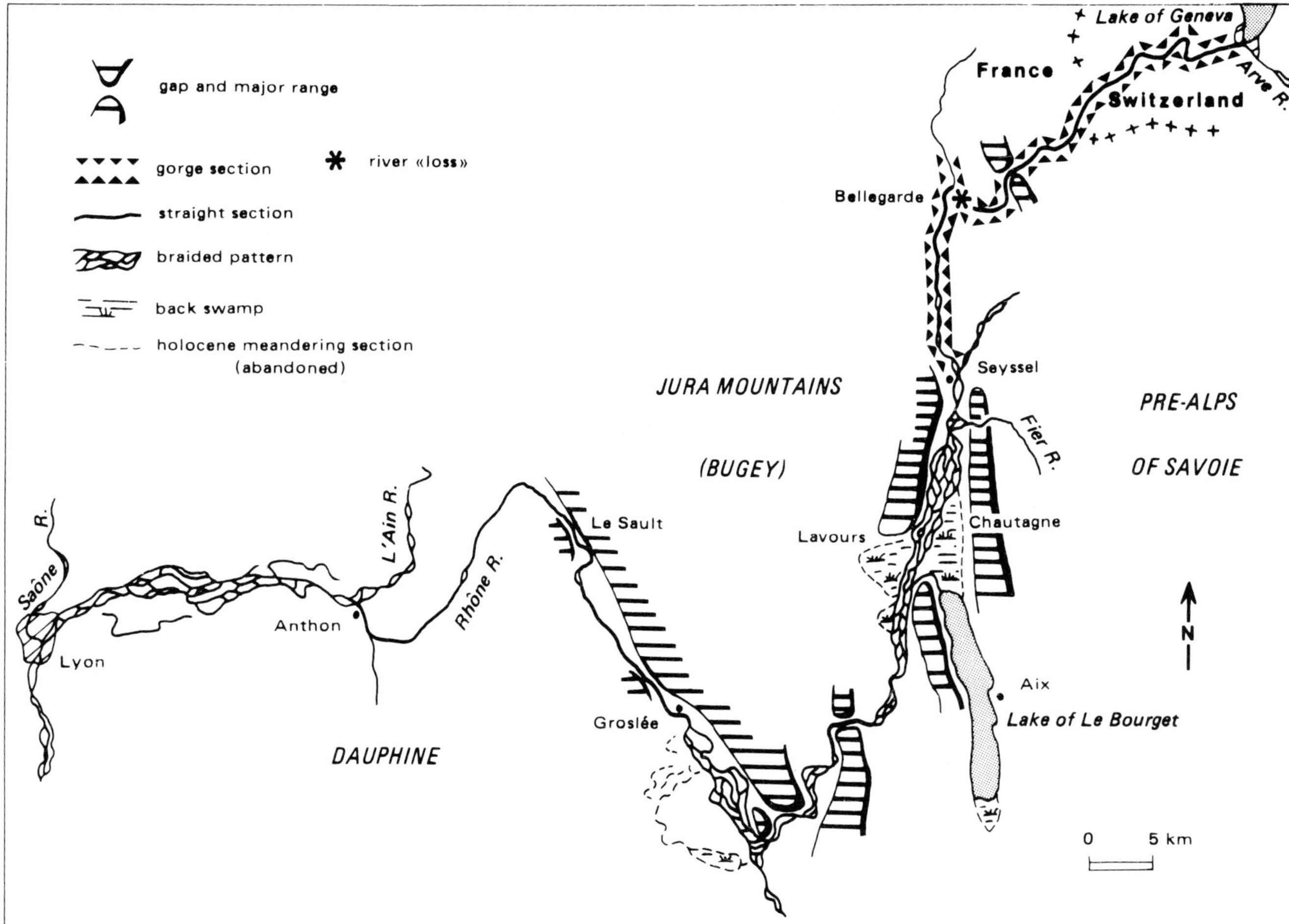

FIGURE 1. The geomorphological characteristics of the Upper Rhône River *c.* 1750

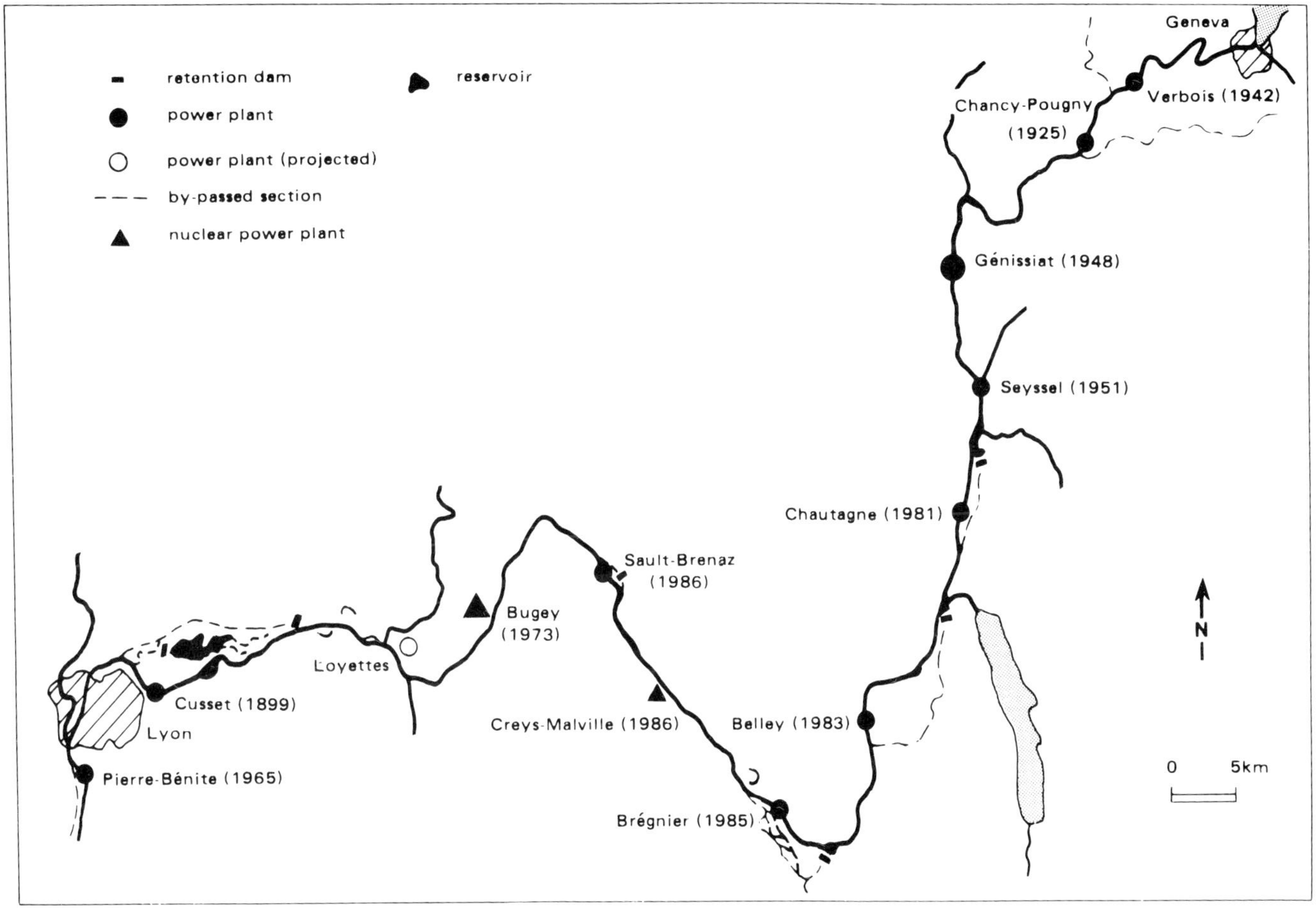

FIGURE 2. The developments on the Rhône River in 1988

gorge sections, braided and anastomosed sections, straight sections, and meandering sections (Figure 1). Furthermore, this diversity can be observed not only in space but also in time; individual sections undergoing a succession of changes during the historical period. Of course the diversity of ecological functioning and evolution of the alluvial plain is correlated with the diversity of structures.

Excluding those areas in which different geomorphological patterns succeeded one another, the complexity of functioning is probably greatest in the braided and anastomosed sections. Between the main channel and the alluvial forests, secondary arms, former channels (some of them completely isolated and others periodically connected with the present river), backwaters and wetlands, can be observed. Furthermore these systems have been used by Man for a long time: by peasant societies before 1750, for navigation during the last century (embankments were necessary in the braided sections), and for hydroelectric power production during the present century (Figure 2). Of course the nature and the intensity of human impacts were very varied, according to the different uses. This chapter uses a range of data including cartographic (Bravard and Bethemont, this volume), palaeoecological (Amoros and Van Urk, this volume) faunistic and floristic evidence from relic channels, and documentary sources, to reconstruct the history of fluvial changes in the French Upper Rhône since 1750.

THE GEOMORPHIC PATTERNS OF RHÔNE RIVER *c.* 1750

The Upper Rhône River, from Lake Geneva downstream to Lyon, is 192 km long, and flows on the western foreland of the Alps. Deprived of its sediment load after leaving Lake Geneva, the Rhône River is supplied with sediments mainly by its alpine tributaries, namely the Arve River immediately downstream from Geneva and the Fier River 45 kms further downstream. After the retreat of the Wurm glacier (i.e. *c.* 18 000 years BP) the river incised to create gorge sections and filled in former lakes, which were found in the overdeepened areas with gravels and sands. The Rhône River prograded in these elongated basins and progressively built up a succession of alluvial plains which aggraded during the Holocene. However, downstream of the Wurmian outer moraine at Anthon (Figure 1), the Rhône River dissected the fluvioglacial deposits. This contrasting evolution occurred because the coarse bedload was stored in the upstream plains. Nevertheless some geomorphic changes were induced by the dynamics of the Ain River, the main tributary upstream of Lyon. Thus, in the year 1750 the Rhône River displayed a wide variety of geomorphic pattern, reflecting its Pleistocene and Holocene history. Four sections can be identified.

First, the upstream course between Geneva and Seyssel is a gorge section deeply incised into morainic deposits and underlying gritstones and limestones. Here, the river either appeared as a torrent with numerous rapids or disappeared locally into the limestone strata.

Secondly, downstream from Seyssel the dominant geomorphic pattern in the 1750s was active braiding created by the interaction of the coarse bedload of the river and its tributaries, the locally steep valley slope averaging 0.007 to 0.001, and the widely varying flow regime ($Qm = 450\ m^3\ s^{-1}$, $Q100 = 4500\ m^3\ s^{-1}$ at Lyon) with summer high flow and winter and spring spates. Braided channels characterized three distinct sections: (1) between Seyssel and Yenne where a Holocene alluvial fan has continuously aggraded; (2) between Le Guiers River confluence and Groslée, due to a natural capture which shortened and steepened the river course, the pattern of which was meandering during Roman Times, and

(3) downstream from the Ain River confluence where the dissection of the glacial outwash deposits supplied a large amount of coarse load.

Thirdly, on the margins of the aggrading braided belt, two large peaty backswamps, at Chautagne and Lavours, developed during the Holocene. During the 18th century they were inundated for several weeks each year, namely in summer, the best period for the growth of the marshy vegetation.

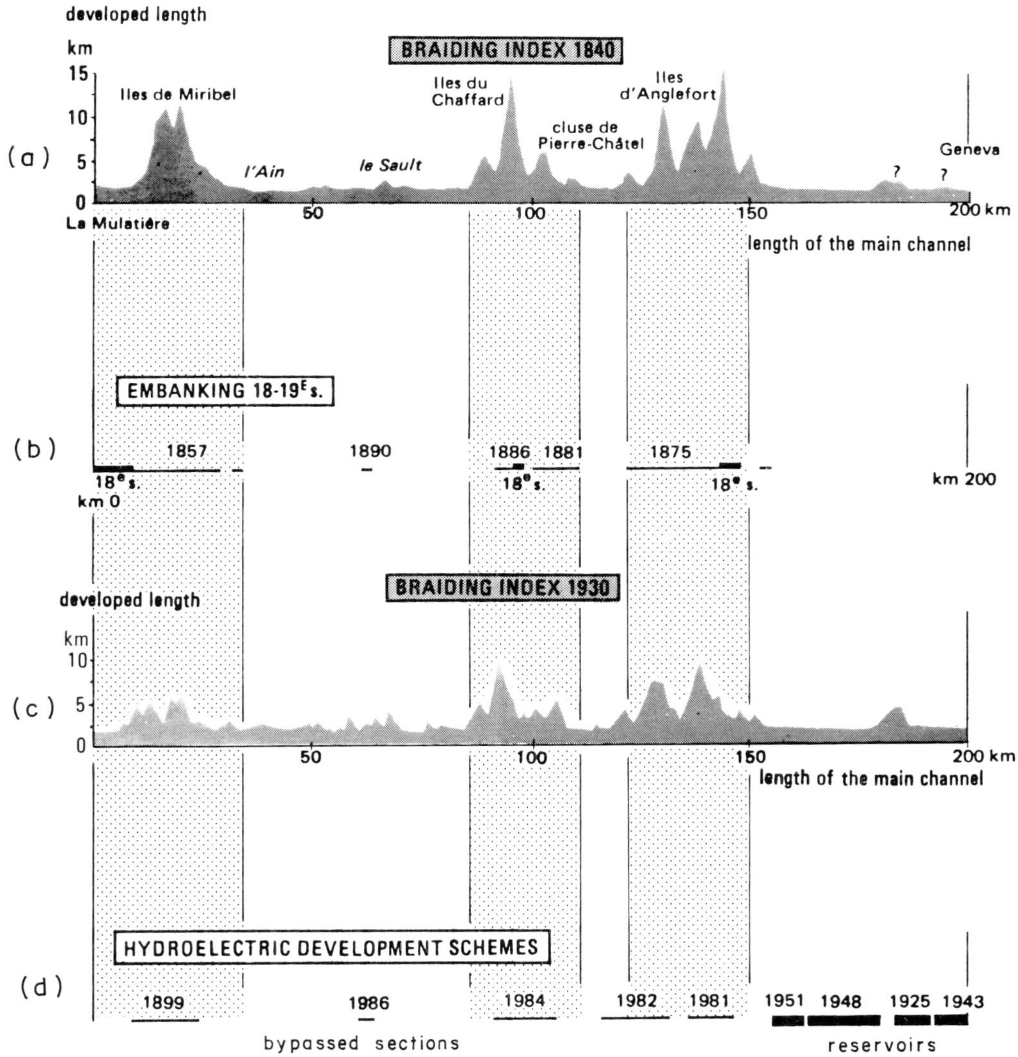

FIGURE 3. Changes of the braided sections resulting from embankments and the development of hydroelectric schemes

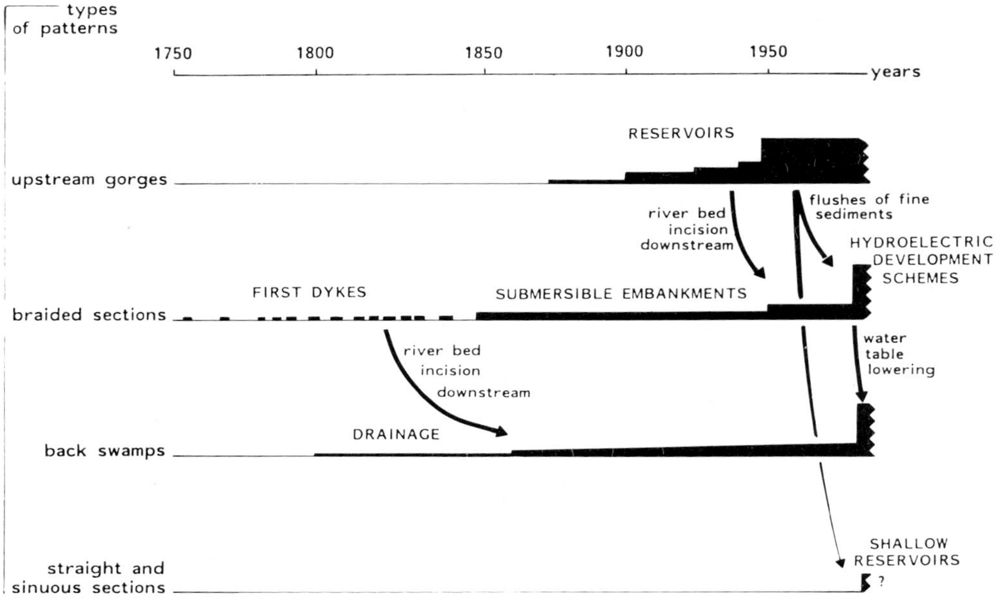

FIGURE 4. The historical development of the different geomorphological sections. The thickness of the black horizontal lines is correlated with the intensity of the changes; interactions between different types of patterns are suggested

Fourthly, between Groslée and Anthon, the river had a straight and, for most of time, unique channel. A few oxbow lakes are preserved in some short meandering sections but the spatial stability of the river during historic times has produced only a few dead arms. Some of the dead arms date back to Roman times in an abandoned valley, south of Groslée; some other ones date back to the modern times (17th and 18th centuries) on the margins of the present course upstream of Groslée and downstream from Anthon.

A major question is to know whether this type of river was 'natural' in 1750. The situation must be considered at two spatial scales. First, at that time, the valley bottom was almost completely cleared because of the demand for tilled land. The backswamps and the gravel bars of the braided belt were devoted to extensive grazing, and the woody islands were clear cut every five years for supplying the riverine populations, the tile crafts and the town of Lyon with fuel. Then, the former rural societies were all the more unable to control the river dynamics: their activities probably enhanced the tendancy to wandering by weakening the banks and decreasing the natural rugosity of the bed. Secondly, the dominant braided pattern may have been influenced or controlled by changes within the drainage area. The climatic deterioration during the Little Ice Age, combined with extensive land clearance in the alpine ranges, changed the two primary controlling variables, namely the discharge and the sediment load.

In the long-term, periods of aggradation and degradation, during the Holocene caused the metamorphosis of channel pattern, evidence of which is to be found along the Rhône River and some of its tributaries. These river responses include the influence of Man through a long history, at least since the Neolithic Times. Indeed, Man is possibly responsible for an increasing tendancy to braiding during the Modern Times.

Thus, the floodplain and channel of the French Upper Rhône, at least, cannot be said to be truly 'natural' even as early as 1750. The geomorphic patterns and the biotopes were clearly influenced by Man, overfishing was already a problem, and the biological communities were affected. Nevertheless all these changes were probably reversible because Man had not yet changed the ability of the fluvial system to recover or to experience geomorphic responses to changes in the watershed.

Major geomorphic and ecological changes date back to the end of the 18th century and affected mainly the braided sections. Embankments were built first (Figure 3b). Except for one of the first European diversion dams, built in 1899 in the vicinity of Lyon, reservoirs were built in the gorges during the first part of the 20th century. More recently, diversion dams have been constructed in the alluvial plains (Figures 2 and 3d). The braided sections have experienced the longest and most diversified development (Figure 4). On their margin the backswamps were affected by some impacts. In contrast, changes in the gorges and, overall, in the straight and sinuous sections are quite recent in an historical timescale.

THE CHANGES IN THE BRAIDED SECTIONS

The ecological situation before 1880

Until the end of the 18th century, due to technical limitations, Man was unable to control this type of wandering gravel bed river, which was characterized by rapid erosion, the construction of broad gravel bars, and the extensive deposition of coarse sand over vegetated islands and the tilled floodplain. At this time the first engineering works were groynes made of wood and limestone blocks in order to divert the flow; the first embankments protected the bridges and the small riverine towns devoted to shipping, such as Seyssel at the head of the navigable river. A well documented section is that located at the Southern end of the Jura Mountains, upstream of Groslée (Figure 1). Here the geomorphic pattern includes an aggrading braided belt and lateral anastomosed arms on the margins of the floodplain, and exemplifies the ecological conditions of the braided sections.

Running water ecosystems of the braided belt

Unfortunately no data are available concerning the macroinvertebrates during this period, but there exist some very rare relic braided sections that have been little disturbed by the embankments of the last century. By referring to the results of investigations performed in these areas before the completion of the recent hydroelectric development schemes, we can propose that the benthic macroinvertebrate populations were probably more rithronic than potamic. The dominant groups were Gammarids, Plecoptera and Ephemeroptera (*Ecdyonurus*, *Heptagenia*, *Baetis*). Fish populations were dominated by Salmonids, trout (*Salmo trutta*) and grayling (*Thymallus thymallus*), accompanied by some small species such as the loach (*Noemacheilus barbatulus*) and the minnow (*Phoxinus phoxinus*). Due to the continuous shifting of the arms in such a braided pattern, all have almost the same fauna; only the anastomosed arms on the margin of the floodplain were deeper and more

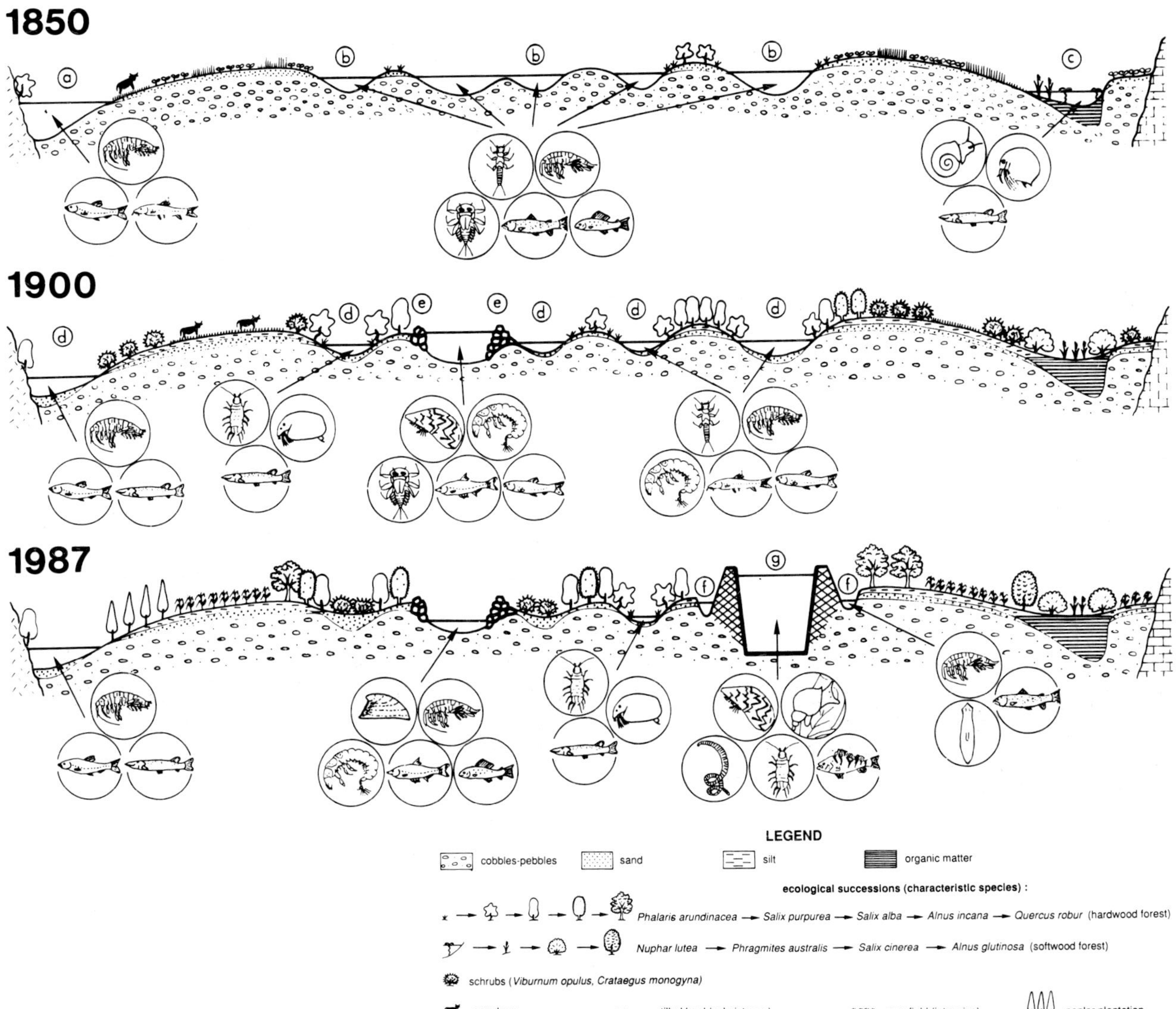
1850
a
b
b
b
c
1900
d
d
e
e
d
d
d
1987
g
f
f
LEGEND
cobbles-pebbles
sand
silt
organic matter
ecological successions (characteristic species) :
Phalaris arundinacea → Salix purpurea → Salix alba → Alnus incana → Quercus robur (hardwood forest)
Nuphar lutea → Phragmites australis → Salix cinerea → Alnus glutinosa (softwood forest)
schrubs (Viburnum opulus, Crataegus monogyna)
meadows
tilled land (subsistence)
corn field (intensive)
poplar plantation

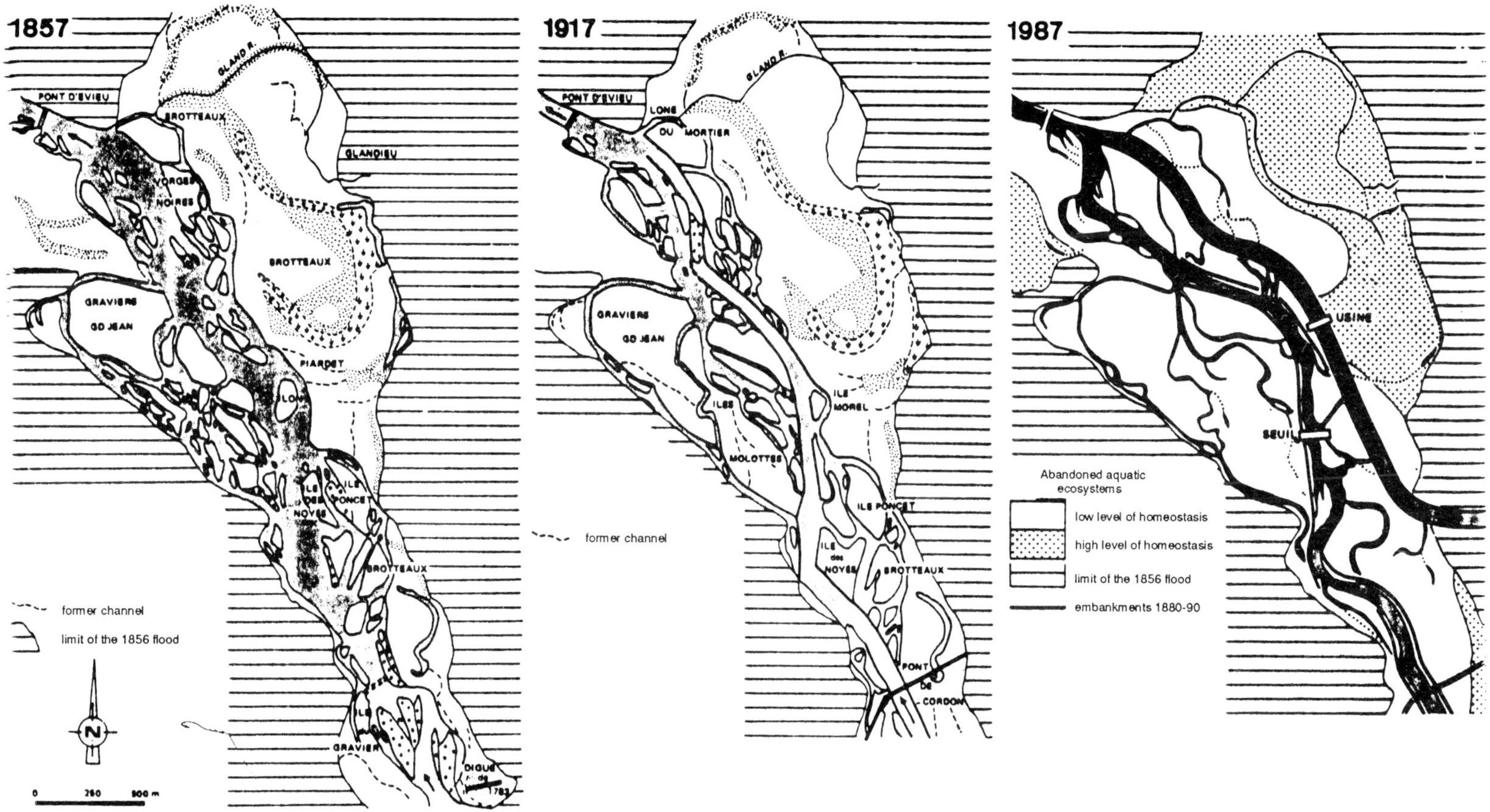

FIGURE 5. Schematic cross-sections of Brégnier–Cordon sector showing the numerous braided channels within the aggrading central area and the anastomosed channels in the lower lateral belts; in 1850 before the civil engineering works; in 1900, after the construction of the submersible embankments; and in 1987, after the hydroelectric equipment construction. (Some animals are plotted for just focusing on rough similarities or differences between aquatic communities. The composition of the populations and their changes are described with details in the text)

lentic and had probably predatory fishes such as the pike (*Esox lucius*) or the perch (*Perca fluviatilis*) (Figure 5).

Standing water ecosystems of the side-arms

The constantly shifting geometry of the braided channels and the destruction of the alluvial plugs at the ends of the side-arms, strongly limited the development of ecological successions within lentic and standing waters. When they occurred, the successions were very short because the former braided channels were rapidly filled with sands (Jacquet, 1986); in such cases, terrestrialization was probably driven by this allogenic process and consequently eutrophication of the dead arms did not occur.

In contrast, the anastomosed side-arms, being more sinuous and with a lower slope, had slower current velocities and their morphology was relatively stable. When such side-arms were cut off, their stability permitted the complete development of ecological successions. After the natural abandonment of the side-arm by the running water of the river, the resulting standing water body was rapidly colonized by a planktonic community characterized by *Bosmina longirostris* (Crustacea, Cladocera). While eutrophication proceeded, the extension of an aquatic plant community (*Myriophyllo–Nupharetum* type) reduced progressively the open-water area. Then, as the organic matter content increased in the superficial sediments, the littoral plant communities, probably dominated by *Carex elata*, established in the shallow areas; with this colonization began the terrestrialization process. The next phase started with the development of *Salix cinerea* bushes, which accelerated the terrestrialization process by increasing the production and accumulation of biomass and by evapotranspiration. The ultimate stage of this ecological succession was a forest community associated with *Alnus glutinosa* and *Fraxinus excelsior*. Whereas this succession was mainly driven by autogenic processes, alluvial deposition during spates accelerated the terrestrialization process. The rate of silting decreased in relation to the distance between the abandoned anastomosed channel and the main stream and consequently the ecological successions evolved at a variety of rates (Bravard *et al.*, 1986).

Thus, prior to embankment during the 1880s, the gradient of rates of successional development and the cooccurrence of former channels abandoned at different times, gave the anastomosed belt a higher diversity of habitats and communities than in the braided area (Figure 5).

Terrestrial biotopes of the floodplain

The fast lateral shifting of the main river channel was a major cause of instability and rejuvenation of the terrestrial biotopes in the vicinity of the braided belt. These ecological conditions fostered the establishment of pioneer plant populations of *Melilotus alba* and *Epilobium dodonaei* and shrubs of *Salix eleagnos*, *S. purpurea*, *Populus nigra* associated with very porous soils.

The flatness of the gravel bars, the speed and the height of flood discharges, and the deficiency of silty sediments, inhibited long plant successions and promoted a complex mosaic and architecture of willow thickets. The populations of *Salix eleagnos*, which characterize large sandy alluvial areas of the present Ain and Rhône Rivers confluence, and *S. daphnoides* were probably more widespread than the populations of *S. alba* and

S. triandra. However, the latter species experienced rapid expansion as soon as civil engineering works allowed the sedimentation of fine sands and silt. On the margins of the floodplain a few hardwood forest of *Fraxinus excelsior*, *Quercus robur* and *Acer pseudoplatanus*, escaped land clearances for pasture, subsistance agriculture and firewood cutting.

From 1880 to 1980: the impact of low-level embankments

Along the Rhône River, engineering works began at the end of 1840 when Lyon tried to promote steam shipping on the Rhône and Saône Rivers upstream of the town. It was then necessary to build stone dykes in order to concentrate low flows to create a unique deep channel. On account of the lack of public funds many dykes were built. These low-level dykes blocked only the upstream end of side channels at low flow; at high discharges they were submerged as water overflowed the dykes rejuvenating the network of minor channels (Bravard, 1987). In the area of Brégnier–Cordon this type of embankment was constructed during the 1880s and their impact lasted until the recent construction of the hydroelectric development schemes. Once again the biological changes have been inferred by analogy to present day or recently existing channels not affected by embanking and by palaeoecological analyses.

Running water ecosystems in the embanked channel and in the side-arms

The management of low flow water levels in braided sectors by using rock-fill low-level embankments created new running water habitats: the embanked channel (navigation channel) and the blocked secondary arms. The latter received seepage water through the dykes during periods of low flow, in addition to flow over these dykes during floods and spates (Figure 5).

In the embanked channel, the flow velocities were fast, and consequently rheophilic and potamic species dominated the benthic macroinvertebrate populations. Among them, must be noticed the flat Ephemeroptera larvae of *Heptagenia sulphurea* and *Rhithrogena*, and Trichoptera larvae of several species of *Hydropsyche*. The stability of the embankments allowed the development of populations of the Zebra Mussel, *Dreissena polymorpha*, all along the dykes below the lowest water level. The numerous interstices between the blocks provided shelters for Gammarids which form the major part of invertebrate biomass. Only rheophilic fishes were able to withstand the strength of the current; four cyprinid species were dominant: the nase (*Chondrostoma nasus*), the dace (*Leuciscus leuciscus*), the chub (*Leuciscus cephalus*) and the barbel (*Barbus barbus*) were present in the vicinity of the dykes where burbots (*Lota lota*) could be caught too (Nelva *et al.*, 1981).

The composition of the macroinvertebrate and fish fauna of the blocked secondary arms was dominated by nearly the same species with some modifications. Among the macroinvertebrates, *Rhithrogena* was rare and *Dreissena* restricted to the upstream dyke of the arm. On the other hand, the diversity was higher, due mainly to the wealth of Plecoptera, Ephemeroptera, and Coleoptera larvae. The fauna was more rhithronic in these blocked secondary arms than in the embanked main channel (Perrin, 1978). Concerning the fish fauna the same situation could be observed: in addition to the four dominant species present in the main channel, several smaller species were more frequent, namely the gudgeon

(*Gobio gobio*), the ruffe (*Acerina cernua*), the blageon (*Leuciscus soufia*), the bleak (*Alburnus alburnus*), the streambleak (*Alburnoides bipunctatus*) and the minnow (*Phoxinus phoxinus*).

From the end of the second world war until 1980, the physicochemical and biological water quality of the Rhône River was periodically affected by the releases from the reservoir built in the upstream gorge section (see part 3.2). Every two or three years, sediments, inorganic or organic matter and pollutants, which were trapped behind the dams, were suddenly flushed downstream. This induced a physicochemical pollution characterized by an increase in suspended matter, an increase in ammonia content, which happened to reach lethal values for fishes, and a depletion of dissolved oxygen content (Roux, 1984). Consequently, siltation was important, namely for the secondary arms situated on the margins of the floodplain, and this accelerated their 'terrestrialization'. The biological effects of these releases differ according to the animal group considered. Among invertebrates, the effects were more severe for organisms having an entirely aquatic life-cycle (e.g. Gammarids) than for those having an aerial phase (e.g. Insects). Nevertheless, the long term effects were small owing to the potential for quick recolonization due to the multiple reproduction cycles exhibited each year by most species (Tinchant, 1979). The long term impacts of these repeated flushings on fishes are more important. The young stages, larvae and juveniles, are more sensitive to pollution than the adults, and it was observed that all size-classes of Salmonids are more sensitive to pollution than Cyprinids. Furthermore, the disappearance of one age-class every three years is more harmful for short lifetime species (e.g. Salmonids) than for long lifetime ones (e.g. Cyprinids). This impact of the releases could explain why rheophilic Cyprinids became more and more abundant between 1950 to 1970 while the populations of Salmonids, namely the graylings, decreased in these braided sections during the same period (Roux, 1984). Furthermore it must be added that, until 1950, anadromous migratory species such as the shad (*Alosa*) were observed and caught regularly in the French Upper Rhône; after the completion of the hydroelectric development schemes on the Rhône downstream from Lyon between 1952 and 1975—most without fish ladders or with inefficient ones—these migratory species disappeared. The situation for the eel (*Anguilla anguilla*) is not exactly the same because fishermen are restocking this species. However, the composition of the populations in these braided sections has been markedly affected by the completion of the hydroelectric stations on the Upper Rhône at the beginning of the nineteen-eighties (see the section *Changes since 1980*).

From the standing water of the dead-arms towards the terrestrial communities

The embankments prevented the erosion of the floodplain without affecting sedimentation processes during periods when floods overflowed the dykes. Hence, the man-induced stability of the braided side-arm allowed the development of ecological successions. During floods and spates, suspended sediment was carried into these arms and deposited behind the dykes. Thereby particular water dynamics were created in these arms i.e. the alternation of running water during high water and semilotic to standing water during low water (Perrin, 1978; Carrel, 1986). These habitats suited the lotic communities associated with *Hydropsychae* larvae (Trichoptera) and planktonic communities characterized by *Bosmina coregoni longispina* (Crustacea, Cladocera); indicators of oligotrophic to mesotrophic waterbodies. The fish populations of these semilotic habitats combined species from the lotic main stream (*Leuciscus cephalus*, *Alburnus alburnus*, for example) with more lentic species (*Perca fluviatilis*, *Esox lucius*).

The silt deposited behind the dykes made the upstream alluvial plug relatively impermeable and consequently changed the flow dynamics. At this stage of the ecological succession, surface water flowed only during spates while in normal conditions the water of the dead-arm was slowly renewed by seepage supply through the dykes, and by river water flowing backward from the downstream end still connected to the main channel (Juget *et al.*, 1979). This successional stage is characterized by *Orectochilus villosus* adults (Insects, Coleoptera), *Bosmina longirostris* (Crustacea, Cladocera) and *Callitriche platicarpa* (aquatic plant) (Amoros, 1980; Castella *et al.*, 1984; Richardot-Coulet *et al.*, 1987). As a result of these ecosystem changes, the lotic fishes were decreasing slightly while the population of standing-water species was expanding (*Tinca tinca*, *Cyprinus carpio*).

The disturbances that occurred at the downstream connection with the main channel induced alluvial deposition and the formation of alluvial plugs. Eventually these plugs completely closed the former braided channels and thereby accelerated the eutrophication and the terrestrialization that had begun during the previous successional stage. Colonization of the isolated dead-arms by marsh communities associated with *Carex elata* and *Senecio paludosus*, later followed by a forest community with *Ulmus minor* and *Lysimachia nummularia*, constituted two of the major subsequent steps in the succession.

The analyses of sediment cores have shown that such a succession was rarely direct or continuous; an oscillating pattern often resulted from the acceleration of successional processes by siltation and retrogression by scouring during spates, and sometimes by the destruction of the alluvial plugs (Jacquet, 1986). The development span of the complete succession might extend from some decades to more than one century depending on the initial depth of the former channel and on the distance to the main channel. About fifty years after the completion of the 1880s river-training program, the braiding index (defined as the developed length of flowing channels per km of river section, at low flow) had markedly decreased; the change is even more obvious in the section called 'Iles de Miribel' which has been embanked between 1848 and 1857 (Figure 2).

In the former anastomosed channels the successions initiated before the embankment underwent slow change (Jacquet *et al.*, 1984). Within the Brégnier–Cordon sector (str. km 98–91) it has been demonstrated that the aquatic ecosystem development was slowed down by raising both the water level and the water table (more than 1 m from 1856 to 1947). These changes occurred in response to the aggradation of the main channel and the water level elevation compensated more or less for the filling of the abandoned channels (Bravard *et al.*, 1986).

One century after the embankment of the river, the braided–anastomosed sector of Brégnier–Cordon supported a high heterogeneity of habitats and communities (Pautou, 1975; Roux, 1982; Castella *et al.*, 1984). The dyke construction had initiated new ecological successions in the blocked braided side-arms. Moreover, the embankment had created a new spatial organisation of alluvial fluxes and hence a diversity of silting conditions. This heterogeneity played an important role in the regulation and the productivity of the whole river hydrosystem. For instance, the semilotic and stagnant dead-arms still connected to the main channel functioned as spawning areas and nurseries for main stream fishes as well as refuges during spates or pollution incidents (Amoros and Roux, 1987; Copp, 1987). Nevertheless, this heterogeneity had to decrease because the ecological successions were

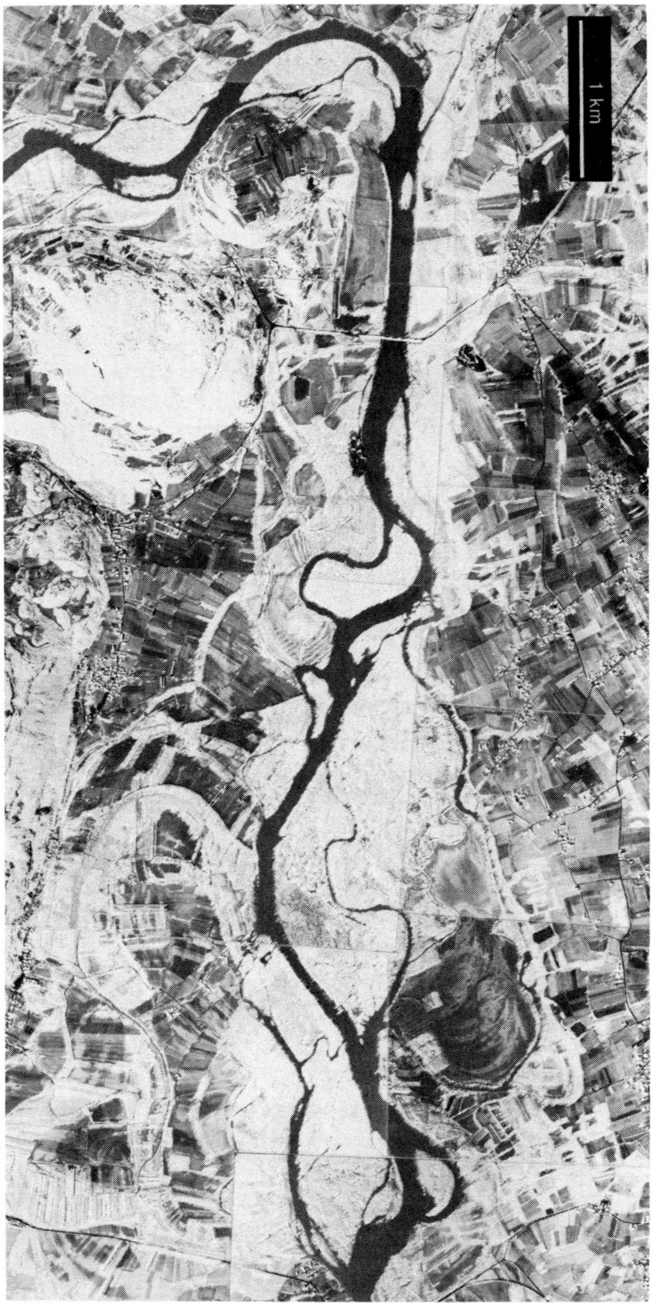
1 km

FIGURE 6. Aerial views of the Brégnier–Cordon area before and during the construction of the hydropower development scheme of Brégnier–Cordon

moving, more or less rapidly, towards the ultimate stages which correspond to terrestrial ecosystems; the embankments that instigated habitat diversity in the braided area, impeded the erosion outside the dykes and thereby the rejuvenation of the successions through the creation of new landforms.

The modification of river dynamics caused bedload deposition and connection of the lowest gravel bars, and increased the deposition of sand on the levées and the siltation in the swales and bottom flats. The areal accretion of islands and the aggradation of terrestrial landforms enhanced the heterogeneity of the floodplain through a spatial gradient of hydrological and pedological conditions. The most evident changes are the following: (1) The contraction of the populations of *Salix* (*S. triandra*, *S. viminalis*, *S. alba*) on the stabilized deposits; (2) the development of hard wood forests (*Fraxinus excelsior* and, to a lesser extent *Quercus robur*) and the development of the populations of *Alnus incana* which are now the most common; and (3) the initiation of new populations of mixed *Alnus incana* and *A. glutinosa*, and *Salix cinerea* in the remote former braided arms. Since the 1920s the decline of agriculture allowed the development of shrubs followed by hardwood forests (Bravard *et al.*, 1986; Pautou *et al.*, 1985; Pautou and Bravard, 1982) (Figure 5).

Changes since 1980

Since 1980 the Compagnie Nationale du Rhône has completed the installation of hydroelectric plants along the Upper Rhône River. A set of diversion dams has been constructed in the former braided plains, namely Chautagne, Belley and Brégnier–Cordon (Figure 2). This type of development scheme diverts up to 700 $m^3 s^{-1}$ into a canal comprising a head race, the power plant and a tail race (Figures 6 and 7). The former river bed, called the by-passed section, receives a 'compensation flow' of 10 to 50 $m^3 s^{-1}$ to be compared with $Qm = 350\ m^3 s^{-1}$, and the flow during spates when the discharge exceeds

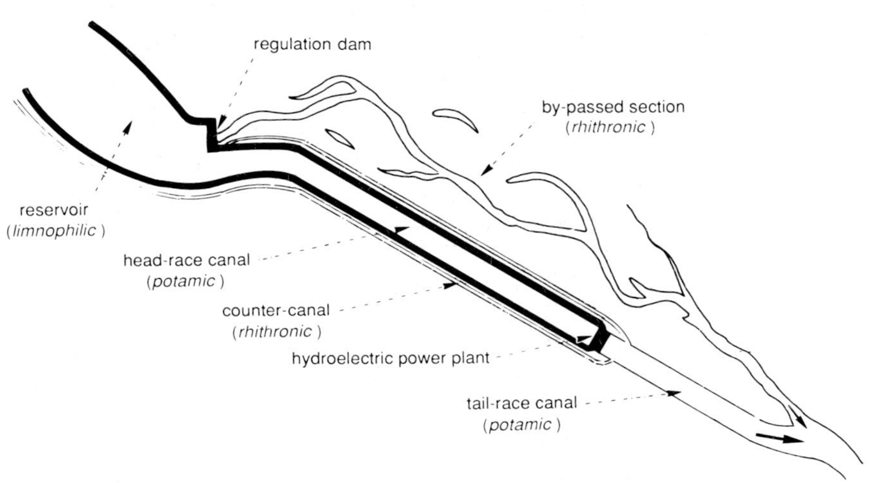

FIGURE 7. Components of the 'Compagnie Nationale du Rhône' hydropower development schemes, now a characteristic of the French Upper Rhône

700 $m^3 s^{-1}$. The Brégnier–Cordon scheme, completed in 1984, has these general characteristics but the minimum flow in the by-passed section has been raised to 80 $m^3 s^{-1}$ in winter and 150 $m^3 s^{-1}$ in summer, and is used also for electricity production at the regulation dam.

In the former braided area the new flow regime imposed a quasipermanent situation of low flow which shortens the period of submersion; the aeration of the soil encourages the mesophilous species to the detriment of mesohygrophilous and hygrophilous species. When not controlled by artificial weirs built in the former embanked main channel, the phreatic water levels as well as their amplitude of variation have been lowered by 1 to 1.5 m. Inputs of suspended matter have decreased because of reservoir trapping but significant sedimentation may occur during floods. These short duration floods are now quite unable to induce the lateral migration of channels and to regenerate the alluvial biotopes; previously, in the sections free of embankments, lateral erosion occurred when the discharge exceeded 400 $m^3 s^{-1}$ (Pautou, 1975).

The running water ecosystems

The condition of the running water ecosystems was drastically affected by the completion of the recent hydroelectric development schemes. In the former embanked channel and in the secondary arms, during most part of the year, the water level and the discharge are now reduced far below the previous lowest waters, but during floods and spates the whole discharge is diverted into these by-passed sections. In contrast, in the artificial, newly created habitats such as the canals and the reservoirs, the water level is nearly the same throughout the year with annually regulated discharge, but with daily level fluctuations (50 to 100 cm) due to the production of peak energy. Furthermore, outside the dykes along the reservoir and along the head-race canal, a counter-canal collects the seepage water. These counter-canals are similar to heavily regulated small lowland streams usually with dense macrophytes. It can be observed that the diversity of aquatic habitats has been enhanced, but the potamic character of the embanked stream is strongly modified (Figure 5).

In these by-passed sections the fauna is more rhithronic: for instance, genuine potamic invertebrate species such as *Heptagenia sulphurea* and some species of *Hydropsyche* are now declining. The same is happening to species that are very sensitive to large discharge variations, such as *Baetis fuscatus*. The by-passed sections have a shallow flow and the transparency of the water is improved. Under these conditions, periphytic algae produce thick sheets of food for scrapers, such as Mollusca (*Ancylus*, *Theodoxus* . . .) among invertebrates, or as the nase (*Chondrostoma nasus*) among fishes. The populations of these scraper species are increasing. As mentioned above, the fish populations are dominated by the nase, accompanied not only by the previously most common species (the chub, the dace and the barbel) but also by the grayling (*Thymallus thymallus*), the blageon (*Leuciscus soufia*) and smaller species such as the gudgeon (*Gobio gobio*), the stone loach (*Neomacheilus barbatulus*), the minnow (*Phoxinus phoxinus*) etc. . . The fish population is becoming similar to the populations of the blocked secondary arms, and probably to the populations of the braided sections before the embankments of the past century. These by-passed sections have experienced a kind of rejuvenation, equivalent to a reduction in stream order. It must be added that the new control policy involving triennial releases from

Verbois Reservoir, include measures to protect the by-passed sections. Indeed, during the releases, the diversion dam situated at the upstream end of the by-passed section is completely closed and the water loaded with suspended matter is conveyed through the artificial diversion canal.

The situation is quite different in the newly created habitats: the canals and impounded reaches have uniform banks, high discharges and large depths. The fauna of the impounded reaches is dominated by limnophilic species, oligochaetae, molluscs, asellids etc, but the fauna of the canals is more unstable due to the variations of the water level. The dominant species of fishes in the impounded reaches are the chub (*Leuciscus cephalus*), the roach (*Rutilus rutilus*), the perch (*Perca fluviatilis*) and the gudgeon (*Gobio gobio*); the species composition in the canals is rather similar but more unstable too. The counter-canals are comparable to small trout streams and shelter the corresponding fauna: Gammarids, Planarians, Baetidae, Chironomids and Hydroptilidae.

Wetlands and floodplain

The new hydrological conditions accelerate the terrestrialization of the former braided channels located downstream from the weirs. Meanwhile the former anastomosed channels should be less impacted because of their higher degree of homeostasis (Rostan *et al.*, 1987). The dead-arms have experienced a reduction in the areal extent of aquatic vegetation, namely the populations of *Myriophyllum spicatum*, of *Phragmites australis* and *Scirpus lacustris*, of the semiaquatic vegetation such as *Carex elata*, *C. riparia* and *Phalaris arundinacea* at the downstream end of the arms. In the canals the regulated discharge and the sedimentation of suspended matter could enhance the development of populations of *Ranunculus fluitans* in the biotopes where water depth is less than 1.50 m with sufficient light penetration.

These changes have reduced the areal extent of the gravelly and sandy biotopes necessary for photophilous species and the populations of *Salix*, *Populus nigra*, *P. alba* and *Robinia pseudacacia*. In the former braided arms the populations of *Alnus glutinosa* and *A. incana*, of *Fraxinus excelsior* and *Prunus padus* extended to the detriment of the populations of *Salix alba* and *S. cinerea* still in place; the hardwood forest extended to the detriment of *Alnus incana* and, on the highest islands, collinean species are infiltrating, namely: *Acer pseudoplatanus*, *A. platanoides*, *A. campestris*, *Juglans regia*, *Corylus avellana*, *Tilia sylvestris* and *Carpinus betulus*. On the most permeable deposits, species of the *Quercus pubescens* forests are spreading. However, this general trend could be slowed by the occurrence of large floods such as the 10-year flood of 1987 (Bravard *et al.*, 1986; Pautou *et al.*, 1985; Pautou, 1983).

In conclusion, it must be emphasized that the braided sections have been most severely affected by river management. First, the steep slopes and the gravelly substrates were very sensitive to flow concentration imposed by major dykes which promoted river bed degradation. In contrast, low-level embankments submerged during high flow have little effect on the long profiles but affect landform rejuvenation in the side-arms; they allow large areas to be inundated and intense siltation of the channels. Secondly the gravelly substrates of the floodplain, inherited from the Holocene periods of valley infill, remain very sensitive to any change of the phreatic level which may control severe changes of soils and terrestrial vegetation.

THE CHANGES IN THE OTHER REACHES OF THE VALLEY

The backswamps

The reclamation of the two large backswamps of Chautagne and Lavours (Figure 1) probably dates back to the middle ages. After cutting the swamp forest, the areas were used for grazing horses and cattle during autumn and winter. In the spring the cattle were moved off the marsh until August to let the grass grow. The hay was mown for making green manure, used to fertilize the vineyards of the surrounding foothills which provided the main financial resources to the population. Until the Revolution of 1789 the cattle were the property of landlords and monasteries (Bravard, 1981, 1986). These flat marshlands displayed a rich diversity of grasses reflecting the wide variety of hydrological and pedological conditions (Pautou, 1975).

The peasants discriminated between different kinds of hay devoted either to cattle, to horses, or to the production of green manure, according to the type of *Carex*. Although some landowners wished to improve the marshes and to convert them into tillage, peasants of these regions fiercely opposed the construction of any embankment which would impede the flooding of the plain. They decided to dig ditches in the early 1800s, not to drain the marshes, but to control the cattle and to share the land between the communities. At this time the marshes were places of stability because spring and summer floods spread silt over them and conditioned their fertility and the production of hay.

Curiously the first changes were induced by land abandonment at the end of the 19th century. After 1860 this region was stricken by a policy of state control upon the range forests, mainly due to the philloxera crisis in vineyards during the 1880s, then by the attraction of workers to industrializing cities, and finally by the death of hundreds of thousands of peasants during World War I. Depopulation began at the end of the 19th century, the need for green manure diminished and the marshes were progressively abandoned. This slow decay of the marshes probably concealed the impacts of the embankments performed in the braided belt of this valley since the end of the 18th century.

In Chautagne, the floodplain was isolated from the braided belt by a high embankment built in the 1780s; on the right bank a straight high embankment was completed in the 1850s for raising up a road and a railway (Figure 8). These engineering works isolated large areas on the sides of the valley and changed the conditions of flooding, namely they decreased their intensity because of induced backwater effects in the Lavours Marsh and in the Chautagne section. The exceptional floods of 1856, i.e. $\simeq Q100$, having destroyed part of the town of Lyon, an 1858 law forbade any high embankment upstream of this town (and any large French town). Nevertheless the upper section experienced bed degradation as a delayed impact of embanking, and so a slow progressive reduction of the intensity of floods. This long trend probably does not reflect any Man induced control upon the floods. The flow regime was changed by the town of Geneva at the outlet of Lake Geneva, after the successive improvements of La Machine dam during the 19th century, but the retention of the flood waters was not affected.

Anyway, the backswamps of the valley were progressively getting drier. Their productivity probably decreased but people did not lack hay any longer at the beginning of the 20th century and did not protest. Direct changes began when the first ditches were improved to create efficient drainage channels, namely in the 1930s when the State bought a large

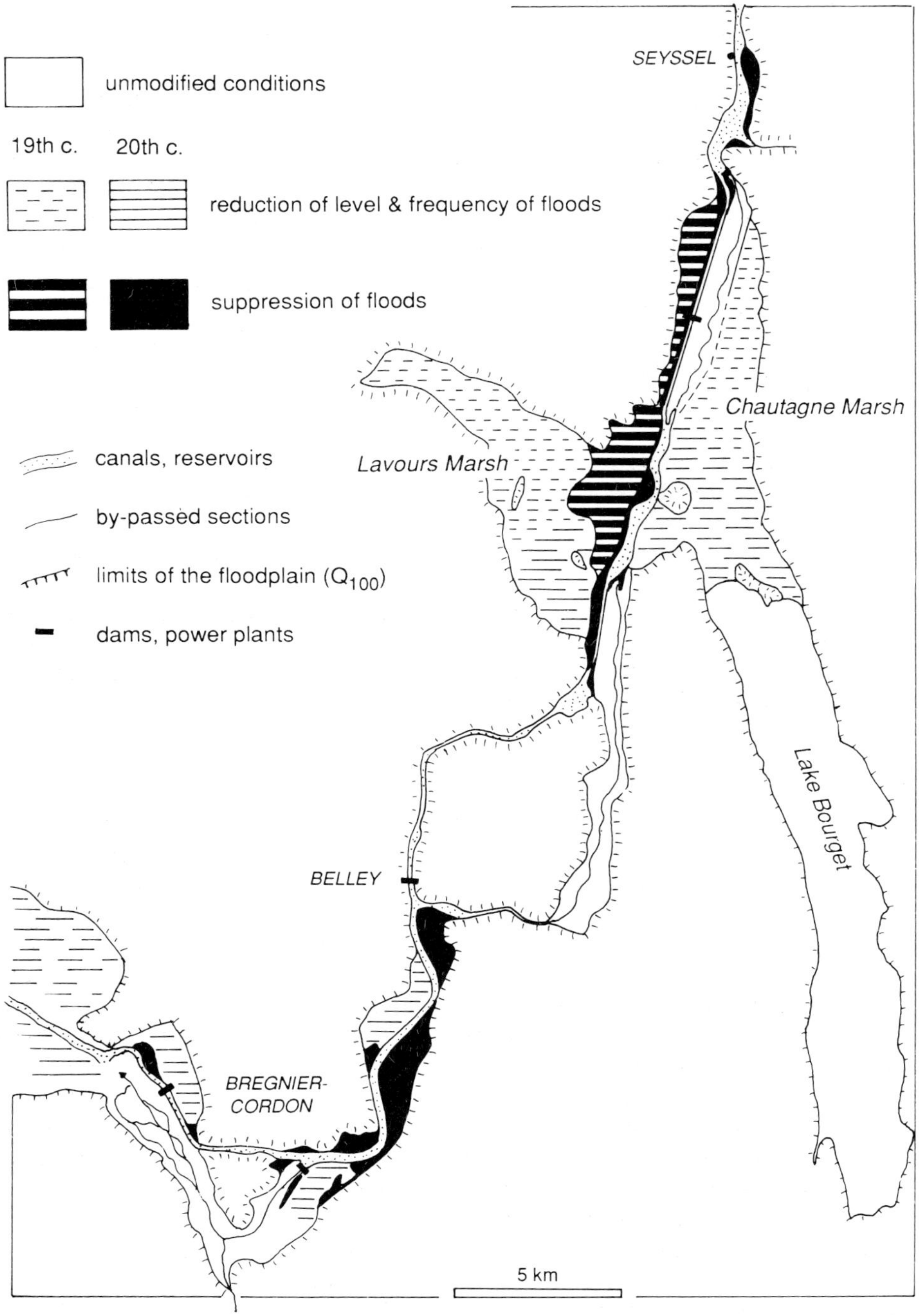

FIGURE 8. The contraction of the Rhône River floodplain since the 18th century

area of the Chautagne marsh in order to plant 700 ha of poplars, and at the end of the 1970s when farmers developed tillage for producing corn.

A second major period of change has begun in the 1980s, due to the harnessing of the River Rhône by the Compagnie Nationale du Rhône, although the 1858 law remains valid. New areas of the floodplain have been completely protected by the embankments of canals and reservoirs, the connection of the Lavours marsh with the river has decreased and many sections of the floodplain have been protected against small floods (*Q*1 to *Q*10) to decrease the risks for agricultural practices. However, within the former braided belts, the natural hydrological regime is maintained along the present bypassed sections (see Figure 8). The large marshes have become progressively perched and abandoned by local populations; since they are no longer integrated in the energetic balance of agricultural lands, the grass species are replaced by phanerophytes:

1. Some hygrophilous and mesohygrophilous species, such as *Schoenus nigricans*, *Molinia cerulea*, *Cladium mariscus*, *Agrostis stolonifera*, *Siglingia decumbens* are superseding species requiring better trophic conditions such as *Carex gracilis*, *C. acutiformis*, *Phragmites australis*.
2. Highly competitive species, such as *Phragmites australis*, *Cladium mariscus*, *Filipendula ulmaria*, *Solidago gigantea* are spreading quickly, to the detriment of rare species such as marsh orchids.
3. Shrubs and thickets of *Alnus glutinosa*, *Salix cinerea*, *Alnus frangula*, *Viburnum opulus*, *Betula verrucosa* and *Fraxinus excelsior* are expanding.

At the moment, new development trends are emerging, i.e. land regrouping for producing corn and oil-seeds, the delimitation of protected natural areas, and the restoration of extensive grazing is being tested through the introduction of hardy highland cattle from Scotland (Marjcrzak and Manneville, 1987).

The gorge sections

The section situated between Geneva and a few kilometres upstream from Seyssel has never been navigable due to natural conditions and was used only for logging until the end of the 19th century. As early as 1871 a dam was built at Bellegarde in order to produce mechanical energy. It initiated the progressive harnessing of the river within the gorges: this section provides a total fall of level of 110 m out of the 200 m for the whole Geneva–Lyon section and did not necessitate the drowning of fields and villages. The gorge section provided an ideal location for a chain of reservoirs: Verbois (Switzerland, completed in 1943), Chancy–Pougny (a Franco-Swiss dam, completed in 1925), Génissiat (1948) and Seyssel (1951), compensating for the artificial peak flows of the latter (Figure 2).

A short section has been kept free from impoundment at the upstream end of the reservoir of Génissiat; it exemplifies the classical changes of this type of environment. The decrease of the flow velocity stabilized the deposits and enabled the sedimentation of fines; the banks are affected by frequent and fast fluctuations of the water level due to the production of daily peak energy, by longer submersions during floods and by a rise of the phreatic level.

The main impacts upon the riverine vegetation are the following: (1) the establishment of mixed populations of *Salix cinerea* and *S. alba* on the deposits of silts; (2) the spatial extension of *Phalaris arundinacea* due to the marked variations of water level; (3) the extension of *Alnus incana* and *Fraxinus excelsior* but the spatial reduction of *Salix alba* and *Impatiens glandulifera*; and (4) the establishment of mixed populations of *Populus nigra*, *Pinus sylvestris* and *Quercus pubescens* on the highest sites (Girel and Doche, 1983).

Before impoundment, the Rhône River was a very swift stream, a typical mountain stream; the fish fauna was dominated by Salmonids, trout (*Salmo trutta*) and grayling (*Thymallus thymallus*), and rheophilic Cyprinids, such as the dace (*Leuciscus leuciscus*), and small species such as the sculpin (*Cottus gobio*) and the minnow (*Phoxinus phoxinus*). After completion of the dams, the fish fauna of the river corresponding to the reservoirs was completely modified, rheophilic species disappeared and now only limnophilic and ubiquitous species are present. The fish populations are dominated by roach (*Rutilus rutilus*), carp (*Cyprinus carpio*), tench (*Tinca tinca*), bream (*Abramis brama*), chub (*Leuciscus cephalus*), pike (*Esox lucius*) and perch (*Perca fluviatilis*). Only in the upstream part of the reservoirs can trout be observed. But the most harmful effect of these dams is due to the releases from the reservoirs which affect periodically and markedly the physicochemical and biological water quality of the Rhône River downstream (see above).

The straight and sinuous sections

The river course between Groslée and Anthon remained free of any embankment until the beginning of the 1980s due to its lateral stability and to the depth of the single channel. After the decline of navigation and former riverine practices, the landscape was restored to a seminatural state; since the beginning of the 20th century the riparian vegetation has colonized the banks and the tow-path.

In sinuous sections, the natural cutting of meanders several centuries ago isolated some oxbow lakes such as 'Le Sauget', upstream Groslée, and 'Le Grand Gravier', downstream from Anthon (Figure 1). Fish populations in these abandoned channels are characteristic of stagnant waters: common species are the roach (*Rutilus rutilus*), the perch (*Perca fluviatilis*), the rudd (*Scardinius erythrophtalmus*), and introduced species such as the pumpkinseed (*Lepomis gibbosus*) and the black bullhead (*Ictalurus melas*).

In the deepest parts of these oxbow lakes hydrophytic plants (*Myriophyllum spicatum* and *Nuphar luteum*) are well represented (Figure 9). The terrestrialization sequence begins with the development of helophyte stands (*Scirpus lacustris*, *Phragmites australis*, *Equisetum fluviatile*, *Typha australis*) (Ain *et al.*, 1973). The roots and the rhizomes of these helophytes produce a sort of mattress which is developing horizontally, reducing the water surface area, and aggrading due to the accumulation of aerial parts. The most developed populations are situated on the convex side, upstream and downstream: *Alnus glutinosa*, *Salix cinerea*. Because of their great depth (>5 m) and their remoteness from the main stream, their ecosystems develop very slowly (the complete terrestrialization spans more than three centuries). Their ecological successions seem similar to those of anastomosed channels and proceed usually in a progressive way even with increasing water level (Jurasz and Amoros, in preparation). These oxbow lake ecosystems can regulate the influence of the main river because of their relatively high degree of homeostasis resulting from the organization of the biocenosis, the biomass production and the organic matter storage (Rostan *et al.*, 1987). Despite the flooding of the Sauget oxbow lake when the Rhône River discharge rises up to $1350\,m^3\,s^{-1}$ ($Qm = 450\,m^3\,s^{-1}$) and despite the consecutive changes of the physicochemical variables (temperature, conductivity, pH, oxygen, ionic concentrations), the consequences of the river inputs are minimized by the biological activity of the ecosystem

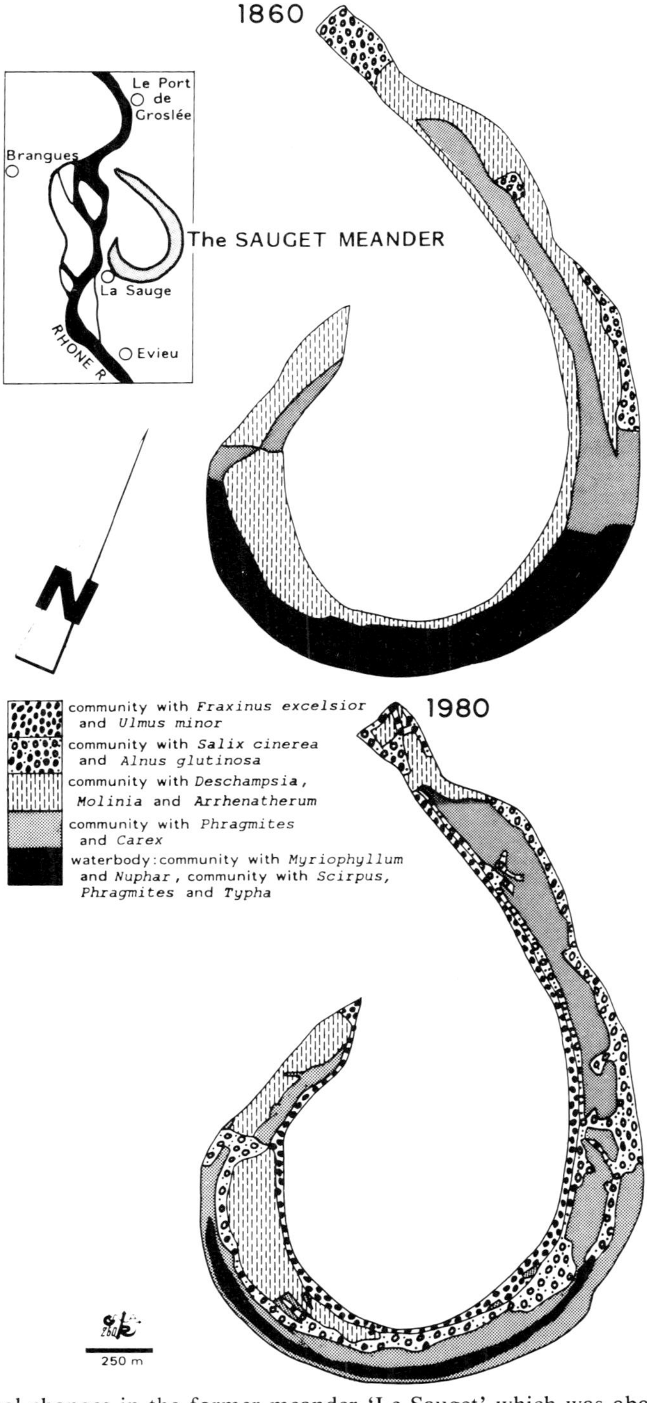

FIGURE 9. Ecological changes in the former meander 'Le Sauget' which was abandoned in 1690

(Carrell and Juget, 1987). Because of this high degree of homeostasis, the indirect impacts of upstream civil engineering works should be negligible in such stagnant waterbodies. Moreover, in spite of the cultivation of the surrounding alluvial plain, they have probably maintained a natural status during several centuries. Vegetation mapping in the oldest abandoned meanders shows that the *Alnus glutinosa* stage occurs in most of them (Pautou, 1975) except in their higher parts that are overgrown by *Fraxinus excelsior* and *Prunus padus*.

The margins of the meanders have been aggraded by silt deposition. For example, on the former island bounded by the Saugey meander, the silt layer is 1 to 2 m thick overlying a 2–4 m thick sand layer. These thick soils are well supplied with capillary underground water and hardwood forests have become well established with *Fraxinus excelsior*, *Quercus robur* and *Ulmus minor* in the abandoned areas. The pristine riverine forest has been completely clear cut, only some small formerly exploited woods remain. The fertile soils of the alluvial plain are intensively tilled and produce corn, soya, tobacco and fruit.

Major changes are presently occurring because of the development of the hydroelectric schemes. The dam at Sault-Brenaz was completed in 1986 and has changed the section from Brégnier to Sault-Brenaz; another dam is planned at Loyettes and this could affect a long section downstream of Le Sault if the construction proceeds. This river section has a low gradient (0.0001 to 0.0003) and is incised into fluvioglacial and Holocene terraces. Thus, the bypassed sections tend to be short for the benefit of a very long, narrow, and rather shallow reservoir. The banks are controlled by embankments in the downstream course but may remain wild in some upstream stretches.

CONCLUSION

The Rhône River represents a good example of a large European river that has been influenced by a range of human influences over several hundred years. In contrast to the small watercourses that were drastically modified by human activities during the last thousand years because available technologies allowed such transformations, in the case of large rivers major impacts are associated with the last three centuries. The most ancient modifications of large rivers were probably military ones when the watercourses acted as state boundary lines, but these modifications were rather slight in comparison with what was to follow. Real impacts are due to civil engineering works implemented to promote and facilitate navigation, to protect towns and lands against floods and to produce hydroelectric energy. The same evolutionary sequence can be observed on the Rhône, the Danube (Bethemont and Bravard, 1986; Pecsi, 1968) and the Rhine (Carbiener, 1983; Friedrich and Muller, 1984; Lelek, in press) Rivers. Indeed, these three large rivers were or still are state boundary rivers, and the same types of engineering works were completed because all three are situated on the external and peripheral alpine area. The main and direct consequences of harnessing the river have been channelization, constriction of the active alluvial valley, and discharge controls; these induced spatial, short-term and long-term impacts.

The short-term impacts of the low-level embankments of the last century increased the structural and functional diversity of the ecosystems in the braided sections: starting from a rhithronic status, the embanked channel became potamic, the secondary arms which were blocked upstream experienced a new type of functioning, the embankments and the dykes offered new biotopes, and new hydrological and sedimentological processes allowed the development of several ecological sequences which were inhibited before. Furthermore the speed of evolution of these sequences was diversified according to their distance from the main channel, the main determinant of the successional rates being the allogenic processes.

In the same braided sections where hydroelectric schemes were developed, the diversity increased once more. So the embanked channel recovered a rhithronic status, the secondary arms became more rhithronic too, some dead arms were rejuvenated, new types of biotopes were created, the counter-canals introduced new small trout streams, and the headraces and the tailraces are potamic as the newly parts of the channel transformed into reservoirs. The diversification of the aquatic biotopes revealed two opposite trends, both a rise and a decrease in the typological levels corresponding respectively to an overstatement of rhithronic and potamic characteristics of the involved ecosystems. But although the absolute number of habitats is increased at present, it must be mentioned that the spatial diversity will probably decrease eventually in most parts of the braided sections. Indeed, due to the lowering of the water table, the alluvial valley is constricted and the extension of the area of active geomorphology reduced. Consequently, long-term impacts will reduce the diversity of habitats and communities: the rejuvenation of ecosystems is prevented by the successive controls on the river, autogenous processes, being the only ones active now, will lead to a simplification in favour of the later stages of the ecological sequences, and finally to a relative uniformity of the braided sections in the long-term.

A range of effects on the downstream river can arise from a major dam such as Génissiat. The area of influence is the perceptible spatial extent of the river changes, in response to localized training works or impoundments. Génissiat dam, for example, has induced downstream impacts through an upstream control: river-bed incision over tens of kilometres, fine sediment retention, and water pollution during some flushes. This dam controls upstream impacts through siltation changes of water quality and fauna; upstream gravel deposition justified excessive extractions in the river-bed which, in turn, induced a damageable regressive erosion.

On this type of river, which experienced several periods of training, it is difficult to determine the links of causality, e.g. the present bypassed channel, in the Chautagne former braided section, has been degraded up to 3 m; this incision is a delayed response to the combined impacts of successive embankments of 1780 and 1840 which concentrated the flow in a 1 km long section, the recent impact of gravel retention upstream from Génissiat dam, and the present impact of the Chautagne hydroelectric station which has changed the flow regime. A section may so react to a combination of impacts which are superimposed and reinforced. This section exemplifies the complex response of river-beds and floodplains: the watertable of the backswamps has been lowered as the direct impact of drainage works as the indirect impact of river-bed degradation then flow shortage. In the same manner, in the braided area of Brégnier–Cordon, considered as an example, ecological changes arise from the interaction of a sequence of human impacts over the historic timescale. These are summarized in Figure 10.

This review of the Rhône River changes exemplifies the interactions between abiotic and biotic factors which may be discussed in a hierarchical framework (Petts, 1984): the dykes and the dams have produced first-order impacts on the flow regime and the sediment load, then changed the channel form, the substrate composition and the aquatic vegetation

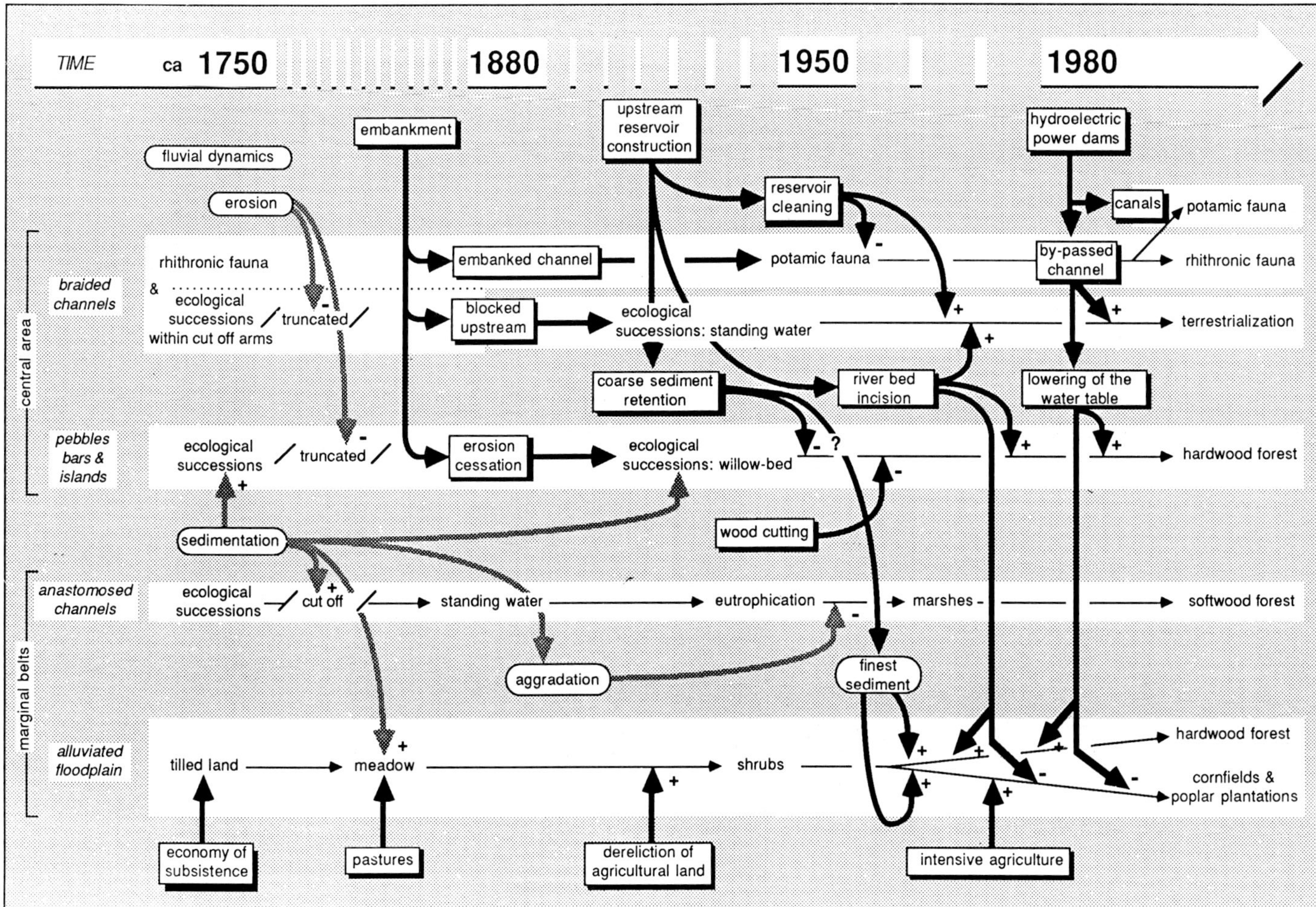

FIGURE 10. Qualitative model of the interactions between natural processes and several human actions responsible for the ecological changes within a braided–anastomosed sector of the French Upper Rhône from 1750 to 1988

(2nd order impacts), and at last invertebrates and fishes (3rd order impacts). This pattern is valid in both the aquatic and terrestrial environments of the floodplain.

The application of palaeoecological studies based on synchronic and diachronic analyses (Amoros *et al.*, 1987) utilizing cartographic and documentary data sources have been used to reconstruct the history of the fluvial hydrosystems on the French Upper Rhône since 1750. Important evidence of the former river, at different times is contained within the floodplain sediments and abandoned channels which remain isolated from more recent changes. Today, the river is 'man-made' but the contemporary hydrosystem reflects the cumulative effects of man's interference over at least two centuries modified by allogenic processes and recolonization from incontrolled tributaries. By understanding the interaction of these complex processes, models are to be developed to improve our understanding of how large alluvial rivers may change in the future.

REFERENCES

Amoros, C. 1980. Structure et fonctionnement des écosystèmes du Haut-Rhône français. 15 Comparaison des peuplements de Cladocères et Copépodes de deux anciens méandres d'âges différents. *Acta Oecologica, Oecol. gener.*, **1**(2), 193–208.

Amoros, C. and Roux, A. L. 1987. *Interactions between water bodies within the floodplains of large rivers: function and development of connectivity*, I.A.L.E. Congress, Münster, July 1987.

Amoros, C., Roux, A. L., Reygrobellet, J. L., Bravard, J. P. and Pautou G. 1987. Methodological research applied to the ecological management of fluvial hydrosystems. *Regulated Rivers*, **1**(1), 17–36.

Bethemont, J. and Bravard, J. P. 1986. Gabcikovo: un grand projet et une controverse. *Rev. Géogr. Lyon*, 19–41.

Bravard, J. P. 1981. *La Chautagne*, Inst. Etud. Univ. Lyon, 182 pp.

Bravard, J. P. 1987. *Le Rhône, Du Léman à Lyon*, La Manufacture, Lyon, 452 pp.

Bravard, J. P., Amoros, C. and Pautou, G. 1986. Impact of civil engineering works on the successions of communities in a fluvial system. A methodological and predictive approach applied to a section of the Upper Rhône River, France. *Oikos*, **47**, 92–111.

Carbiener, R. 1983. Le grand Ried central d'Alsace: écologie et évolution d'une zone humide d'origine fluviale rhénane. *Bull. Ecol.*, **14**(4), 249–277.

Carrel, G. 1986. *Caractérisation physico-chimique du Haut-Rhône français et de ses annexes; incidence sur la croissance des populations d'alevins*, Thèse doctorat, Lyon I, 185 pp.

Carrel, G. and Juget, J. 1987. La Morte du Sauget, un ancien méandre du Rhône: bilan hydrologique et biogéochimique. *Schweiz. Z. Hydrol*, **49**(1), 102–125.

Castella, E., Richardot-Coulet, M., Roux, C. and Richoux, P. 1984. Macroinvertebrates as 'describers' of morphological and hydrological types of aquatic ecosystems abandoned by the Rhône River. *Hydrobiologia*, **119**(3), 219–225.

Copp, G. H. 1987. *Le rôle et le fonctionnement des milieux aquatiques du Haut-Rhône français comme sites de reproduction et de nurserie pour les poissons du fleuve*, Thèse Doctorat, Univ. Lyon I, 97 pp.

Friedrich, G. and Muller, D. 1984. Rhine. In Whitton, B. A. (Ed.), *Ecology of European Rivers*, Blackwell, London, 265–315.

Girel, J. and Doche, B. 1983. Influence des activités humaines sur la genèse, l'évolution et la disparition de groupements végétaux alluviaux (Exemple du site de l'Etournel, dans la vallée du Haut-Rhône entre Bellegarde et Genève). *Rev. Géogr. Alp.*, **LXXI**, 343–351.

Jacquet-Labrot, C. 1986. *Evolution des anciens chenaux du Rhône et de l'Ain: écologie rétrospective basée sur les restes de Cladocères*, Thèse Doctorat, Univ. Lyon I, 194 pp.

Jacquet, C., Amoros, C. and Auda, Y. 1984. Evolution des écosystèmes aquatiques abandonnés par les fleuves: recherches méthodologiques sur l'utilisation des restes de Cladocères en écologie rétrospective. *Archiv. f. Hydrobiol.*, **102**(1), 73–89.

Juget, J., Yi, B. J., Roux, C., Richoux, P., Richardot-Coulet, M., Reygrobellet, J. L. and Amoros, C. 1979. Structure et fonctionnement des écosystèmes du Haut-Rhône français. VII. Le complexe hydrographique de la Lône des Pêcheurs (un ancien méandre du Rhône). *Schweiz. Z. Hydrol.* **41**(2) 395–417.

Lelek, A. in press. The Rhine River and some of its tributaries under human impact in the last two centuries. *LARS* (in press).

Majchrzak, Y. and Maneville, O. 1987. *Gestion des zones naturelles humides: méthodologie de suivi du pâturage par des Bovins Highland Cattle dans la Réserve Naturelle du marais de Lavours (Béon, Ain, France*), Congrès Soc. Sav., Lyon.

Nelva, A., Pattee, E., Perrin, J. F., Persat, H. and Roux, A. L. 1981. Structure et fonctionnement des écosystèmes du Haut-Rhône français. 25 — Premières observations sur les populations piscicoles dans le secteur de Brégnier Cordon. *Verh. intern. Verein. Limnol.*, **21**, 1276–1282.

Pautou, G. 1975. *Contribution à l'étude écologique de la plaine alluviale du Rhône entre Seyssel et Lyon*, Thèse Univ. de Grenoble, 375 pp.

Pautou, G. 1983. Répercussions des aménagements hydroélectriques sur le dynamisme de la végétation. *Rev. Géogr. Alpine*, **LXXI**, 4, 331–342.

Pautou, G. and Bravard, J. P. 1982. L'incidence des activités humaines sur la dynamique de l'eau et l'évolution de la végétation dans la vallée dy Haut-Rhône français. *Rev. Géogr. Lyon*, **57**(3), 63–79.

Pautou, G., Decamps, H., Amoros, C. and Bravard, J. P. 1985. Successions végétales dans les couloirs fluviaux: l'exemple de la plaine alluviale du Haut-Rhône français. *Bull. Ecol.*, **16**(3), 203–212.

Pecsi M. 1968. Evolution of the floodplain levels of the Danube and their principal bearings on the geography of agriculture. (*Ac. Sc. Hungarica, Inst. Geogr.), Fold. Kozl.*, **3**, 215–222.

Perrin, J. F. 1978. *Signification écologique des peuplements benthiques du Haut-Rhône français*, Thèse Doct. Spéc, Lyon, 171 pp.

Petts, G. E. 1984. *Impounded Rivers — Perspective for Ecological Management*, Wiley, Chichester, U.K., 326 pp.

Richardot-Coulet, M., Castella, E. and Castella, C. 1987. Classification and succession of former channels of the French Upper Rhône Alluvial plain using Mollusca. *Regulated Rivers*, **1**, 111–127.

Rostan, J. C., Amoros, C. and Juget, J. 1987. The organic content of the surficial sediment: a method for the study of ecosystem development in abandoned river channels. *Hydrobiologia*, **148**, 45–62.

Roux, A. L. (ouvrage collectif publié sous la direction de:) 1982. *Cartographie polythématique appliquée à la gestion écologique des eaux. Etude d'un hydrosystème fluvial: le Haut-Rhône français*, Edit. C.N.R.S., Centre région. Publ. Lyon, 116 pp.

Roux A. L. 1984. Impacts of emptying and cleaning reservoirs on the downstream Rhône physico-chemical and biological water quality. 2nd Intern. Symp. on Regulated Streams, Oslo août 1982. In Lillehammer, A. and Saltven, S. J. (Eds), *Regulated Rivers*, Oslo, 61–70.

Tinchant, A. 1979. *Effets de la vidange du barrage de Génissiat sur la faune des macroinvertébrés benthiques*, D.E.A. Univ. Lyon, 26 pp.

Index